AN INTRODUCTION TO STATISTICAL SIGNAL PROCESSING

This book describes the essential tools and techniques of statistical signal processing. At every stage theoretical ideas are linked to specific applications in communications and signal processing using a range of carefully chosen examples. The book begins with a development of basic probability, random objects, expectation, and second-order moment theory followed by a wide variety of examples of the most popular random process models and their basic uses and properties. Specific applications to the analysis of random signals and systems for communicating, estimating, detecting, modulating, and other processing of signals are interspersed throughout the book. Hundreds of homework problems are included and the book is ideal for graduate students of electrical engineering and applied mathematics. It is also a useful reference for researchers in signal processing and communications.

ROBERT M. GRAY received his Ph.D. from the University of Southern California, and is the Lucent Technologies Professor of Engineering and Professor and Vice Chair of Electrical Engineering at Stanford University. He has written over 200 scientific papers in areas including information theory, applied probability, signal processing, speech and image processing and coding, ergodic theory, and the theory of Toeplitz matrices. He is a fellow of the IEEE and the Institute of Mathematical Statistics.

LEE D. DAVISSON received his Ph.D. from Princeton University, and is Emeritus Professor and Chair of electrical engineering at the University of Maryland, College Park. He is the author or coauthor of three books and over 100 scientific papers. He is a fellow of the IEEE.

AN INTRODUCTION TO STATISTICAL SIGNAL PROCESSING

ROBERT M. GRAY
Information Systems Laboratory
Department of Electrical Engineering
Stanford University

LEE D. DAVISSON
Department of Electrical Engineering and Computer Science
University of Maryland

PUBLISHED BY THE PRESS SYNDICATE OF THE UNIVERSITY OF CAMBRIDGE
The Pitt Building, Trumpington Street, Cambridge, United Kingdom

CAMBRIDGE UNIVERSITY PRESS
The Edinburgh Building, Cambridge CB2 2RU, UK
40 West 20th Street, New York, NY 10011–4211, USA
477 Williamstown Road, Port Melbourne, VIC 3207, Australia
Ruiz de Alarcón 13, 28014 Madrid, Spain
Dock House, The Waterfront, Cape Town 8001, South Africa

http://www.cambridge.org

First published 2004

Printed in the United Kingdom at the University Press, Cambridge

Typeface Times 11/14 pt *System* LaTeX 2ε [TB]

A catalog record for this book is available from the British Library

Library of Congress Cataloging in Publication data

Gray, Robert M.
An introduction to statistical signal processing / Robert M. Gray and Lee D. Davisson.
p. cm.
Includes bibliographical references and index.
ISBN 0-521-83860-6
1. Signal processing – Statistical methods. I. Davisson, Lee D. II. Title.

TK5102.5.G67 2004
621.382′2–dc22 2004045892

ISBN 0 521 83860 6 hardback

Contents

Preface

The origins of this book lie in our earlier book *Random Processes: A Mathematical Approach for Engineers* (Prentice Hall, 1986). This book began as a second edition to the earlier book and the basic goal remains unchanged – to introduce the fundamental ideas and mechanics of random processes to engineers in a way that accurately reflects the underlying mathematics, but does not require an extensive mathematical background and does not belabor detailed general proofs when simple cases suffice to get the basic ideas across. In the years since the original book was published, however, it has evolved into something bearing little resemblance to its ancestor. Numerous improvements in the presentation of the material have been suggested by colleagues, students, teaching assistants, and reviewers, and by our own teaching experience. The emphasis of the book shifted increasingly towards examples and a viewpoint that better reflected the title of the courses we taught using the book for many years at Stanford University and at the University of Maryland: *An Introduction to Statistical Signal Processing.* Much of the basic content of this course and of the fundamentals of random processes can be viewed as the analysis of statistical signal processing systems: typically one is given a probabilistic description for one random object, which can be considered as an *input signal.* An operation is applied to the input signal (*signal processing*) to produce a new random object, the *output signal.* Fundamental issues include the nature of the basic probabilistic description, and the derivation of the probabilistic description of the output signal given that of the input signal and the particular operation performed. A perusal of the literature in statistical signal processing, communications, control, image and video processing, speech and audio processing, medical signal processing, geophysical signal processing, and classical statistical areas of time series analysis, classification and regression, and pattern recognition shows a wide variety of probabilistic models for input processes and for operations on those processes, where the operations might be deterministic or random, natural or artificial, linear or nonlinear, digital or analog, or beneficial or harmful. An introductory course focuses on the fundamentals underlying the analysis of such systems: the theories of probability, random processes, systems, and signal processing.

When the original book went out of print, the time seemed ripe to convert the manuscript from the prehistoric troff format to the widely used LaTeX format and to undertake a serious revision of the book in the process. As the revision became more extensive, the title changed to match the course name and content. We reprint the original preface to provide some of the original motivation for the book, and then close this preface with a description of the goals sought during the many subsequent revisions.

Preface to Random Processes: An Introduction for Engineers

> Nothing in nature is random ... A thing appears random only through the incompleteness of our knowledge.
>
> *Spinoza, Ethics I*

> I do not believe that God rolls dice.
>
> *attributed to Einstein*

Laplace argued to the effect that given complete knowledge of the physics of an experiment, the outcome must always be predictable. This metaphysical argument must be tempered with several facts. The relevant parameters may not be measurable with sufficient precision due to mechanical or theoretical limits. For example, the uncertainty principle prevents the simultaneous accurate knowledge of both position and momentum. The deterministic functions may be too complex to compute in finite time. The computer itself may make errors due to power failures, lightning, or the general perfidy of inanimate objects. The experiment could take place in a remote location with the parameters unknown to the observer; for example, in a communication link, the transmitted message is unknown a priori, for if it were not, there would be no need for communication. The results of the experiment could be reported by an unreliable witness – either incompetent or dishonest. For these and other reasons, it is useful to have a theory for the analysis and synthesis of processes that behave in a random or unpredictable manner. The goal is to construct mathematical models that lead to reasonably accurate prediction of the long-term average behavior of random processes. The theory should produce good estimates of the average behavior of real processes and thereby correct theoretical derivations with measurable results.

In this book we attempt a development of the basic theory and applications of random processes that uses the language and viewpoint of rigorous mathematical treatments of the subject but which requires only a typical bachelor's degree level of electrical engineering education including elementary discrete and continuous time linear systems theory, elementary probability, and transform theory and applications. Detailed proofs are presented only when within the scope of this background. These simple proofs, however, often provide the groundwork for "handwaving" justifications of more general and complicated results that are semi-rigorous in that they can be made rigorous by the appropriate delta-epsilontics of real analysis or measure theory. A primary goal of this approach is thus to use intuitive arguments that accurately reflect the underlying mathematics and which will hold up under scrutiny if the student continues to more advanced courses. Another goal is to enable the student who might not continue to more advanced courses to be able to read and generally follow the modern literature on applications of random processes to information and communication theory, estimation and detection, control, signal processing, and stochastic systems theory.

Revisions

Through the years the original book has continually expanded to roughly double its original size to include more topics, examples, and problems. The material has been significantly reorganized in its grouping and presentation. Prerequisites and preliminaries have been moved to the appendices. Major additional material has been added on jointly Gaussian vectors, minimum mean squared error estimation, linear and affine least squared error estimation, detection and classification, filtering, and, most recently, mean square calculus and its applications to the analysis of continuous time processes. The index has been steadily expanded to ease navigation through the book. Numerous errors reported by reader email have been fixed and suggestions for clarifications and improvements incorporated.

This book is a work in progress. Revised versions will be made available through the World Wide Web page http://ee.stanford.edu/~gray/sp.html. The material is copyrighted by Cambridge University Press, but is freely available as a pdf file to any individuals who wish to use it provided only that the contents of the entire text remain intact and together. Comments, corrections, and suggestions should be sent to rmgray@stanford.edu. Every effort will be made to fix typos and take suggestions into account on at least an annual basis.

Acknowledgements

We repeat our acknowledgements of the original book: to Stanford University and the University of Maryland for the environments in which the book was written, to the John Simon Guggenheim Memorial Foundation for its support of the first author during the writing in 1981–2 of the original book, to the Stanford University Information Systems Laboratory Industrial Affiliates Program which supported the computer facilities used to compose this book, and to the generations of students who suffered through the ever changing versions and provided a stream of comments and corrections.

Thanks are also due to Richard Blahut and anonymous referees for their careful reading and commenting on the original book. Thanks are due to the many readers who have provided corrections and helpful suggestions through the Internet since the revisions began being posted. Particular thanks are due to Yariv Ephraim for his continuing thorough and helpful editorial commentary.

Thanks also to Sridhar Ramanujam, Raymond E. Rogers, Isabel Milho, Zohreh Azimifar, Dan Sebald, Muzaffer Kal, Greg Coxson, Mihir Pise, Mike Weber, Munkyo Seo, James Jacob Yu, and several anonymous reviewers for Cambridge University Press. Thanks also to Philip Meyler, Lindsay Nightingale, and Joseph Bottrill of Cambridge University Press for their help in the production of the final version of the book.

Thanks to Ian Lee, Michael Gutmann, Frédéric Vrins, André Isidio de Melo, Philippe Bonnet, and to the champion typo finder, Ron Aloysius.

Lastly, the first author would like to acknowledge his debt to his professors who taught him probability theory and random processes, especially Al Drake and Wilbur B. Davenport Jr. at MIT and Tom Pitcher at USC.

Glossary

$\{\,\}$	a collection of points satisfying some property, e.g. $\{r : r \leq a\}$ is the collection of all real numbers less than or equal to a value a
$[\,]$	an interval of real points including the end points, e.g. for $a \leq b$ $[a, b] = \{r : a \leq r \leq b\}$. Called a *closed interval*
$(\,)$	an interval of real points excluding the end points, e.g. for $a \leq b$ $(a, b) = \{r : a < r < b\}$. Called an *open interval.* Note this is empty if $a = b$
$(\,]$, $[\,)$	denote intervals of real points including one endpoint and excluding the other, e.g. for $a \leq b$ $(a, b] = \{r : a < r \leq b\}$, $[a, b) = \{r : a \leq r < b\}$
$\emptyset$	the empty set, the set that contains no points.
$\forall$	for all
Ω	the sample space or universal set, the set that contains all of the points
$\#(F)$	the number of elements in a set F
$\stackrel{\Delta}{=}$	equal by definition
exp	the exponential function, $\exp(x) \stackrel{\Delta}{=} e^x$, used for clarity when x is complicated
$\mathcal{F}$	sigma-field or event space
$\mathcal{B}(\Omega)$	Borel field of Ω, that is, the sigma-field of subsets of the real line generated by the intervals or the Cartesian product of a collection of such sigma-fields
iff	if and only if
l.i.m.	limit in the mean
$o(u)$	function of u that goes to zero as $u \to 0$ faster than u

P	probability measure
P_X	distribution of a random variable or vector X
p_X	probability mass function (pmf) of a random variable X
f_X	probability density function (pdf) of a random variable X
F_X	cumulative distribution function (cdf) of a random variable X
$E(X)$	expectation of a random variable X
$M_X(ju)$	characteristic function of a random variable X
$\oplus$	addition modulo 2
$1_F(x)$	indicator function of a set F: $1_F(x) = 1$ if $x \in F$ and 0 otherwise
Φ	Φ-function (Eq. (2.78))
Q	complementary Phi function (Eq. (2.79))
$\mathcal{Z}_k$	$\triangleq \{0, 1, 2, \ldots, k-1\}$
$\mathcal{Z}_+$	$\triangleq \{0, 1, 2, \ldots\}$, the collection of nonnegative integers
$\mathcal{Z}$	$\triangleq \{\ldots, -2, -1, 0, 1, 2, \ldots\}$, the collection of all integers

1
Introduction

A random or stochastic process is a mathematical model for a phenomenon that evolves in time in an unpredictable manner from the viewpoint of the observer. The phenomenon may be a sequence of real-valued measurements of voltage or temperature, a binary data stream from a computer, a modulated binary data stream from a modem, a sequence of coin tosses, the daily Dow–Jones average, radiometer data or photographs from deep space probes, a sequence of images from a cable television, or any of an infinite number of possible sequences, waveforms, or signals of any imaginable type. It may be unpredictable because of such effects as interference or noise in a communication link or storage medium, or it may be an information-bearing signal, deterministic from the viewpoint of an observer at the transmitter but random to an observer at the receiver.

The theory of random processes quantifies the above notions so that one can construct mathematical models of real phenomena that are both tractable and meaningful in the sense of yielding useful predictions of future behavior. Tractability is required in order for the engineer (or anyone else) to be able to perform analyses and syntheses of random processes, perhaps with the aid of computers. The "meaningful" requirement is that the models must provide a reasonably good approximation of the actual phenomena. An oversimplified model may provide results and conclusions that do not apply to the real phenomenon being modeled. An overcomplicated one may constrain potential applications, render theory too difficult to be useful, and strain available computational resources. Perhaps the most distinguishing characteristic between an average engineer and an outstanding engineer is the ability to derive effective models providing a good balance between complexity and accuracy.

Random processes usually occur in applications in the context of environments or systems which *change* the processes to produce other processes.

The intentional operation on a signal produced by one process, an "input signal," to produce a new signal, an "output signal," is generally referred to as *signal processing*, a topic easily illustrated by examples.

- A time-varying voltage waveform is produced by a human speaking into a microphone or telephone. The signal can be modeled by a random process. This signal might be modulated for transmission, then it might be digitized and coded for transmission on a digital link. Noise in the digital link can cause errors in reconstructed bits, the bits can then be used to reconstruct the original signal within some fidelity. All of these operations on signals can be considered as signal processing, although the name is most commonly used for manmade operations such as modulation, digitization, and coding, rather than the natural possibly unavoidable changes such as the addition of thermal noise or other changes out of our control.
- For digital speech communications at very low bit rates, speech is sometimes converted into a model consisting of a simple linear filter (called an autoregressive filter) and an input process. The idea is that the parameters describing the model can be communicated with fewer bits than can the original signal, but the receiver can synthesize the human voice at the other end using the model so that it sounds very much like the original signal. A system of this type is called a *vocoder*.
- Signals including image data transmitted from remote spacecraft are virtually buried in noise added to them on route and in the front end amplifiers of the receivers used to retrieve the signals. By suitably preparing the signals prior to transmission, by suitable filtering of the received signal plus noise, and by suitable decision or estimation rules, high quality images are transmitted through this very poor channel.
- Signals produced by biomedical measuring devices can display specific behavior when a patient suddenly changes for the worse. Signal processing systems can look for these changes and warn medical personnel when suspicious behavior occurs.
- Images produced by laser cameras inside elderly North Atlantic pipelines can be automatically analyzed to locate possible anomalies indicating corrosion by looking for locally distinct random behavior.

How are these signals characterized? If the signals are random, how does one find stable behavior or structures to describe the processes? How do operations on these signals change them? How can one use observations based on random signals to make intelligent decisions regarding future behavior? All of these questions lead to aspects of the theory and application of random processes.

Courses and texts on random processes usually fall into either of two general and distinct categories. One category is the common engineering approach, which involves fairly elementary probability theory, standard undergraduate Riemann calculus, and a large dose of "cookbook" formulas – often with insufficient attention paid to conditions under which the formulas are valid. The results are often justified by nonrigorous and occasionally mathematically inaccurate handwaving or intuitive plausibility arguments that may not reflect the actual underlying mathematical structure and may not be supportable by a precise proof. While intuitive arguments can be extremely valuable in providing insight into deep theoretical results, they can be a handicap if they do not capture the essence of a rigorous proof.

A development of random processes that is insufficiently mathematical leaves the student ill prepared to generalize the techniques and results when faced with a real-world example not covered in the text. For example, if one is faced with the problem of designing signal processing equipment for predicting or communicating measurements being made for the first time by a space probe, how does one construct a mathematical model for the physical process that will be useful for analysis? If one encounters a process that is neither stationary nor ergodic (terms we shall consider in detail), what techniques still apply? Can the law of large numbers still be used to construct a useful model?

An additional problem with an insufficiently mathematical development is that it does not leave the student adequately prepared to read modern literature such as the many *Transactions of the IEEE* and the journals of the European Association for Signal, Speech, and Image Processing (EURASIP). The more advanced mathematical language of recent work is increasingly used even in simple cases because it is precise and universal and focuses on the structure common to all random processes. Even if an engineer is not directly involved in research, knowledge of the current literature can often provide useful ideas and techniques for tackling specific problems. Engineers unfamiliar with basic concepts such as *sigma-field* and *conditional expectation* will find many potentially valuable references shrouded in mystery.

The other category of courses and texts on random processes is the typical mathematical approach, which requires an advanced mathematical background of real analysis, measure theory, and integration theory. This approach involves precise and careful theorem statements and proofs, and uses far more care to specify precisely the conditions required for a result to hold. Most engineers do not, however, have the required mathematical background, and the extra care required in a completely rigorous development severely limits the number of topics that can be covered in a typical course – in particular, the applications that are so important to engineers tend to be neglected. In addition, too much time is spent with the formal details, obscuring the often simple and elegant ideas behind a proof. Often little, if any, physical motivation for the topics is given.

This book attempts a compromise between the two approaches by giving the basic theory and a profusion of examples in the language and notation of the more advanced mathematical approaches. The intent is to make the crucial concepts clear in the traditional elementary cases, such as coin flipping, and thereby to emphasize the mathematical structure of all random processes in the simplest possible context. The structure is then further developed by numerous increasingly complex examples of random processes that have proved useful in systems analysis. The complicated examples are constructed from the simple examples by signal processing, that is, by using a simple process as an input to a system whose output is the more complicated process. This has the double advantage of describing the action of the system, the actual signal processing, and the interesting random process which is thereby produced. As one might suspect, signal processing also can be used to produce simple processes from complicated ones.

Careful proofs are usually constructed only in elementary cases. For example, the fundamental theorem of expectation is proved only for discrete random variables, where it is proved simply by a change of variables in a sum. The continuous analog is subsequently given without a careful proof, but with the explanation that it is simply the integral analog of the summation formula and hence can be viewed as a limiting form of the discrete result. As another example, only weak laws of large numbers are proved in detail in the mainstream of the text, but the strong law is treated in detail for a special case in a starred section. Starred sections are used to delve into other relatively advanced results, for example the use of mean square convergence ideas to make rigorous the notion of integration and filtering of continuous time processes.

By these means we strive to capture the spirit of important proofs without undue tedium and to make plausible the required assumptions and constraints. This, in turn, should aid the student in determining when certain tools do or do not apply and what additional tools might be necessary when new generalizations are required.

A distinct aspect of the mathematical viewpoint is the "grand experiment" view of random processes as being a probability measure on sequences (for discrete time) or waveforms (for continuous time) rather than being an infinity of smaller experiments representing individual outcomes (called random variables) that are somehow glued together. From this point of view random variables are merely special cases of random processes. In fact, the grand experiment viewpoint was popular in the early days of applications of random processes to systems and was called the "ensemble" viewpoint in the work of Norbert Wiener and his students. By viewing the random process as a whole instead of as a collection of pieces, many basic ideas, such as stationarity and ergodicity, that characterize the dependence on time of probabilistic descriptions and the relation between time averages and probabilistic averages are much easier to define and study. This also permits a more complete discussion of processes that violate such probabilistic regularity requirements yet still have useful relations between time and probabilistic averages.

Even though a student completing this book will not be able to follow the details in the literature of many proofs of results involving random processes, the basic results and their development and implications should be accessible, and the most common examples of random processes and classes of random processes should be familiar. In particular, the student should be well equipped to follow the gist of most arguments in the various *Transactions of the IEEE* dealing with random processes, including the *IEEE Transactions on Signal Processing*, *IEEE Transactions on Image Processing*, *IEEE Transactions on Speech and Audio Processing*, *IEEE Transactions on Communications*, *IEEE Transactions on Control*, and *IEEE Transactions on Information Theory*, and the EURASIP/Elsevier journals such as *Image Communication*, *Speech Communication*, and *Signal Processing*.

It also should be mentioned that the authors are electrical engineers and, as such, have written this text with an electrical engineering flavor. However, the required knowledge of classical electrical engineering is slight, and engineers in other fields should be able to follow the material presented.

This book is intended to provide a one-quarter or one-semester course that develops the basic ideas and language of the theory of random processes and provides a rich collection of examples of commonly encountered processes, properties, and calculations. Although in some cases these examples may seem somewhat artificial, they are chosen to illustrate the way engineers should think about random processes. They are selected for simplicity and conceptual content rather than to present the method of solution to some particular application. *Sections that can be skimmed or omitted for the shorter one-quarter curriculum are marked with a star* ($\star$). Discrete time processes are given more emphasis than in many texts because they are simpler to handle and because they are of increasing practical importance in digital systems. For example, linear filter input/output relations are carefully developed for discrete time; then the continuous time analogs are obtained by replacing sums with integrals. The mathematical details underlying the continuous time results are found in a starred section.

Most examples are developed by beginning with simple processes. These processes are filtered or modulated to obtain more complicated processes. This provides many examples of typical probabilistic computations on simple processes and on the output of operations on simple processes. Extra tools are introduced as needed to develop properties of the examples.

The prerequisites for this book are elementary set theory, elementary probability, and some familiarity with linear systems theory (Fourier analysis, convolution, discrete and continuous time linear filters, and transfer functions). The elementary set theory and probability may be found, for example, in the classic text by Al Drake [18] or in the current MIT basic probability text by Bertsekas and Tsitsiklis [3]. The Fourier and linear systems material can by found in numerous texts, including Gray and Goodman [33]. Some of these basic topics are reviewed in this book in Appendix A. These results are considered prerequisite as the pace and density of material would likely be overwhelming to someone not already familiar with the fundamental ideas of probability such as probability mass and density functions (including the more common named distributions), computing probabilities, derived distributions, random variables, and expectation. It has long been the authors' experience that the students having the most difficulty with this material are those with little or no experience with elementary probability.

Organization of the book

Chapter 2 provides a careful development of the fundamental concept of probability theory – a probability space or experiment. The notions of sample space, event space, and probability measure are introduced and illustrated by examples. Independence and elementary conditional probability are developed in some detail. The ideas of signal processing and of random variables are introduced briefly as functions or operations on the output of an experiment. This in turn allows mention of the idea of expectation at an early stage as a generalization of the description of probabilities by sums or integrals.

Chapter 3 treats the theory of measurements made on experiments: random variables, which are scalar-valued measurements; random vectors, which are a vector or finite collection of measurements; and random processes, which can be viewed as sequences or waveforms of measurements. Random variables, vectors, and processes can all be viewed as forms of signal processing: each operates on "inputs," which are the sample points of a probability space, and produces an "output," which is the resulting sample value of the random variable, vector, or process. These output points together constitute an output sample space, which inherits its own probability measure from the structure of the measurement and the underlying experiment. As a result, many of the basic properties of random variables, vectors, and processes follow from those of probability spaces. Probability distributions are introduced along with probability mass functions, probability density functions, and cumulative distribution functions. The basic derived distribution method is described and demonstrated by example. A wide variety of examples of random variables, vectors, and processes are treated. Expectations are introduced briefly as a means of characterizing distributions and to provide some calculus practice.

Chapter 4 develops in depth the ideas of expectation – averages of random objects with respect to probability distributions. Also called probabilistic averages, statistical averages, and ensemble averages, expectations can be thought of as providing simple but important parameters describing probability distributions. A variety of specific averages are considered, including mean, variance, characteristic functions, correlation, and covariance. Several examples of unconditional and conditional expectations and their properties and applications are provided. Perhaps the most important application is to the statement and proof of laws of large numbers or ergodic theorems, which relate long-term sample-average behavior of random processes to expectations. In this chapter laws of large numbers are proved for simple, but important, classes of random processes. Other important applications of expectation arise in performing and analyzing signal processing applications such as detecting, classifying, and estimating data. Minimum mean squared nonlinear and linear estimation of scalars and vectors is treated in some detail, showing the fundamental connections among conditional expectation, optimal estimation, and second-order moments of random variables and vectors.

Chapter 5 concentrates on the computation and applications of second-order moments – the mean and covariance – of a variety of random processes. The primary example is a form of derived distribution problem: if a given random process with known second-order moments is put into a linear system what are the second-order moments of the resulting output random process? This problem is treated for linear systems represented by convolutions and for linear modulation systems. Transform techniques are shown to provide a simplification in the computations, much like their ordinary role in elementary linear systems theory. Mean square convergence is revisited and several of its applications to the analysis of continuous time random processes are collected under the heading of mean square calculus. Included are a careful definition of integration and filtering of random processes, differentiation of random processes, and sampling and orthogonal expansions of random processes. In all of these examples the behavior of the second-order moments determines the applicability of the results. The chapter closes with a development of several results from the theory of linear least squares estimation. This provides an example of both the computation and the application of second-order moments.

In Chapter 6 a variety of useful models of sometimes complicated random processes are developed. A powerful approach to modeling complicated random processes is to consider linear systems driven by simple random processes. Chapter 5 used this approach to compute second-order moments, this chapter goes beyond moments to develop a complete description of the output processes. To accomplish this, however, one must make additional assumptions on the input process and on the form of the linear filters. The general model of a linear filter driven by a memoryless process is used to develop several popular models of discrete time random processes. Analogous continuous time random process models are then developed by direct description of their behavior. The principal class of random processes considered is the class of independent increment processes, but other processes with similar definitions but quite different properties are also introduced. Among the models considered are autoregressive processes, moving-average processes, ARMA (autoregressive moving-average) processes, random walks, independent increment processes, Markov processes, Poisson and Gaussian processes, and the random telegraph wave process. We also briefly consider an example of a nonlinear system where the output random processes can at least be partially described – the exponential function of a Gaussian or Poisson process which models phase or frequency modulation. We close with examples of a type of "doubly stochastic" process – a compound process formed by adding a random number of other random effects.

Appendix A sketches several prerequisite definitions and concepts from elementary set theory and linear systems theory using examples to be encountered elsewhere in the book. The first subject is crucial at an early stage and should be reviewed before proceeding to Chapter 2. The second subject is not required until Chapter 5, but it serves as a reminder of material with which the student should already be familiar. Elementary probability is not reviewed, as our basic development includes elementary probability presented in a rigorous manner that sets the stage for more advanced probability. The review of prerequisite material in the appendix serves to collect together some notation and many definitions that will be used throughout the book. It is, however, only a brief review and cannot serve as a substitute for a complete course on the material. This chapter can be given as a first reading assignment and either skipped or skimmed briefly in class; lectures can proceed from an introduction, perhaps incorporating some preliminary material, directly to Chapter 2.

Appendix B provides some scattered definitions and results needed in the book that detract from the main development, but may be of interest for background or detail. These fall primarily in the realm of calculus and range from the evaluation of common sums and integrals to a consideration of different definitions of integration. Many of the sums and integrals should be prerequisite material, but it has been the authors' experience that many students have either forgotten or not seen many of the standard tricks. Hence several of the most important techniques for probability and signal processing applications are included. Also in this appendix some background information on limits of double sums and the Lebesgue integral is provided.

Appendix C collects the common univariate probability mass functions and probability density functions along with their second-order moments for reference.

The book concludes with Appendix D suggesting supplementary reading, providing occasional historical notes, and delving deeper into some of the technical issues raised in the book. In that section we assemble references on additional background material as well as on books that pursue the various topics in more depth or on a more advanced level. We feel that these comments and references are supplementary to the development and that less clutter results by putting them in a single appendix rather than strewing them throughout the text. The section is intended as a guide for further study, not as an exhaustive description of the relevant literature, the latter goal being beyond the authors' interests and stamina.

Each chapter is accompanied by a collection of problems, many of which have been contributed by collegues, readers, students, and former students. It is important when doing the problems to justify any "yes/no" answers. If an answer is "yes," prove it is so. If the answer is "no," provide a counterexample.

2
Probability

2.1 Introduction

The theory of random processes is a branch of probability theory and probability theory is a special case of the branch of mathematics known as measure theory. Probability theory and measure theory both concentrate on functions that assign real numbers to certain sets in an abstract space according to certain rules. These set functions can be viewed as measures of the size or weight of the sets. For example, the precise notion of area in two-dimensional Euclidean space and volume in three-dimensional space are both examples of measures on sets. Other measures on sets in three dimensions are mass and weight. Observe that from elementary calculus we can find volume by integrating a constant over the set. From physics we can find mass by integrating a mass density or summing point masses over a set. In both cases the set is a region of three-dimensional space. In a similar manner, probabilities will be computed by integrals of densities of probability or sums of "point masses" of probability.

Both probability theory and measure theory consider only nonnegative real-valued set functions. The value assigned by the function to a set is called the *probability* or the *measure* of the set, respectively. The basic difference between probability theory and measure theory is that the former considers only set functions that are normalized in the sense of assigning the value of 1 to the entire abstract space, corresponding to the intuition that the abstract space contains every possible outcome of an experiment and hence should happen with certainty or probability 1. Subsets of the space have some uncertainty and hence have probability less than 1.

Probability theory begins with the concept of a *probability space*, which is a collection of three items:

1. An *abstract space* Ω, as encountered in Appendix A, called a *sample space*, which

contains all distinguishable *elementary outcomes* or results of an experiment. These points might be names, numbers, or complicated signals.

2. An *event space* or *sigma-field* $\mathcal{F}$ consisting of a collection of subsets of the abstract space which we wish to consider as possible events and to which we wish to assign a probability. We require that the event space have an algebraic structure in the following sense: any finite or countably infinite sequence of set-theoretic operations (union, intersection, complementation, difference, symmetric difference) on events must produce other events.
3. A *probability measure* P – an assignment of a number between 0 and 1 to every event, that is, to every set in the event space. A probability measure must obey certain rules or *axioms* and will be computed by integrating or summing, analogously to area, volume, and mass computations.

This chapter is devoted to developing the ideas underlying the triple $(\Omega, \mathcal{F}, P)$, which is collectively called a *probability space* or an *experiment*. Before making these ideas precise, however, several comments are in order.

First of all, it should be emphasized that a probability space is composed of three parts; an abstract space is only one part. Do not let the terminology confuse you: "space" has more than one usage. Having an abstract space model all possible distinguishable outcomes of an experiment should be an intuitive idea since it simply gives a precise mathematical name to an imprecise English description. Since subsets of the abstract space correspond to collections of elementary outcomes, it should also be possible to assign probabilities to such sets. It is a little harder to see, but we can also argue that we should focus on the sets and not on the individual points when assigning probabilities since in many cases a probability assignment known only for points will not be very useful. For example, if we spin a pointer and the outcome is known to be equally likely to be any number between 0 and 1, then the probability that any particular point such as 0.3781984637 or exactly $1/\pi$ occurs is zero because there is an uncountable infinity of possible points, none more likely than the others[1].Hence knowing only that the probability of each and every point is zero, we would be hard pressed to make any meaningful inferences about the probabilities of other events such as the outcome being between $1/2$ and $3/4$. Writers of fiction (including Patrick O'Brian in his Aubrey–Maturin series) have made much of the fact that extremely unlikely events often occur. One can say that zero probability events occur virtually all the time since the a-prioriprobability

[1] A set is countably infinite if it can be put into one-to-one correspondence with the nonnegative integers and hence can be counted. For example, the set of positive integers is countable and the set of all rational numbers is countable. The set of all irrational numbers and the set of all real numbers are both uncountable. See Appendix A for a discussion of countably infinite vs. uncountably infinite spaces.

that the Universe will be exactly in a particular configuration at 13:15 Coordinated Universal Time (also known as Greenwich Mean Time) is zero, yet the Universe will indeed be in some configuration at that time.

The difficulty inherent in this example leads to a less natural aspect of the probability space triumvirate – the fact that we must specify an event space or collection of subsets of our abstract space to which we wish to assign probabilities. In the example it is clear that taking the individual points and their *countable* combinations is not enough (see also Problem 2.3). On the other hand, why not just make the event space the class of *all* subsets of the abstract space? Why require the specification of which subsets are to be deemed sufficiently important to be blessed with the name "event"? In fact, this concern is one of the principal differences between elementary probability theory and advanced probability theory (and the point at which the student's intuition frequently runs into trouble). When the abstract space is finite or even countably infinite, one can consider all possible subsets of the space to be events, and one can build a useful theory. When the abstract space is uncountably infinite, however, as in the case of the space consisting of the real line or the unit interval, one cannot build a useful theory without constraining the subsets to which one will assign a probability. Roughly speaking, this is because probabilities of sets in uncountable spaces are found by integrating over sets, and some sets are simply too nasty to be integrated over. Although it is difficult to show, for such spaces there does not exist a reasonable and consistent means of assigning probabilities to all subsets without contradiction or without violating desirable properties. In fact, it is so difficult to show that such "non-probability-measurable" subsets of the real line exist that we will not attempt to do so in this book. The reader should at least be aware of the problem so that the need for specifying an event space is understood. It also explains why the reader is likely to encounter phrases like "measurable sets" and "measurable functions" in the literature – some things are unmeasurable!

Thus a probability space must make explicit not just the elementary outcomes or "finest-grain" outcomes that constitute our abstract space; it must also specify the collections of sets of these points to which we intend to assign probabilities. Subsets of the abstract space that do not belong to the event space will simply not have probabilities defined. The algebraic structure that we have postulated for the event space will ensure that if we take (countable) unions of events (corresponding to a logical "or") or intersections of events (corresponding to a logical "and"), then the resulting sets are also events and hence will have probabilities. In fact, this is one of the main functions of probability theory: given a probabilistic description of a collection of events, find the probability of some new event formed by set-theoretic operations on the given events.

Up to this point the notion of *signal processing* has not been mentioned. It enters at a fundamental level if one realizes that each individual point $\omega \in \Omega$ produced in an experiment can be viewed as a *signal*: it might be a single voltage conveying the value of a measurement, a vector of values, a sequence of values, or a waveform, any one of which can be interpreted as a *signal* measured in the environment or received from a remote transmitter or extracted from a physical medium that was previously recorded. *Signal processing* in general is the performing of some operation on the signal. In its simplest yet most general form this consists of applying some function or mapping or operation g to the signal or input ω to produce an output $g(\omega)$, which might be intended to guess some hidden parameter, extract useful information from noise, or enhance an image, or might be any simple or complicated operation intended to produce a useful outcome. If we have a probabilistic description of the underlying experiment, then we should be able to derive a probabilistic description of the outcome of the signal processor. This is the core problem of derived distributions, one of the fundamental tools of both probability theory and signal processing. In fact, this idea of defining functions on probability spaces is the foundation for the definition of random variables, random vectors, and random processes, which will inherit their basic properties from the underlying probability space, thereby yielding new probability spaces. Much of the theory of random processes and signal processing consists of developing the implications of certain operations on probability spaces: beginning with some probability space we form new ones by operations called variously mappings, filtering, sampling, coding, communicating, estimating, detecting, averaging, measuring, enhancing, predicting, smoothing, interpolating, classifying, analyzing, or other names denoting linear or nonlinear operations. Stochastic systems theory is the combination of systems theory with probability theory. The essence of stochastic systems theory is the connection of a system to a probability space. Thus a precise formulation and a good understanding of probability spaces are prerequisites to a precise formulation and correct development of examples of random processes and stochastic systems.

Before proceeding to a careful development, several of the basic ideas are illustrated informally with simple examples.

2.2 Spinning pointers and flipping coins

Many of the basic ideas at the core of this text can be introduced and illustrated by two very simple examples, the continuous experiment of spinning a pointer inside a circle and the discrete experiment of flipping a coin.

A uniform spinning pointer

Suppose that Nature (or perhaps Tyche, the Greek goddess of chance) spins a pointer in a circle as depicted in Figure 2.1. When the pointer stops it can

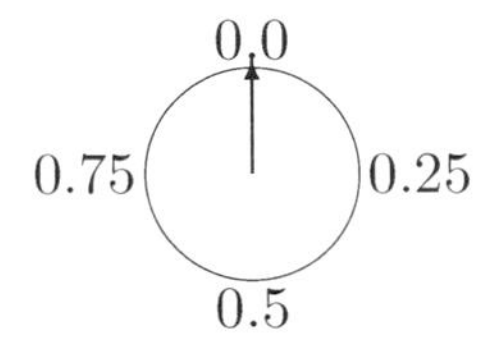

Figure 2.1 The spinning pointer

point to any number in the unit interval $[0, 1) \triangleq \{r : 0 \leq r < 1\}$. We call $[0, 1)$ the *sample space* of our experiment and denote it by a capital Greek omega, Ω. What can we say about the probabilities or chances of particular events or outcomes occurring as a result of this experiment? The sorts of events of interest are things like "the pointer points to a number between 0.0 and 0.5" (which one would expect should have probability 0.5 if the wheel is indeed fair) or "the pointer does not lie between 0.75 and 1" (which should have a probability of 0.75). Two assumptions are implicit here. The first is that an "outcome" of the experiment or an "event" to which we can assign a probability is simply a subset of $[0, 1)$. The second assumption is that the probability of the pointer landing in any particular interval of the sample space is proportional to the length of the interval. This should seem reasonable if we indeed believe the spinning pointer to be "fair" in the sense of not favoring any outcomes over any others. The bigger a region of the circle, the more likely the pointer is to end up in that region. We can formalize this by stating that for any interval $[a, b] = \{r : a \leq r \leq b\}$ with $0 \leq a \leq b < 1$ we have that the probability of the event "the pointer lands in the interval $[a, b]$" is

$$P([a, b]) = b - a. \tag{2.1}$$

We do not have to restrict interest to intervals in order to define probabilities consistent with (2.1). The notion of the length of an interval can be made precise using calculus and simultaneously extended to any subset of $[0, 1)$ by defining the probability $P(F)$ of a set $F \subset [0, 1)$ as

$$P(F) \triangleq \int_F f(r)\, dr, \tag{2.2}$$

where $f(r) = 1$ for all $r \in [0, 1)$. With this definition it is clear that for any $0 \leq a < b \leq 1$ that

$$P([a, b]) = \int_a^b f(r)\, dr = b - a. \tag{2.3}$$

We could also arrive at effectively the same model by considering the sample space to be the entire real line, $\Omega = \Re \triangleq (-\infty, \infty)$ and defining the probability density function (pdf) to be

$$f(r) = \begin{cases} 1 & \text{if } r \in [0, 1) \\ 0 & \text{otherwise} \end{cases}. \tag{2.4}$$

The integral can also be expressed without specifying limits of integration by using the indicator function of a set

$$1_F(r) \triangleq \begin{cases} 1 & \text{if } r \in F \\ 0 & \text{if } r \notin F \end{cases} \tag{2.5}$$

as

$$P(F) \triangleq \int 1_F(r) f(r)\, dr. \tag{2.6}$$

Other implicit assumptions have been made here. The first is that probabilities must satisfy some *consistency* properties. We cannot arbitrarily define probabilities of distinct subsets of $[0, 1)$ (or, more generally, $\Re$) without regard to the implications of probabilities for other sets; the probabilities must be consistent with each other in the sense that they do not contradict each other. For example, if we have two formulas for computing probabilities of a common event, as we have with (2.1) and (2.2) for computing the probability of an interval, then both formulas must give the same numerical result – as they do in this example.

The second implicit assumption is that the integral exists in a well-defined sense, that it can be evaluated using calculus. As surprising as it may seem to readers familiar only with typical engineering-oriented developments of Riemann integration, the integral of (2.2) is in fact not well defined for all subsets of $[0, 1)$. But we leave this detail for later and assume for the moment that we only encounter sets for which the integral (and hence the probability) is well defined.

The function $f(r)$ is called a *probability density function* or *pdf* since it is a nonnegative point function that is integrated to compute total probability of a set, just as a mass density function is integrated over a region to compute the mass of a region in physics. Since in this example $f(r)$ is constant over a region, it is called a *uniform pdf*.

The formula (2.2) for computing probability has many implications, three of which merit comment at this point.

- Probabilities are nonnegative:

$$P(F) \geq 0 \text{ for any } F. \tag{2.7}$$

This follows since integrating a nonnegative argument yields a nonnegative result.

- The probability of the entire sample space is 1:

$$P(\Omega) = 1. \tag{2.8}$$

This follows since integrating 1 over the unit interval yields 1, but it has the intuitive interpretation that the probability that "something happens" is 1.

- The probability of the union of *disjoint* or *mutually exclusive* regions is the sum of the probabilities of the individual events:

$$\text{If } F \cap G = \emptyset, \text{ then } P(F \cup G) = P(F) + P(G). \tag{2.9}$$

This follows immediately from the properties of integration:

$$\begin{aligned} P(F \cup G) &= \int_{F \cup G} f(r)\,dr = \int_F f(r)\,dr + \int_G f(r)\,dr \\ &= P(F) + P(G). \end{aligned}$$

An alternative proof of the third property follows by observing that since F and G are disjoint, $1_{F \cup G}(r) = 1_F(r) + 1_G(r)$ and hence linearity of integration implies that

$$\begin{aligned} P(F \cup G) &= \int 1_{F \cup G}(r) f(r)\,dr = \int (1_F(r) + 1_G(r)) f(r)\,dr \\ &= \int 1_F(r) f(r)\,dr + \int 1_G(r) f(r)\,dr \\ &= P(F) + P(G). \end{aligned}$$

This property is often called the *additivity* property of probability. The second proof makes it clear that additivity of probability is an immediate result of the linearity of integration, i.e., that the integral of the sum of two functions is the sum of the two integrals.

Repeated application of additivity for two events shows that for any finite collection $\{F_k;\ k = 1, 2, \ldots, K\}$ of disjoint events, i.e., events with the property that $F_k \bigcap F_j = \emptyset$ for all $k \neq j$, we have that

$$P\left(\bigcup_{k=1}^{K} F_k\right) = \sum_{k=1}^{K} P(F_k), \tag{2.10}$$

showing that additivity is equivalent to *finite additivity*, the extension of the additivity property from two to a finite collection of sets. Since additivity is a special case of finite additivity and it implies finite additivity, the two notions are equivalent and we can use them interchangeably.

These three properties of nonnegativity, normalization, and additivity are fundamental to the definition of the general notion of probability and will form three of the four axioms needed for a precise development. It is tempting to call an assignment P of numbers to subsets of a sample space a *probability measure* if it satisfies these three properties, but we shall see that a fourth condition, which is crucial for having well-behaved limits and asymptotics, will be needed to complete the definition. Pending this fourth condition, (2.2) defines a probability measure. In fact, this definition is complete in the simple case where the sample space Ω has only a finite number of points since in that case limits and asympotics become trivial. A sample space together with a probability measure provide a mathematical model for an experiment. This model is often called a *probability space*, but for the moment we shall stick to the less intimidating word of *experiment*.

Simple properties

Several simple properties of probabilities can be derived from what we have so far. As particularly simple, but still important, examples, consider the following.

Assume that P is a set function defined on a sample space Ω that satisfies the properties of equations (2.7)–(2.9). Then

(a) $P(F^c) = 1 - P(F)$.
(b) $P(F) \leq 1$.
(c) Let $\emptyset$ be the null or empty set, then $P(\emptyset) = 0$.
(d) If $\{F_i;\ i = 1, 2, \ldots, K\}$ is a finite partition of Ω, i.e., if $F_i \cap F_k = \emptyset$

when $i \neq k$ and $\bigcup_{i=1} F_i = \Omega$, then

$$P(G) = \sum_{i=1}^{K} P(G \cap F_i) \tag{2.11}$$

for any event G.

Proof

(a) $F \cup F^c = \Omega$ implies $P(F \cup F^c) = 1$ (Property (2.8)). $F \cap F^c = \emptyset$ implies $1 = P(F \cup F^c) = P(F) + P(F^c)$ (Property (2.9)).
(b) $P(F) = 1 - P(F^c) \leq 1$ (Property (2.7) and (a) above).
(c) By Property (2.8) and (a) above, $P(\Omega^c) = P(\emptyset) = 1 - P(\Omega) = 0$.
(d) $P(G) = P(G \cap \Omega) = P(G \cap (\bigcup_i F_i)) = P(\bigcup_i (G \cap F_i)) = \sum_i P(G \cap F_i)$. □

Observe that although the null or empty set $\emptyset$ has probability 0, the converse is not true in that a set need not be empty just because it has zero probability. In the uniform fair wheel example the set $F = \{1/n : n = 1, 2, 3, \ldots\}$ is not empty, but it does have probability zero. This follows roughly because for any finite N $P(\{1/n : n = 2, 3, \ldots, N\}) = 0$ (since the integral of 1 over a finite set of points is zero) and therefore the limit as $N \to \infty$ must also be zero, a "continuity of probability" idea that we shall later make rigorous.

A single coin flip

The original example of a spinning wheel is continuous in that the sample space consists of a continuum of possible outcomes, all points in the unit interval. Sample spaces can also be discrete, as is the case of modeling a single flip of a "fair" coin with heads labeled "1" and tails labeled "0", i.e., heads and tails are equally likely. The sample space in this example is $\Omega = \{0, 1\}$ and the probability for any event or subset of Ω can be defined in a reasonable way by

$$P(F) = \sum_{r \in F} p(r), \tag{2.12}$$

or, equivalently,

$$P(F) = \sum 1_F(r) p(r), \tag{2.13}$$

where now $p(r) = 1/2$ for each $r \in \Omega$. The function p is called a *probability mass function* or *pmf* because it is summed over points to find total probability, just as point masses are summed to find total mass in physics. Be cautioned that P is defined for sets and p is defined only for points in the sample space. This can be confusing when dealing with one-point or singleton sets, for example

$$P(\{0\}) = p(0)$$
$$P(\{1\}) = p(1).$$

This may seem too much work for such a little example, but keep in mind that the goal is a formulation that will work for far more complicated and interesting examples. This example is different from the spinning wheel in that the sample space is discrete instead of continuous and that the probabilities of events are defined by sums instead of integrals, as one should expect when doing discrete mathematics. It is easy to verify, however, that the basic properties (2.7)–(2.9) hold in this case as well (since sums behave like integrals), which in turn implies that the simple properties (a)–(d) also hold.

A single coin flip as signal processing

The coin flip example can also be derived in a very different way that provides our first example of signal processing. Consider again the spinning pointer so that the sample space is Ω and the probability measure P is described by (2.2) using a uniform pdf as in (2.4). Performing the experiment by spinning the pointer will yield some real number $r \in [0, 1)$. Define a measurement q made on this outcome by

$$q(r) = \begin{cases} 1 & \text{if } r \in [0, 0.5] \\ 0 & \text{if } r \in (0.5, 1) \end{cases}. \tag{2.14}$$

This function can also be defined somewhat more economically in terms of an indicator function as

$$q(r) = 1_{[0,0.5]}(r). \tag{2.15}$$

This is an example of a *quantizer*, an operation that maps a continuous value into a discrete value. Quantization is an example of *signal processing* since it is a function or mapping defined on an input space, here $\Omega = [0, 1)$ or $\Omega = \Re$, producing a value in some output space. In this example $\Omega_q = \{0, 1\}$. The dependence of a function on its input space or *domain of definition* Ω and its output space or *range* Ω_q,is often denoted by $q : \Omega \to \Omega_q$. Although introduced as an example of simple signal processing, the usual name for a real-valued function defined on the sample space of a probability space is a *random variable*. We shall see in the next chapter that there is an extra technical condition on functions to merit this name, but that is a detail that can be postponed.

The output space Ω_q can be considered as a new sample space, the space corresponding to the possible values seen by an observer of the output of the quantizer (an observer who might not have access to the original space). If we know both the probability measure on the input space and the function, then in theory we should be able to describe the probability measure that the output space inherits from the input space. Since the output space is discrete, it should be described by a pmf, say p_q. Since there are only two points, we need only find the value of $p_q(1)$ (or $p_q(0)$ since $p_q(0) + p_q(1) = 1$). An output of 1 is seen if and only if the input sample point lies in $[0, 0.5]$, so it follows easily that $p_q(1) = P([0, 0.5]) = \int_0^{0.5} f(r)\, dr = 0.5$, exactly the value assumed for the fair coin flip model. The pmf p_q implies a probability measure P_q on the output space Ω_q defined by

$$P_q(F) = \sum_{\omega \in F} p_q(\omega),$$

where the subscript q distinguishes the probability measure P_q on the output space from the probability measure P on the input space. Note that we can define any other binary quantizer corresponding to an "unfair" or biased coin by changing the 0.5 to some other value.

This simple example makes several fundamental points that will evolve in depth in the course of this material. First, it provides an example of *signal processing* and the first example of a *random variable*, which is essentially just a mapping of one sample space into another. Second, it provides an example of a *derived distribution*: given a probability space described by Ω and P and a function (random variable) q defined on this space, we have derived a new probability space describing the outputs of the function with sample space Ω_q and probability measure P_q. Third, it is an example of a common phenomenon that quite different models can result in identical sample spaces and probability measures. Here the coin flip could be modeled in a *directly given* fashion by just describing the sample space and the probability measure, or it could be modeled in an indirect fashion as a function (signal processing, random variable) on another experiment. This suggests, for example, that to study coin flips empirically we could either actually flip a fair coin, or we could spin a fair wheel and quantize the output. Although the second method seems more complicated, it is in fact extremely common since most random number generators (or pseudo-random number generators) strive to produce random numbers with a uniform distribution on $[0, 1)$ and all other probability measures are produced by further signal processing. We have seen how to do this for a simple coin flip. In fact any pdf or pmf can be generated in this way. (See Problem 3.7.) The generation of uniform random numbers is both a science and an art. Most function roughly as follows. One begins with a floating point number in $(0, 1)$ called the *seed*, say a, and uses another positive floating point number, say b, as a multiplier. A sequence x_n is then generated recursively as $x_0 = a$ and $x_n = b \times x_{n-1} \mod (1)$ for $n = 1, 2, \ldots$, that is, the fractional part of $b \times x_{n-1}$. If the two numbers a and b are suitably chosen then x_n should appear to be uniform. (Try it!) In fact, since there are only a finite number (albeit large) of possible numbers that can be represented on a digital computer, this algorithm must eventually repeat and hence x_n must be a periodic sequence. As a result such a sequence of numbers is a *pseudo-random* sequence and not a genuine sequence of random numbers. The goal of designing a good pseudo-random number generater is to make the period as long as possible and to make the sequences produced look as much as possible like a random sequence in the sense that statistical tests for independence are fooled. If one wanted to generate a truly random generator, one might use some natural phenomenon such as thermal noise, treated near the end of the book – measure the voltage across a heated resistor and let the random action of molecules in motion produce a random measurement.

Abstract versus concrete

It may seem strange that the axioms of probability deal with apparently abstract ideas of measures instead of corresponding to physical intuition. Physical intuition says that the probability tells you something about the fraction of times specific events will occur in a sequence of trials, such as the relative frequency of a pair of dice summing to seven in a sequence of many rolls, or a decision algorithm correctly detecting a single binary symbol in the presence of noise in a transmitted data file. Such real-world behavior can be quantified by the idea of a relative frequency; that is, suppose the output of the nth trial of a sequence of trials is x_n and we wish to know the relative frequency that x_n takes on a particular value, say a. Then given an infinite sequence of trials $x = \{x_0, x_1, x_2, \ldots\}$ we could define the relative frequency of a in x by

$$r_a(x) = \lim_{n\to\infty} \frac{\text{number of } k \in \{0, 1, \ldots, n-1\} \text{ for which } x_k = a}{n}. \tag{2.16}$$

For example, the relative frequency of heads in an infinite sequence of fair coin flips should be 0.5, and the relative frequency of rolling a pair of fair dice and having the sum be 7 in an infinite sequence of rolls should be $1/6$ since the pairs $(1, 6), (6, 1), (2, 5), (5, 2), (3, 4), (4, 3)$ are equally likely and form 6 of the possible 36 pairs of outcomes. Thus one might suspect that to make a rigorous theory of probability requires only a rigorous definition of probabilities as such limits and a reaping of the resulting benefits. In fact much of the history of theoretical probability consisted of attempts to accomplish this, but unfortunately it does not work. Such limits might not exist, or they might exist and not converge to the same thing for different repetitions of the same experiment. Even when the limits do exist there is no guarantee they will behave as intuition would suggest when one tries to do calculus with probabilities, that is, to compute probabilities of complicated events from those of simple related events. Attempts to get around these problems uniformly failed and probability was not put on a rigorous basis until the axiomatic approach was completed by Kolmogorov. (A discussion of some of the contributions of Kolmogorov may be found in the Kolmogorov memorial issue of the *Annals of Probability*, **17**, 1989. His contributions to information theory, a shared interest area of the authors, are described in [11].) The axioms do, however, capture certain intuitive aspects of relative frequencies. Relative frequencies are nonnegative, the relative frequency of the entire set of possible outcomes is one, and relative frequencies are additive in the sense that the relative frequency of the symbol a or the symbol b occurring, $r_{a\cup b}(x)$, is clearly $r_a(x) + r_b(x)$. Kolmogorov realized that beginning with simple axioms could lead to rigorous limiting results of the type needed, whereas there was no way to begin with the limiting results as part of the axioms. In fact it is the fourth axiom, a limiting version of additivity, that plays the key role in making the asymptotics work.

2.3 Probability spaces

We now turn to a more thorough development of the ideas introduced in the previous section.

A *sample space* Ω is an abstract space, a nonempty collection of points or members or elements called *sample points* (or *elementary events* or *elementary outcomes*).

An *event space* (or *sigma-field* or *sigma-algebra*) $\mathcal{F}$ of a sample space Ω is a nonempty collection of subsets of Ω called *events* with the following three properties:

$$\text{If } F \in \mathcal{F}, \text{ then also } F^c \in \mathcal{F}, \tag{2.17}$$

that is, if a given set is an event, then its complement must also be an event. Note that any particular subset of Ω may or may not be an event (review the quantizer example).

$$\text{If for some finite } n, F_i \in \mathcal{F}, \ i = 1, 2, \ldots, n, \text{ then also}$$

$$\bigcup_{i=1}^{n} F_i \in \mathcal{F}, \tag{2.18}$$

that is, a finite union of events must also be an event.

$$\text{If } F_i \in \mathcal{F}, \ i = 1, 2, \ldots, \text{ then also}$$

$$\bigcup_{i=1}^{\infty} F_i \in \mathcal{F}, \tag{2.19}$$

that is, a *countable* union of events must also be an event.

We shall later see alternative ways of describing (2.19), but this form is the most common.

Equation (2.18) can be considered as a special case of (2.19) since, for example, given a finite collection F_i, $i = 1, \ldots, N$, we can construct an infinite sequence of sets with the same union. For example, given F_k, $k = 1, \ldots, N$, construct an infinite sequence G_n with the same union by choosing $G_n = F_n$ for $n = 1, \ldots, N$ and $G_n = \emptyset$ otherwise. It is convenient, however, to consider the finite case separately. If a collection of sets satisfies only (2.17) and (2.18) but not (2.19), then it is called a *field* or *algebra* of sets. For this reason, in elementary probability theory one often refers to "set algebra" or to the "algebra of events." (Don't worry about why (2.19) might not be satisfied.) Both (2.17) and (2.18) can be considered as "closure" properties; that is, an event space must be closed under complementation and unions in the sense that performing a sequence of complementations or unions of events must yield a set that is also in the collection, i.e., a set that is also an event. Observe also that (2.17), (2.18), and (A.11) imply that

$$\Omega \in \mathcal{F}, \tag{2.20}$$

that is, the whole sample space considered as a set must be in $\mathcal{F}$; that is, it must be an event. Intuitively, Ω is the "certain event," the event that "something happens." Similarly, (2.20) and (2.17) imply that

$$\emptyset \in \mathcal{F}, \tag{2.21}$$

and hence the empty set must be in $\mathcal{F}$, corresponding to the intuitive event "nothing happens."

A few words about the different nature of membership in Ω and $\mathcal{F}$ is in order. If the set F is a subset of Ω, then we write $F \subset \Omega$. If the subset F is also in the event space, then we write $F \in \mathcal{F}$. Thus we use set inclusion when considering F as a subset of an abstract space, and element inclusion when considering F as a member of the event space and hence as an event. Alternatively, the elements of Ω are points, and a collection of these points is a subset of Ω; but the elements of $\mathcal{F}$ are sets – subsets of Ω – and not points. A student should ponder the different natures of abstract spaces of points and event spaces consisting of sets until the reasons for set inclusion in the former and element inclusion in the latter space are clear. Consider especially the difference between an element of Ω and a subset of Ω that consists of a single point. The latter *might* or *might not* be an element of $\mathcal{F}$, the former is *never* an element of $\mathcal{F}$. Although the difference might seem to be merely semantics, the difference is important and should be thoroughly understood.

A *measurable space* $(\Omega, \mathcal{F})$ is a pair consisting of a sample space Ω and an event space or sigma-field $\mathcal{F}$ of subsets of Ω. The strange name "measurable space" reflects the fact that we can assign a measure such as a probability measure to such a space and thereby form a probability space or probability measure space.

A *probability measure* P on a measurable space $(\Omega, \mathcal{F})$ is an assignment of a real number $P(F)$ to every member F of the sigma-field (that is, to every event) such that P obeys the following rules, which we refer to as the *axioms of probability.*

Axiom 2.1

$$P(F) \geq 0 \text{ for all } F \in \mathcal{F} \tag{2.22}$$

i.e., no event has negative probability.

Axiom 2.2

$$P(\Omega) = 1 \tag{2.23}$$

i.e., the probability of "everything" is one.

Axiom 2.3 *If F_i, $i = 1, \ldots, n$ are disjoint, then*

$$P\left(\bigcup_{i=1}^{n} F_i\right) = \sum_{i=1}^{n} P(F_i). \tag{2.24}$$

Axiom 2.4 *If F_i, $i = 1, \ldots$ are disjoint, then*

$$P\left(\bigcup_{i=1}^{\infty} F_i\right) = \sum_{i=1}^{\infty} P(F_i). \tag{2.25}$$

Note that nothing has been said to the effect that probabilities must be sums or integrals, but the first three axioms should be recognizable from the three basic properties of non-negativity, normalization, and additivity encountered in the simple examples given in the introduction to this chapter where probabilities were defined by an integral of a pdf over a set or a sum of a pmf over a set. The axioms capture these properties in a general form and will be seen to include more general constructions, including multidimensional integrals and combinations of integrals and sums. The fourth axiom can be viewed as an extra technical condition that must be included in order to get various limits to behave. Just as Property (2.19) of an event space will later be seen to have an alternative statement in terms of limits of sets, the fourth axiom of probability, Axiom 2.4, will be shown to have an alternative form in terms of explicit limits, a form providing an important continuity property of probability. Also as in the event space properties, the fourth axiom implies the third.

As with the defining properties of an event space, for the purposes of discussion we have listed separately the finite special case (2.24) of the general condition (2.25). The finite special case is all that is required for elementary discrete probability. The general condition is required to get a useful theory for continuous probability. A good way to think of these conditions is that they essentially describe probability measures as set functions defined by either summing or integrating over sets, or by some combination thereof. Hence much of probability theory is simply calculus, especially the evaluation of sums and integrals.

To emphasize an important point: a function P which assigns numbers to elements of an event space of a sample space is a probability measure *if and only if* it satisfies all of the four axioms!

A *probability space* or *experiment* is a triple $(\Omega, \mathcal{F}, P)$ consisting of a sample space Ω, an event space $\mathcal{F}$ of subsets of Ω, and a probability measure P defined for all members of $\mathcal{F}$.

Before developing each idea in more detail and providing several examples of each piece of a probability space, we pause to consider two simple examples of the complete construction. The first example is the simplest possible probability space and is commonly referred to as the *trivial probability space*. Although useless for application, the model does serve a purpose by showing that a well-defined model need not be interesting. The second example is essentially the simplest nontrivial probability space, a slight generalization of the fair coin flip permitting an unfair coin.

Examples

[2.0] Let Ω be any abstract space and let $\mathcal{F} = \{\Omega, \emptyset\}$; that is, $\mathcal{F}$ consists of exactly two sets – the sample space (everything) and the empty set (nothing). This is called the trivial event space. This is a model of an experiment where only two events are possible: "something happens" or "nothing happens" – not a very interesting description. There is only one possible probability measure for this measurable space: $P(\Omega) = 1$ and $P(\emptyset) = 0$. (*Why?*) This probability measure meets the required rules that define a probability measure; they can be directly verified since there are only two possible events. Equations (2.22) and (2.23) are obvious. Equations (2.24) and (2.25) follow since the only possible values for F_i are Ω and $\emptyset$. At most one of the F_i can be Ω. If one of the F_i is Ω, then both sides of the equality are 1. Otherwise, both sides are 0.

[2.1] Let $\Omega = \{0, 1\}$. Let $\mathcal{F} = \{\{0\}, \{1\}, \Omega = \{0, 1\}, \emptyset\}$. Since $\mathcal{F}$ contains *all* of the subsets of Ω, the properties (2.17) through (2.19) are trivially satisfied, and hence it is an event space. (There is one other possible event space that could be defined for Ω in this example. *What is it?*) Define the set function P by

$$P(F) = \begin{cases} 1 - p & \text{if } F = \{0\} \\ p & \text{if } F = \{1\} \\ 0 & \text{if } F = \emptyset \\ 1 & \text{if } F = \Omega \end{cases},$$

where $p \in [0, 1]$ is a fixed parameter. (If $p = 0$ or $p = 1$ the space becomes trivial.) It is easily verified that P satisfies the axioms of probability and hence is a probability measure. Therefore $(\Omega, \mathcal{F}, P)$ is a probability space. Note that we had to give the value of $P(F)$ for *all* events F, a construction that would clearly be absurd for large sample spaces. Note also that the choice of $P(F)$ is not unique for the given measurable space $(\Omega, \mathcal{F})$; we could have chosen any value in $[0, 1]$ for $P(\{1\})$ and used the axioms to complete the definition.

The preceding example is the simplest nontrivial example of a probability space and provides a rigorous mathematical model for applications such as the binary transmission of a single bit or for the flipping of a single biased coin once. It therefore provides a complete and rigorous mathematical model for the single coin flip of the introduction.

We now develop in more detail properties and examples of the three components of probability spaces: sample spaces, event spaces, and probability measures.

2.3.1 Sample spaces

Intuitively, a sample space is a listing of all conceivable finest-grain, distinguishable outcomes of an experiment to be modeled by a probability space. Mathematically it is just an abstract space.

Examples

[2.2] A finite space $\Omega = \{a_k;\ k = 1, \ldots, K\}$. Specific examples are the binary space $\{0, 1\}$ and the finite space of integers $\mathcal{Z}_k \triangleq \{0, 1, \ldots, k-1\}$.

[2.3] A countably infinite space $\Omega = \{a_k;\ k = 0, 1, \ldots\}$, for some sequence $\{a_k\}$. Specific examples are the space of all nonnegative integers $\{0, 1, \ldots\}$, which we denote by $\mathcal{Z}_+$, and the space of all integers $\{\ldots, -2, -1, 0, 1, 2, \ldots\}$, which we denote by $\mathcal{Z}$. Other examples are the space of all rational numbers, the space of all even integers, and the space of all periodic sequences of integers.

Both Examples [2.2] and [2.3] are called *discrete* spaces. Spaces with finite or countably infinite numbers of elements are called discrete spaces.

[2.4] An interval of the real line $\Re$, for example, $\Omega = (a, b)$. We might consider an open interval (a, b), a closed interval $[a, b]$, a half-open interval $[a, b)$ or $(a, b]$, or even the entire real line $\Re$ itself. (See Appendix A for details on these different types of intervals.)

Spaces such as Example [2.4] that are not discrete are said to be *continuous*. In some cases it is more accurate to think of spaces as being a mixture of discrete and continuous parts, e.g., the space $\Omega = (1, 2) \cup \{4\}$ consisting of a continuous interval and an isolated point. Such spaces can usually be handled by treating the discrete and continuous components separately.

[2.5] A space consisting of k-dimensional vectors with coordinates taking values in one of the previously described spaces. A useful notation for such vector spaces is a *product space*. Let A denote one of the abstract spaces previously considered. Define the Cartesian product A^k by

$$A^k = \{ \text{all vectors } \mathbf{a} = (a_0, a_1, \ldots, a_{k-1}) \text{ with } a_i \in A \}.$$

Thus, for example, $\Re^k$ is k-dimensional Euclidean space. $\{0,1\}^k$ is the space of all binary k-tuples, that is, the space of all k-dimensional binary vectors. As particular examples, $\{0,1\}^2 = \{00, 01, 10, 11\}$ and $\{0,1\}^3 = \{000, 001, 010, 011, 100, 101, 110, 111\}$, $[0,1]^2$ is the unit square in the plane, and $[0,1]^3$ is the unit cube in three-dimensional Euclidean space.

Alternative notations for a Cartesian product space are

$$\prod_{i \in \mathcal{Z}_k} A_i = \prod_{i=0}^{k-1} A_i,$$

where again the A_i are all replicas or copies of A, that is, where $A_i = A$, all i. Other notations for such a finite-dimensional Cartesian product are

$$\times_{i \in \mathcal{Z}_k} A_i = \times_{i=0}^{k-1} A_i = A^k.$$

This and other product spaces will prove to be useful ways of describing abstract spaces which model sequences of elements from another abstract space.

Observe that a finite-dimensional vector space constructed from a discrete space is also discrete since if one can count the number of possible values that one coordinate can assume, then one can count the number of possible values that a finite number of coordinates can assume.

[2.6] A space consisting of infinite sequences drawn from one of the Examples [2.2] through [2.4]. Points in this space are often called *discrete time signals*. This is also a product space. Let A be a sample space and let A_i be replicas or copies of A. We will consider both one-sided and two-sided infinite products to model sequences with and without a finite origin, respectively. Define the two-sided space

$$\prod_{i \in \mathcal{Z}} A_i = \{ \text{all sequences } \{a_i;\, i = \ldots, -1, 0, 1, \ldots\};\, a_i \in A_i \},$$

and the one-sided space

$$\prod_{i \in \mathcal{Z}_+} A_i = \{ \text{all sequences } \{a_i;\, i = 0, 1, \ldots\};\, a_i \in A_i \}.$$

These two spaces are also denoted by $\prod_{i=-\infty}^{\infty} A_i$ or $\times_{i=-\infty}^{\infty} A_i$ and $\prod_{i=0}^{\infty} A_i$ or $\times_{i=0}^{\infty} A_i$, respectively.

The two spaces under discussion are often called *sequence spaces.* Even if the original space A is discrete, the sequence space constructed from A will be continuous. For example, suppose that $A_i = \{0,1,2,3,4,5,6,7,8,9\}$ for all integers i. Then $\times_{i=0}^{\infty} A_i$ is the space of all semi-infinite (one-sided) decimal sequences, which is equivalent to the space of all real numbers in the unit interval $[0,1)$. This follows since if $\omega \in \Omega$, then $\omega = (\omega_0, \omega_1, \omega_2, \ldots)$, which can be written as $0.\omega_0\omega_1\omega_2\ldots$, and can represent any real number in the unit interval by the decimal expansion $\sum_{i=0}^{\infty} \omega_i 10^{-i-1}$. This space contains the decimal representations of all of the real numbers in the unit interval, an uncountable infinity of numbers. Similarly, there is an uncountable infinity of one-sided binary sequences because one can express all points in the unit interval in the binary number system as sequences to the right of the "decimal" point (Problem A.11).

[2.7] Let A be one of the sample spaces of Examples [2.2] through [2.4]. Form a new abstract space consisting of all waveforms or functions of time with values in A, for example all real-valued time functions or *continuous time signals.* This space is also modeled as a product space. For example, the infinite two-sided space for a given A is

$$\prod_{t \in \Re} A_t = \{\text{ all waveforms } \{x(t);\, t \in (-\infty,\infty)\};\ x(t) \in A, \text{ all } t\},$$

with a similar definition for one-sided spaces and for time functions on a finite time interval.

Note that we indexed sequences (discrete time signals) using subscripts, as in x_n, and we indexed waveforms (continuous time signals) using parentheses, as in $x(t)$. In fact, the notations are interchangeable; we could denote waveforms as $\{x(t);\, t \in \Re\}$ or as $\{x_t;\, t \in \Re\}$. The notation using subscripts for sequences and parentheses for waveforms is the most common, and we will usually stick to it. Yet another notation for discrete time signals is $x[n]$, a common notation in the digital signal processing literature. It is worth remembering that vectors, sequences, and waveforms are all just indexed collections of numbers; the only difference is the index set: finite for vectors, countably infinite for sequences, and continuous for waveforms.

⋆ *General product spaces*

All of the product spaces we have described can be viewed as special cases of the general product space defined next.

Let $\mathcal{I}$ be an index set such as a finite set of integers $\mathcal{Z}_k$, the set of all integers $\mathcal{Z}$, the set of all nonnegative integers $\mathcal{Z}_+$, the real line $\Re$, or the nonnegative reals $[0, \infty)$. Given a family of spaces $\{A_t;\ t \in \mathcal{I}\}$, define the product space

$$A^{\mathcal{I}} = \prod_{t \in \mathcal{I}} A_t = \{\,\text{all }\ \{a_t;\ t \in \mathcal{I}\};\ a_t \in A_t,\ \text{ all } t\}.$$

The notation $\times_{t \in \mathcal{I}} A_t$ is also used for the same thing. Thus product spaces model spaces of vectors, sequences, and waveforms whose coordinate values are drawn from some fixed space. This leads to two notations for the space of all k-dimensional vectors with coordinates in A : A^k and $A^{\mathcal{Z}_k}$. This shorter and simpler notation is usually more convenient.

2.3.2 Event spaces

Intuitively, an event space is a collection of subsets of the sample space or groupings of elementary events which we shall consider as physical events and to which we wish to assign probabilities. Mathematically, an event space is a collection of subsets that is closed under certain set-theoretic operations; that is, performing certain operations on events or members of the event space must give other events. Thus, for example, if in the example of a single voltage measurement we have $\Omega = \Re$ and we are told that the set of all voltages greater than 5 volts, $\{\omega : \omega \geq 5\}$, is an event, that is, it is a member of a sigma-field $\mathcal{F}$ of subsets of $\Re$, then necessarily its complement $\{\omega : \omega < 5\}$ must also be an event, that is, a member of the sigma-field $\mathcal{F}$. If the latter set is not in $\mathcal{F}$ then $\mathcal{F}$ cannot be an event space! Observe that no problem arises if the complement physically cannot happen – events that "cannot occur" can be included in $\mathcal{F}$ and then assigned probability zero when choosing the probability measure P. For example, even if you know that the voltage does not exceed 5 volts, if you have chosen the real line $\Re$ as your sample space, then you must include the set $\{r : r > 5\}$ in the event space if the set $\{r : r \leq 5\}$ is an event. The impossibility of a voltage greater than 5 is then expressed by assigning $P(\{r : r > 5\}) = 0$.

Although the definition of a sigma-field requires only that the class be closed under complementation and countable unions, these requirements immediately yield additional closure properties. The countably infinite version of DeMorgan's "laws" of elementary set theory (see Appendix A) requires that if F_i, $i = 1, 2, \ldots$, are all members of a sigma-field, then so is

$$\bigcap_{i=1}^{\infty} F_i = \left(\bigcup_{i=1}^{\infty} F_i^c \right)^c .$$

It follows by similar set-theoretic arguments that any countable sequence of any of the set-theoretic operations (union, intersection, complementation, difference, symmetric difference) performed on events must yield other events. Observe, however, that there is no guarantee that *uncountable* operations on events will produce new events; they may or may not. For example, if we are told that $\{F_r;\, r \in [0,1]\}$ is a family of events, then it is not necessarily true that $\bigcup_{r\in[0,1]} F_r$, is an event (see Problem 2.3 for an example).

The requirement that a finite sequence of set-theoretic operations on events yields other events is an intuitive necessity and is easy to verify for a given collection of subsets of an abstract space: It is intuitively necessary that logical combinations (*and* and *or* and *not*) of events corresponding to physical phenomena should also be events to which a probability can be assigned. If you know the probability of a voltage being greater than zero and you know the probability that the voltage is not greater than 5 volts, then you should also be able to determine the probability that the voltage is greater than zero but not greater than 5 volts. It is easy to verify that finite sequences of set-theoretic combinations yield events because the finiteness of elementary set theory usually yields simple proofs.

A natural question arises in regard to (2.17) and (2.18): why not try to construct a useful probability theory on the more general notion of a field rather than a sigma-field? The response is that it unfortunately does not work. Probability theory requires many results involving limits, and such asymptotic results require the infinite relations of (2.19) and (2.25). In some special cases, such as single coin flipping or single die rolling, the simpler finite results suffice because there are only a finite number of possible outcomes, and hence limiting results become trivial – any finite field is automatically a sigma-field. If, however, one flips a coin forever, then there is an uncountable infinity of possible outcomes, and the asymptotic relations become necessary. Let Ω be the space of all one-sided binary sequences. Suppose that you consider the smallest field formed by all finite set-theoretic operations on the individual one-sided binary sequences, that is, on singleton sets in the sequence space. Then many countably infinite sets of binary sequences (say the set of all periodic sequences) are not events since they cannot be expressed as finite sequences of set-theoretic operations on the singleton sets. Obviously, the sigma-field formed by including countable set-theoretic operations does not have this defect. This is why sigma-fields must be used rather than fields.

Limits of sets

The condition (2.19) can be related to a condition on limits by defining the notion of a limit of a sequence of sets. This notion will prove useful when interpreting the axioms of probability. Consider a sequence $F_n, n = 1, 2, \ldots$, of sets with the property that each set contains its predecessor, that is, that $F_{n-1} \subset F_n$ for all n. Such a sequence of sets is said to be *nested* and *increasing*. For example, the sequence $F_n = [1, 2 - 1/n)$ of subsets of the real line is increasing. The sequence $(-n, a)$ is also increasing. Intuitively, the first example increases to a limit of $[1, 2)$ in the sense that every point in the set $[1, 2)$ is eventually included in one of the F_n. Similarly, the sequence in the second example increases to $(-\infty, a)$. Formally, the limit of an increasing sequence of sets can be defined as the union of all of the sets in the sequence since the union contains all of the points in all of the sets in the sequence and does not contain any points not contained in at least one set (and hence an infinite number of sets) in the sequence:

$$\lim_{n\to\infty} F_n = \bigcup_{n=1}^{\infty} F_n.$$

Figure 2.2(a) illustrates such a sequence in a Venn diagram. Thus the limit

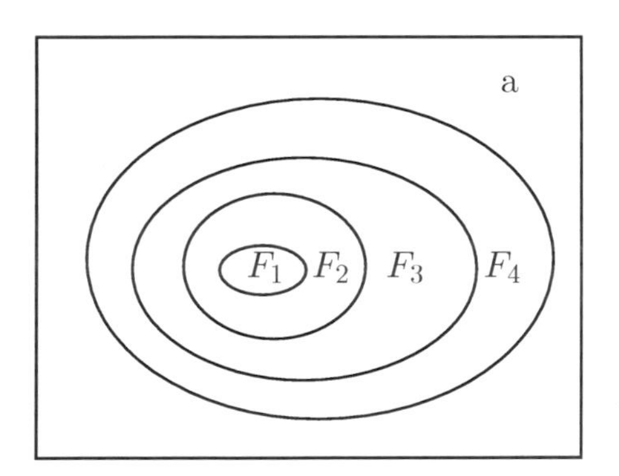

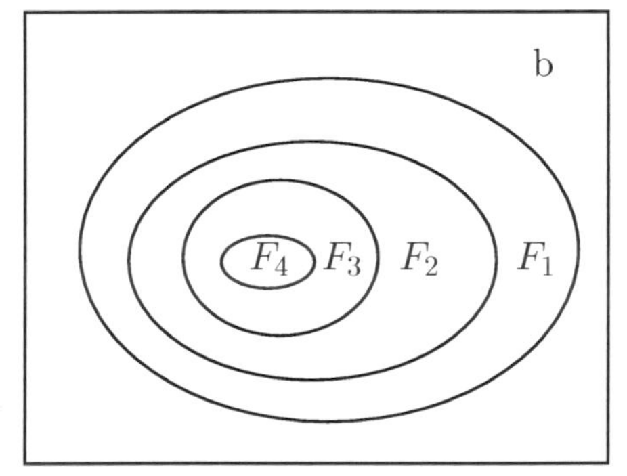

Figure 2.2 (a) Increasing sets, (b) decreasing sets

of the sequence of sets $[1, 2 - 1/n)$ is indeed the set $[1, 2)$, as desired, and the limit of $(-n, a)$ is (∞, a). If F is the limit of a sequence of increasing sets F_n, then we write $F_n \uparrow F$.

Similarly, suppose that F_n, $n = 1, 2, \ldots$ is a *decreasing* sequence of nested sets in the sense that $F_n \subset F_{n-1}$ for all n as illustrated by the Venn diagram in Figure 2.2(b). For example, the sequences of sets $[1, 1 + 1/n)$ and

$(1 - 1/n, 1 + 1/n)$ are decreasing. Again we have a natural notion of the limit of this sequence: both these sequences of sets collapse to the point of singleton set $\{1\}$ – the point in common to all the sets. This suggests a formal definition based on the countably infinite intersection of the sets.

Given a decreasing sequence of sets F_n, $n = 1, 2, \ldots$, we define the limit of the sequence by

$$\lim_{n \to \infty} F_n = \bigcap_{n=1}^{\infty} F_n,$$

that is, a point is in the limit of a decreasing sequence of sets *if and only if* it is contained in all the sets of the sequence. If F is the limit of a sequence of decreasing sets F_n, then we write $F_n \downarrow F$.

Thus, given a sequence of increasing or decreasing sets, the limit of the sequence can be defined in a natural way: the union of the sets of the sequence or the intersection of the sets of the sequence, respectively.

Say that we have a sigma-field $\mathcal{F}$ and an increasing sequence of sets F_n $n = 1, 2, \ldots$, of sets in the sigma-field. Since the limit of the sequence is defined as a union and since the union of a countable number of events must be an event, then the limit must be an event. For example, if we are told that the sets $[1, 2 - 1/n)$ are all events, then the limit $[1, 2)$ must also be an event. If we are told that all finite intervals of the form (a, b), where a and b are finite, are events, then the semi-infinite interval $(-\infty, b)$ must also be an event, since it is the limit of the sequence of sets $(-n, b)$ and $n \to \infty$.

By a similar argument, if we are told that each set in a decreasing sequence F_n is an event, then the limit must be an event, since it is an intersection of a countable number of events. Thus, for example, if we are told that all finite intervals of the form (a, b) are events, then the points of singleton sets must also be events, since a point $\{a\}$ is the limit of the decreasing sequence of sets $(a - 1/n, a + 1/n)$.

If a class of sets is only a field rather than a sigma-field, that is, if it satisfies only (2.17) and (2.18), then there is no guarantee that the class will contain *all* limits of sets. Hence, for example, knowing that a class of sets contains all half-open intervals of the form $(a, b]$ for a and b finite does not ensure that it will also contain points or singleton sets! In fact, it is straightforward to show that the collection of all such half-open intervals together with the complements of such sets and all finite unions of the intervals and complements forms a field. The singleton sets, however, are not in the field! (See Problem 2.6.)

Thus if we tried to construct a probability theory based on only a field, we might have probabilities defined for events such as (a, b) meaning "the output voltage of a measurement is between a and b" and yet not have probabilities defined for a singleton set $\{a\}$ meaning "the output voltage is exactly a." By requiring that the event space be a sigma-field instead of only a field, we are assured that all such limits are indeed events.

It is a straightforward exercise to show that given (2.17) and (2.18), Property (2.19) is equivalent to the following:

If $F_n \in \mathcal{F}$; $n = 1, 2, \ldots$, is a decreasing sequence or an increasing sequence, then

$$\lim_{n\to\infty} F_n \in \mathcal{F}. \tag{2.26}$$

We have already seen that (2.19) implies (2.26). For example, if (2.26) is true and G_n is an arbitrary sequence of events, then define the increasing sequence

$$F_n = \bigcup_{i=1}^{n} G_i.$$

Obviously $F_{n-1} \subset F_n$, and then (2.26) implies (2.19), since

$$\bigcup_{i=1}^{\infty} G_i = \bigcup_{n=1}^{\infty} F_n = \lim_{n\to\infty} F_n \in \mathcal{F}.$$

As we have noted, for a given sample space the selection of an event space is not unique; it depends on the events to which it is desired to assign probabilities and also on analytical limitations on the ability to assign probabilities. We begin with two examples that represent the extremes of event spaces – one possessing the minimum quantity of sets and the other possessing the maximum. We then study event spaces useful for the sample space examples of the preceding section.

Examples

[2.8] Given a sample space Ω, then the collection $\{\Omega, \emptyset\}$ is a sigma-field. This is just the trivial event space already treated in Example [2.0]. Observe again that this is the smallest possible event space for any given sample space because no other event space can have fewer elements.

[2.9] Given a sample space Ω, then the collection of *all subsets of* Ω is a sigma-field. This is true since any countable sequence of set-theoretic operations on subsets of Ω must yield another subset of Ω and hence must be in the collection of *all* possible subsets. The collection of all subsets of a space is called the *power set* of the space. Observe that this is the largest possible event space for the given sample space, because it contains every possible subset of the sample space.

This sigma-field is a useful event space for the sample spaces of Examples [2.2] and [2.3], that is, for sample spaces that are discrete. We shall always take our event space as the power set when dealing with a discrete sample space (except possibly for a few perverse homework problems). A discrete sample space with n elements has a power set with 2^n elements (Problem 2.5). For example, the power set of the binary sample space $\Omega = \{0, 1\}$ is the collection $\{\{0\}, \{1\}, \Omega = \{0, 1\}, \emptyset\}$, a list of all four possible subsets of the space.

Unfortunately, the power set is too large to be useful for continuous spaces. To treat the reasons for this is beyond the scope of a book at this level, but we can say that it is not possible in general to construct interesting probability measures on the power set of a continuous space. There are special cases where we can construct particular probability measures on the power set of a continuous space by mimicking the construction for a discrete space (see, e.g., Problems 2.5, 2.7, and 2.10). Truly continuous experiments cannot, however, be rigorously defined for such a large event space because integrals cannot be defined over all events in such spaces.

Although both of the preceding examples can be used to provide event spaces for the special case of $\Omega = \Re$, the real line, neither leads to a useful probability theory in that case. In the next example we consider another event space for the real line that is more useful and, in fact, is used almost always for $\Re$ and higher-dimensional Euclidean spaces. First, however, we need to treat the idea of *generating* an event space from a collection of important events. Intuitively, given a collection of important sets $\mathcal{G}$ that we require to be events, the event space $\sigma(\mathcal{G})$ *generated* by $\mathcal{G}$ is the smallest event space $\mathcal{F}$ to which all the sets in $\mathcal{G}$ belong. That is, $\sigma(\mathcal{G})$ is an event space, it contains all the sets in $\mathcal{G}$, and no smaller collection of sets satisfies these two conditions.

Regardless of the details, it is worth emphasizing the key points of this discussion.

- The notion of a generated sigma-field allows one to describe an event space for the real line, called the Borel field, that contains all physically important events and that will lead to a useful calculus of probability. It is usually not important to understand the detailed structure of this event space past the facts that it
  - is indeed an event space, and
  - contains all the important events such as *intervals*.
- The notion of a generated sigma-field can be used to extend the event space of

the real line to event spaces of vectors, sequences, and waveforms taking on real values. Again the detailed structure is usually not important past the fact that it
– is indeed an event space, and
– contains all the important events such as those described by requiring any finite collection of coordinate values to lie within intervals.

⋆*Generating event spaces*

Any useful event space for the real line should include as members all intervals of the form (a, b) since we certainly wish to consider events of the form "the output voltage is between 3 and 5 volts." Furthermore, we obviously require it satisfy the defining properties for an event space; that is, require a collection of subsets of Ω that satisfy Properties (2.17) through (2.19). A means of accomplishing both of these goals in a relatively simple fashion is to define our event space as the *smallest* sigma-field that contains the desired subsets, to wit, the intervals and all of their *countable* set-theoretic combinations (bewildering as it may seem, this is not the same as all subsets of $\Re$). Of course, although a sigma-field that is based on the intervals is most useful, it is also possible to consider other starting points. These considerations motivate the following general definition.

Given a sample space Ω (such as the real line $\Re$) and an arbitrary class $\mathcal{G}$ of subsets of Ω – usually the class of all open intervals of the form (a, b) when $\Omega = \Re$ – define $\sigma(\mathcal{G})$, the sigma-field *generated* by the class $\mathcal{G}$, to be the smallest sigma-field containing all of the sets in $\mathcal{G}$, where by "smallest" we mean that if $\mathcal{F}$ is any sigma-field and it contains $\mathcal{G}$, then it contains $\sigma(\mathcal{G})$. (See any book on measure theory, e.g. [1].)

For example, as noted before, we might require that a sigma-field of the real line contain all intervals; then it would also have to contain at least all complements of intervals and all countable unions and intersections of intervals and all countable complements, unions, and intersections of these results, ad infinitum. This technique will be used several times to specify useful event spaces in complicated situations such as continuous simple spaces, sequence spaces, and function spaces. We are now ready to provide the proper, most useful event space for the real line.

[2.10] Given the real line $\Re$, the *Borel field* (or, more accurately, the *Borel sigma-field*) is defined as the sigma-field generated by all the open intervals of the form (a, b). The members of the Borel field are called *Borel sets*. We shall denote the Borel field by $\mathcal{B}(\Re)$, and hence

$$\mathcal{B}(\Re) = \sigma \text{ (all open intervals)}.$$

Since $\mathcal{B}(\Re)$ is a sigma-field and since it contains all of the open intervals, it must also consider limit sets of the form

$$(-\infty, b) = \lim_{n\to\infty} (-n, b)\,,$$
$$(a, \infty) = \lim_{n\to\infty} (a, n)\,,$$

and

$$\{a\} = \lim_{n\to\infty} (a - 1/n,\, a + 1/n),$$

that is, the Borel field must include semi-infinite open intervals and the singleton sets or individual points. Furthermore, since the Borel field is a sigma-field it must contain differences. Hence it must contain semi-infinite half-open sets of the form

$$(-\infty, b] = (-\infty, \infty) - (b, \infty),$$

and since it must contain unions of its members, it must contain half-open intervals of the form

$$(a, b] = (a, b) \cup \{b\} \text{ and } [a, b) = (a, b) \cup \{a\}.$$

In addition, it must contain all closed intervals and all finite or countable unions and complements of intervals of any of the preceding forms. Roughly speaking, the Borel field contains all subsets of the real line that can be constructed by countable sequences of operations on intervals. It is a deep and difficult result of measure theory that the Borel field of the real line is in fact different from the power set of the real line; that is, there exist subsets of the real line that are not in the Borel field. Although we will not describe such a subset, we can guarantee that these "unmeasurable" sets have no physical importance, that they are very hard to construct, and that an engineer will never encounter such a subset in practice. It may, however, be necessary to demonstrate that some weird subset is in fact an event in this sigma-field. This is typically accomplished by showing that it is the limit of simple Borel sets.

In some cases we wish to deal not with a sample space that is the entire real line, but with one that is some subset of the real line. In this case we define the Borel field as the Borel field of the real line "cut down" to the smaller space.

Given that the sample space, Ω, is a Borel subset of the real line $\Re$, the *Borel field* of Ω, denoted $\mathcal{B}(\Omega)$, is defined as the collection of all sets of the form $F \cap \Omega$, for $F \in \mathcal{B}(\Re)$; that is, the intersection of Ω with all of the Borel sets of $\Re$ forms the class of Borel sets of Ω.

It can be shown (Problem 2.4) that, given a discrete subset A of the real line, the Borel field $\mathcal{B}(A)$ is identical to the power set of A. Thus, for the first three examples of sample spaces, the Borel field serves as a useful event space since it reduces to the intuitively appealing class of all subsets of the sample space.

The remaining examples of sample spaces are all product spaces. The construction of event spaces for such product spaces – that is, spaces of vectors, sequences, or waveforms – is more complicated and less intuitive than the constructions for the preceding event spaces. In fact, there are several possible techniques of construction, which in some cases lead to different event spaces. We wish to convey an understanding of the structure of such event spaces, but we do not wish to dwell on the technical difficulties that can be encountered. Hence we shall study only one of the possible constructions – the simplest possible definition of a product sigma-field – by making a direct analogy to a product sample space. This definition will suffice for most systems studied herein, but it has shortcomings. At this time we mention one particular weakness: the event space that we shall define may not be big enough when studying the theory of continuous time random processes.

[2.11] Given an abstract space A, a sigma-field $\mathcal{F}$ of subsets of A, an index set $\mathcal{I}$, and a product sample space of the form

$$A^{\mathcal{I}} = \prod_{t \in \mathcal{I}} A_t,$$

where the A_t are all replicas of A, the product sigma-field

$$\mathcal{F}^{\mathcal{I}} = \prod_{t \in \mathcal{I}} \mathcal{F}_t,$$

is defined as the sigma-field generated by all "one-dimensional" sets of the form

$$\{\{a_t;\, t \in \mathcal{I}\} : a_t \in F \;\text{ for }\; t = s \;\text{ and }\; a_t \in A_t \;\text{ for }\; t \neq s\}$$

for some $s \in \mathcal{I}$ and some $F \in \mathcal{F}$; that is, the product sigma-field is the sigma-field generated by all "one-dimensional" events formed by collecting all of the vectors or sequences or waveforms with one coordinate constrained to lie in a one-dimensional event and with the other coordinates unrestricted. The product sigma-field must contain *all* such events; that is, for all possible indices s and all possible events F.

Thus, for example, given the one-dimensional abstract space $\Re$, the real line along with its Borel field, Figure 2.3(a)–(c) depicts three examples of one-dimensional sets in $\Re^2$, the two-dimensional Euclidean plane. Note, for example, that the unit circle $\{(x, y) : x^2 + y^2 \leq 1\}$ is *not* a one-dimensional set since it requires simultaneous constraints on two coordinates.

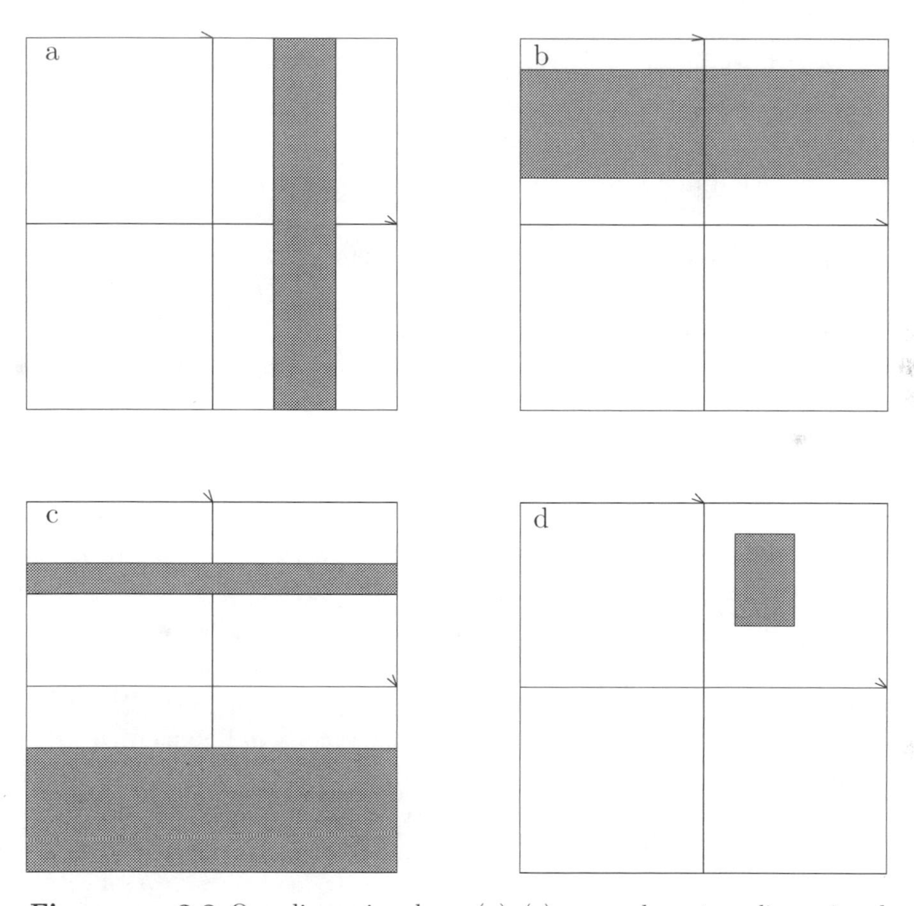

Figure 2.3 One-dimensional (a)–(c) and two-dimensional events (d) in two-dimensional space. (a)$\{(x_0, x_1) : x_0 \in (1, 3)\}$, (b)$\{(x_0, x_1) : x_1 \in (3, 6)\}$, (c)$\{(x_0, x_1) : x_1 \in (4, 5) \cup (-\infty, -2)\}$, (d)$\{(x_0, x_1) : x_0 \in (1, 3); x_1 \in (3, 6)\}$.

More generally, for a fixed finite k the product sigma-field $\mathcal{B}(\Re)^{\mathcal{Z}_k}$ (or simply $\mathcal{B}(\Re)^k$) of k-dimensional Euclidean space $\Re^k$ is the smallest sigma-field containing all one-dimensional events of the form

$$\{\mathbf{x} = (x_0, x_1, \ldots, x_{k-1}) : x_i \in F\}$$

for some $i = 0, 1, \ldots, k - 1$ and some Borel set F of $\Re$. The two-dimensional example Figure 2.3(a) has this form with $k = 2, i = 0$, and $F = (1, 3)$. This one-dimensional set consists of all values in the infinite rectangle between 1 and 3 in the x_0 direction and between $-\infty$ and ∞ in the x_1 direction.

To summarize, we have defined a space A with event space $\mathcal{F}$, and an index set $\mathcal{I}$ such as $\mathcal{Z}_+, \mathcal{Z}, \Re$, or [0,1), and we have formed the product space $A^{\mathcal{I}}$ and the associated product event space $\mathcal{F}^{\mathcal{I}}$. We know that this event space contains all one-dimensional events by construction. We next consider what other events must be in $\mathcal{F}^{\mathcal{I}}$ by virtue of its being an event space.

After the one-dimensional events that pin down the value of a single coordinate of the vector or sequence or waveform, the next most general kinds of events are finite-dimensional sets that separately pin down the values at a finite number of coordinates. Let $\mathcal{K}$ be a finite collection of members of $\mathcal{I}$ and hence $\mathcal{K} \subset \mathcal{I}$. Say that $\mathcal{K}$ has K members, which we shall denote as $\{k_i;\ i = 0, 1, \ldots, K-1\}$. These K numbers can be thought of as a collection of sample times such as $\{1, 4, 8, 156, 1027\}$ for a sequence or $\{1.5, 9.07, 40.0, 41.2, 41.3\}$ for a waveform. We assume for convenience that the sample times are ordered in increasing fashion. Let $\{F_{k_i};\ i = 0, 1, \ldots, K-1\}$ be a collection of members of $\mathcal{F}$. Then a set of the form $\{\{x_t;\ t \in \mathcal{I}\} : x_{k_i} \in F_{k_i};\ i = 0, 1, \ldots, K-1\}$ is an example of a finite-dimensional set. Note that it collects all sequences or waveforms such that a finite number of coordinates are constrained to lie in one-dimensional events. An example of two-dimensional sets of this form in two-dimensional space is illustrated in Figure 2.3(d). Observe that when the one-dimensional sets constraining the coordinates are intervals, then the two-dimensional sets are rectangles. Analogous to the two-dimensional example, finite-dimensional events having separate constraints on each coordinate are called *rectangles.* Observe, for example, that a circle or sphere in Euclidean space is not a rectangle because it cannot be defined using separate constraints on the coordinates; the constraints on each coordinate depend on the values of the others – e.g., in two dimensions for a circle of radius one we require that $x_0^2 \le 1 - x_1^2$.

Note that Figure 2.3(d) is just the intersection of examples (a) and (b) of Figure 2.3. In general we can express finite-dimensional rectangles as intersections of one-dimensional events as follows

$$\{\{x_t;\ t \in \mathcal{I}\} : x_{k_i} \in F_{k_i};\ i = 0, 1, \ldots, K-1\} = \bigcap_{i=0}^{K-1} \{\{x_t;\ t \in \mathcal{I}\} : x_{k_i} \in F_i\},$$

that is, a set constraining a finite number of coordinates to each lie in one-dimensional events or sets in $\mathcal{F}$ is the intersection of a collection of one-dimensional events. Since $\mathcal{F}^{\mathcal{I}}$ is a sigma-field and since it contains the one-dimensional events, it must contain such finite intersections, and hence it must contain such finite-dimensional events.

By concentrating on events that can be represented as the finite intersection of one-dimensional events we do not mean to imply that all events in the product event space can be represented in this fashion – the event space will also contain all possible limits of finite unions of such rectangles, complements of such sets, and so on. For example, the unit circle in two dimensions is not a rectangle, but it can be considered as a limit of unions of rectangles and hence is in the event space generated by the rectangles. (See Problem 2.36.)

The moral of this discussion is that the product sigma-field for spaces of sequences and waveforms must contain (but not consist exclusively of) all sets that are described by requiring that the outputs of coordinates for a finite number of events lie in sets in the one-dimensional event space $\mathcal{F}$.

We shall further explore such product event spaces when considering random processes, but the key points remain:

1. a product event space is a sigma-field
2. it contains all "one-dimensional events" consisting of subsets of the product sample space formed by grouping together all vectors or sequences or waveforms having a single fixed coordinate lying in a one-dimensional event. In addition, it contains all rectangles or finite-dimensional events consisting of all vectors or sequences or waveforms having a finite number of coordinates constrained to lie in one-dimensional events.

2.3.3 Probability measures

The defining axioms of a probability measure as given in equations (2.22) through (2.25) correspond generally to intuitive notions, at least for the first three properties. The first property requires that a probability be a nonnegative number. In a purely mathematical sense, this is an arbitrary restriction, but it is in accord with the long history of intuitive and combinatorial developments of probability. Probability measures share this property with other measures such as area, volume, weight, and mass.

The second defining property corresponds to the notion that the probability that *something* will happen or that an experiment will produce one of its possible outcomes is one. This, too, is mathematically arbitrary but is a convenient and historical assumption. (From childhood we learn about things that are "100% certain" – obviously we could as easily take 1 or π (but *not* infinity – *why?*) to represent certainty.)

The third property, "additivity" or "finite additivity," is the key one. In English it reads that the probability of occurrence of a finite collection of events having no points in common must be the sum of the probabilities of the separate events. More generally, the basic assumption of measure theory is that *any* measure – probabilistic or not – such as weight, volume, mass, and area should be additive: the mass of a group of disjoint regions of matter should be the sum of the separate masses; the weight of a group of objects should be the sum of the individual weights. Equation (2.24) only pins down this property for finite collections of events. The additional restriction of (2.25), called *countable additivity,* is a limiting or asymptotic or infinite version, analogous to (2.19) for set algebra. This again leads to the rhetorical questions of why the more complicated, more restrictive, and less intuitive infinite version is required. In fact, it was the addition of this limiting property that provided the fundamental idea for Kolmogorov's development of modern probability theory in the 1930s.

The response to the rhetorical question is essentially the same as that for the asymptotic set algebra property: countably infinite properties are required to handle asymptotic and limiting results. Such results are crucial because we often need to evaluate the probabilities of complicated events that can only be represented as a limit of simple events. (This is analogous to the way that integrals are obtained as limits of finite sums.)

Note that it is *countable* additivity that is required. Uncountable additivity cannot be defined sensibly. This is easily seen in terms of the fair wheel mentioned at the beginning of the chapter. If the wheel is spun, any particular number has probability zero. On the other hand, the probability of the event made up of all of the *uncountable* numbers between 0 and 1 is obviously one. If you consider defining the probability of all the numbers between 0 and 1 to be the uncountable sum of the individual probabilities, you see immediately the essential contradiction that results.

Since countable additivity has been added to the axioms proposed in the introduction, the formula (2.11) used to compute probabilities of events broken up by a partition immediately extends to partitions with a countable number of elements; that is, if F_k, $k = 1, 2, \ldots$, forms a partition of Ω into disjoint events ($F_n \cap F_k = \emptyset$ if $n \neq k$ and $\bigcup_k F_k = \Omega$), then for any event G

$$P(G) = \sum_{k=1}^{\infty} P(G \cap F_k). \tag{2.27}$$

Limits of probabilities

At times we are interested in finding the probability of the limit of a sequence of events. To relate the countable additivity property of (2.25) to limiting properties, recall the discussion of the limiting properties of events given earlier in this chapter in terms of increasing and decreasing sequences of events. Say we have an increasing sequence of events F_n, $n = 0, 1, 2, \ldots$, $F_{n-1} \subset F_n$, and let F denote the limit set, that is, the union of all of the F_n. We have already argued that the limit set F is itself an event. Intuitively, since the F_n converge to F, the probabilities of the F_n should converge to the probability of F. Such convergence is called a *continuity* property of probability and is very useful for evaluating the probabilities of complicated events as the limit of a sequence of probabilities of simpler events. We shall show that countable additivity implies such continuity. To accomplish this, define the sequence of sets $G_0 = F_0$ and $G_n = F_n - F_{n-1}$ for $n = 1, 2, \ldots$ The G_n are disjoint and have the same union as do the F_n (see Figure 2.2(a) as a visual aid). Thus we have from countable additivity that

$$\begin{aligned} P\left(\lim_{n\to\infty} F_n\right) &= P\left(\bigcup_{k=0}^{\infty} F_k\right) = P\left(\bigcup_{k=0}^{\infty} G_k\right) \\ &= \sum_{k=0}^{\infty} P(G_k) = \lim_{n\to\infty} \sum_{k=0}^{n} P(G_k), \end{aligned}$$

where the last step simply uses the definition of an infinite sum. Since $G_n = F_n - F_{n-1}$ and $F_{n-1} \subset F_n$, $P(G_n) = P(F_n) - P(F_{n-1})$ and hence

$$\sum_{k=0}^{n} P(G_k) = P(F_0) + \sum_{k=1}^{n} (P(F_n) - P(F_{n-1})) = P(F_n).$$

This is an example of what is called a "telescoping sum" where each term cancels the previous term and adds a new piece, i.e.,

$$\begin{aligned} P(F_n) = &+ P(F_n) - P(F_{n-1}) \\ &+ P(F_{n-1}) - P(F_{n-2}) \\ &+ P(F_{n-2}) - P(F_{n-3}) \\ &\vdots \\ &+ P(F_1) - P(F_0) \\ &+ P(F_0). \end{aligned}$$

Combining these results completes the proof of the following statement.

If F_n is a sequence of increasing events, then

$$P\left(\lim_{n\to\infty} F_n\right) = \lim_{n\to\infty} P(F_n), \tag{2.28}$$

that is, the probability of the limit of a sequence of increasing events is the limit of the probabilities.

Note that the sequence of probabilities on the right-hand side of (2.28) is increasing with increasing n. Thus, for example, probabilities of semi-infinite intervals can be found as a limit as $P((-\infty, a]) = \lim_{n\to\infty} P((-n, a])$. A similar argument can be used to show that one can also interchange the limit with the probability measure given a sequence of decreasing events; that is,

If F_n is a sequence of decreasing events, then

$$P\left(\lim_{n\to\infty} F_n\right) = \lim_{n\to\infty} P(F_n). \tag{2.29}$$

that is, the probability of the limit of a sequence of decreasing events is the limit of the probabilities.

Note that the sequence of probabilities on the right-hand side of (2.29) is decreasing with increasing n. Thus, for example, the probabilities of points can be found as a limit of probabilities of intervals, $P(\{a\}) = \lim_{n\to\infty} P((a - 1/n, a + 1/n))$.

It can be shown (see Problem 2.21) that, given (2.22) through (2.24), the three conditions (2.25), (2.28), and (2.29) are equivalent; that is, any of the three could serve as the fourth axiom of probability.

Property (2.28) is called *continuity from below*, and (2.29) is called *continuity from above*. The designations "from below" and "from above" relate to the direction from which the respective sequences of probabilities approach their limit. These continuity results are the basis for using integral calculus to compute probabilities, since integrals can be expressed as limits of sums.

2.4 Discrete probability spaces

We now provide several examples of probability measures on our examples of sample spaces and sigma-fields and thereby give complete examples of probability spaces.

The first example formalizes the description of a probability measure as a sum of a pmf as first mentioned in the introductory section.

[2.12] Let Ω be a finite set and let $\mathcal{F}$ be the power set of Ω. Suppose that we have a function $p(\omega)$ that assigns a real number to each sample point

ω in such a way that

$$p(\omega) \geq 0, \text{ all } \omega \in \Omega \tag{2.30}$$

and

$$\sum_{\omega \in \Omega} p(\omega) = 1. \tag{2.31}$$

Define the set function P by

$$P(F) = \sum_{\omega \in F} p(\omega) = \sum_{\omega \in \Omega} 1_F(\omega) p(\omega), \text{ all } F \in \mathcal{F} \tag{2.32}$$

where $1_F(\omega)$ is the indicator function of the set F, 1 if $\omega \in F$ and 0 otherwise.

For simplicity we drop the $\omega \in \Omega$ underneath the sum; that is, when no range of summation is explicit, it should be assumed the sum is over all possible values. Thus we can abbreviate (2.32) to

$$P(F) = \sum 1_F(\omega) p(\omega), \text{ all } F \in \mathcal{F}. \tag{2.33}$$

The set function P is easily verified to be a probability measure: it obviously satisfies Axioms 2.1 and 2.2. It is finitely and countably additive from the properties of sums. In particular, given a sequence of disjoint events, only a finite number can be distinct (since the power set of a finite space has only a finite number of members). To be disjoint, the balance of the sequence must equal $\emptyset$. The probability of the union of these sets will be the finite sum of the $p(\omega)$ over the points in the union. This equals the sum of the probabilities of the sets in the sequence. Example [2.1] is a special case of Example [2.12], as is the coin flip example of the introductory section.

A function $p(\omega)$ satisfying (2.30) and (2.31) is called a *probability mass function* or *pmf*. Analogous to a positive-valued point mass function in physics which is summed to find a total mass in a volume, the pmf is summed over region to find the probability of that region.

Expectations

The summation (2.33) used to define probability measures for a discrete space is a special case of a more general weighted sum, which we pause to define and consider. Suppose that g is a real-valued function defined on Ω, i.e., $g : \Omega \to \Re$ assigns a real number $g(\omega)$ to every $\omega \in \Omega$. We could consider more general complex-valued functions, but for the moment it is simpler to stick to real-valued functions. Also, we could consider mappings of Ω into subsets of $\Re$, but it is convenient for the moment to let the range of g be the entire real line. Recall that in the introductory section we considered such a function to be an example of signal processing and called it a *random variable*. Given a pmf p, define the *expectation* of g (with respect to p) as [21]

$$E(g) = \sum g(\omega)p(\omega). \tag{2.34}$$

With this definition, (2.33) with $g(\omega) = 1_F(\omega)$ yields

$$P(F) = E(1_F), \tag{2.35}$$

showing that the probability of an event is the expectation of the indicator function of the event. Mathematically, we can think of expectation as a *generalization* of the idea of probability since probability is the special case of expectation that results when the only functions allowed are indicator functions.

Expectations are also called *probabilistic averages* or *statistical averages*. For the time being, probabilities are the most important examples of expectation. We shall see many examples, however, so it is worthwhile to mention a few of the most important. Suppose that the sample space is a subset of the real line, e.g., $\mathcal{Z}$ or $\mathcal{Z}_n$. One of the most commonly encountered expectations is the *mean* or *first moment*

$$m = \sum \omega p(\omega), \tag{2.36}$$

where $g(\omega) = \omega$, the *identity function*. A more general idea is the kth moment defined by

$$m^{(k)} = \sum (\omega)^k p(\omega), \tag{2.37}$$

so that $m = m^{(1)}$. After the mean, the most commonly encountered moment in practice is the second moment,

$$m^{(2)} = \sum (\omega)^2 p(\omega). \tag{2.38}$$

Moments can be thought of as parameters describing a pmf, and some computations involving signal processing will turn out to depend only on certain moments.

A slight variation on kth (or kth order) moments is the so-called *centralized* moment formed by subtracting the mean before taking the power:

$$\sum (\omega - m)^k p(\omega), \tag{2.39}$$

[1] This is not in fact the fundamental definition of expectation that will be introduced in Chapter 4, but it will be seen to be equivalent.

but the only such moment commonly encountered in practice is the *variance*

$$\sigma^2 = \sum (\omega - m)^2 p(\omega). \tag{2.40}$$

The variance and the second moment are easily related as

$$\begin{aligned}
\sigma^2 &= \sum (\omega - m)^2 p(\omega) \\
&= \sum (\omega^2 - 2\omega m + m^2) p(\omega) \\
&= \sum \omega^2 p(\omega) - 2m \sum \omega p(\omega) + m^2 \sum p(\omega) \\
&= m^{(2)} - 2m^2 + m^2 = m^{(2)} - m^2.
\end{aligned} \tag{2.41}$$

Probability mass functions

It is important to observe that the probability mass function is defined only for *points* in the sample space, whereas a probability measure is defined for events, sets which belong to an event space. Intuitively, the probability of a set is given by the sum of the probabilities of the points as given by the pmf. Obviously it is much easier to describe the probability function than the probability measure since it need only be specified for points. The axioms of probability then guarantee that the probability function can be used to compute the probability measure. Note that given one probability form, we can always determine the other. In particular, given the pmf p, we can construct P using (2.32). Given P, we can find the corresponding pmf p from the formula

$$p(\omega) = P(\{\omega\}).$$

We list below several of the most common examples of pmf's. The reader should verify that they are all indeed valid pmf's, that is, that they satisfy (2.30) and (2.31).

The binary pmf. $\Omega = \{0, 1\}$; $p(0) = 1 - p$, $p(1) = p$, where p is a parameter in $(0, 1)$.

A uniform pmf. $\Omega = \mathcal{Z}_n = \{0, 1, \ldots, n-1\}$ and $p(k) = 1/n$; $k \in \mathcal{Z}_n$.

The binomial pmf. $\Omega = \mathcal{Z}_{n+1} = \{0, 1, \ldots, n\}$ and

$$p(k) = \binom{n}{k} p^k (1-p)^{n-k}; \; k \in \mathcal{Z}_{n+1},$$

where

$$\binom{n}{k} = \frac{n!}{k!(n-k)!}$$

is the binomial coefficient (read as "n choose k").

The binary pmf is a probability model for coin flipping with a biased coin or for a single sample of a binary data stream. A uniform pmf on $\mathcal{Z}_6$ can model the roll of a fair die. Observe that it would not be a good model for ASCII data since, for example, the letters t and e and the symbol for space have a higher probability than other letters. The binomial pmf is a probability model for the number of heads in n successive independent flips of a biased coin, as will later be seen.

The same construction provides a probability measure on countably infinite spaces such as $\mathcal{Z}$ and $\mathcal{Z}_+$. It is no longer as simple to prove countable additivity, but it should be fairly obvious that it holds and, at any rate, it follows from standard results in elementary analysis for convergent series. Hence we shall only state the following example without proving countable additivity, but bear in mind that it follows from the properties of infinite summations.

[2.13] Let Ω be a space with a countably infinite number of elements and let $\mathcal{F}$ be the power set of Ω. Then if $p(\omega)$; $\omega \in \Omega$ satisfies (2.30) and (2.31), the set function P defined by (2.32) is a probability measure.

Two common examples of pmf's on countably infinite sample spaces follow. The reader should test their validity.

The geometric pmf. $\Omega = \{1, 2, 3, \ldots\}$ and $p(k) = (1-p)^{k-1}p$; $k = 1, 2, \ldots$, where $p \in (0, 1)$ is a parameter.

The Poisson pmf. $\Omega = \mathcal{Z}_+ = \{0, 1, 2, \ldots\}$ and $p(k) = (\lambda^k e^{-\lambda})/k!$, where λ is a parameter in $(0, \infty)$. (Keep in mind that $0! \triangleq 1$.)

We will later see the origins of several of these pmf's and their applications. For example, both the binomial and the geometric pmf will be derived from the simple binary pmf model for flipping a single coin. For the moment they should be considered as common important examples. Various properties of these pmf's and a variety of calculations involving them are explored in the problems at the end of the chapter.

Computational examples

The various named pmf's provide examples for computing probabilities and other expectations. Although much of this is prerequisite material, it does not hurt to collect several of the more useful tricks that arise in evaluating sums. The binary pmf on its own is too simple to provide much interest, so first consider the uniform pmf on $\mathcal{Z}_n$. This is trivially a valid pmf since it is nonnegative and sums to 1. The probability of any set is simply

$$P(F) = \frac{1}{n} \sum 1_F(\omega) = \frac{\#(F)}{n},$$

where $\#(F)$ denotes the number of elements or points in the set F. The mean is given by

$$m = \frac{1}{n} \sum_{k=0}^{n-1} k = \frac{n-1}{2}, \tag{2.42}$$

a standard formula easily verified by induction, as detailed in Appendix B. The second moment is given by

$$m^{(2)} = \frac{1}{n} \sum_{k=0}^{n-1} k^2 = \frac{(2n-1)(n-1)}{6}, \tag{2.43}$$

as can also be verified by induction. The variance can be found by combining (2.43), (2.42), and (2.41).

The binomial pmf is more complicated. The first issue is to prove that it sums to one and hence is a valid pmf (it is obviously nonnegative). This is accomplished by recalling the binomial theorem from high-school algebra:

$$(a+b)^n = \sum_{k=0}^{n} \binom{n}{k} a^n b^{n-k} \tag{2.44}$$

and setting $a = p$ and $b = 1 - p$ to write

$$\sum_{k=0}^{n} p(k) = \sum_{k=0}^{n} \binom{n}{k} p^k (1-p)^{n-k}$$
$$= (p + 1 - p)^n = 1.$$

Finding moments is trickier here, and we shall later develop a much easier way to do this using exponential transforms. Nonetheless, it provides useful practice to compute an example sum, if only to demonstrate later how much work can be avoided! Finding the mean requires evaluation of the sum

$$\begin{aligned} m &= \sum_{k=0}^{n} k \binom{n}{k} p^k (1-p)^{n-k} \\ &= \sum_{k=1}^{n} \frac{n!}{(n-k)!(k-1)!} p^k (1-p)^{n-k}. \end{aligned}$$

The trick here is to recognize that the sum looks very much like the terms in the binomial theorem, but a change of variables is needed to get the binomial theorem to simplify things. Changing variables by defining $l = k - 1$, the sum becomes

$$m = \sum_{l=0}^{n-1} \frac{n!}{(n-l-1)!l!} p^{l+1} (1-p)^{n-l-1},$$

which will very much resemble the binomial theorem with $n - 1$ replacing n if we factor out a p and an n:

$$\begin{aligned} m &= np \sum_{l=0}^{n-1} \frac{(n-1)!}{(n-1-l)!l!} p^l (1-p)^{n-1-l} \\ &= np(p+1-p)^{n-1} = np. \end{aligned} \tag{2.45}$$

The second moment is messier, so its evaluation is postponed until simpler means are developed.

The geometric pmf is handled using the geometric progression, usually treated in high-school algebra and summarized in Appendix B. From (B.4) in Appendix B we have for any real a with $|a| < 1$

$$\sum_{k=0}^{\infty} a^k = \frac{1}{1-a}, \tag{2.46}$$

which proves that the geometric pmf indeed sums to 1.

Evaluation of the mean of the geometric pmf requires evaluation of the sum

$$m = \sum_{k=1}^{\infty} kp(k) = \sum_{k=1}^{\infty} kp(1-p)^{k-1}.$$

One may have access to a book of tables which includes this sum, but a useful trick can be used to evaluate the sum from the well-known result for summing a geometric series. The trick involves differentiating the usual geometric progression sum, as detailed in Appendix B, where it is shown for any $q \in (0,1)$ that

$$\sum_{k=0}^{\infty} kq^{k-1} = \frac{1}{(1-q)^2}. \tag{2.47}$$

Set $q = 1 - p$ in (2.47) and the formula for m yields

$$m = \frac{1}{p}. \tag{2.48}$$

A similar idea works for the second moment. From (B.7) of Appendix B the second moment is given by

$$m^{(2)} = \sum_{k=1}^{\infty} k^2 p(1-p)^{k-1} = p\left(\frac{2}{p^3} - \frac{1}{p^2}\right) \tag{2.49}$$

and hence from (2.41) the variance is

$$\sigma^2 = \frac{1-p}{p^2}. \tag{2.50}$$

As an example of a probability computation using a geometric pmf, suppose that $(\Omega, \mathcal{F}, P)$ is a discrete probability space with $\Omega = \mathcal{Z}_+$, $\mathcal{F}$ the power set of Ω, and P the probability measure induced by the geometric pmf with parameter p. Find the probabilities of the event $F = \{k : k \geq 10\}$ and $G = \{k : k \text{ is odd}\}$. Alternatively note that $F = \{10, 11, 12, \ldots\}$ and $G = \{1, 3, 5, 7, \ldots\}$ (we consider only odd numbers in the sample space, that is, only positive odd numbers). We have that

$$\begin{aligned}
P(F) &= \sum_{k \in F} p(k) = \sum_{k=10}^{\infty} p(1-p)^{k-1} \\
&= \frac{p}{1-p} \sum_{k=10}^{\infty} (1-p)^k = \frac{p}{1-p}(1-p)^{10} \sum_{k=10}^{\infty} (1-p)^{k-10} \\
&= p(1-p)^9 \sum_{k=0}^{\infty} (1-p)^k = (1-p)^9,
\end{aligned}$$

where the suitable form of the geometric progression has been derived from the basic form (B.4). Although we have concentrated on the calculus, this problem could be interpreted as a solution to a word problem. For example, suppose you arrive at the post office and you know that the probability of k people being in line is a geometric distribution with $p = 1/2$. What is the probability that there are at least ten people in line? From the solution just obtained the answer is $(1 - 0.5)^9 = 2^{-9}$.

To find the probability of an odd outcome, we proceed in the same general fashion to write

$$\begin{aligned} P(G) &= \sum_{k \in G} p(k) = \sum_{k=1,3,\ldots} p(1-p)^{k-1} \\ &= p \sum_{k=0,2,4,\ldots} (1-p)^k = p \sum_{k=0}^{\infty} [(1-p)^2]^k \\ &= \frac{p}{1-(1-p)^2} = \frac{1}{2-p}. \end{aligned}$$

Thus in the English example of the post office lines, the probability of finding an odd number of people in line is 2/3.

Lastly we consider the Poisson pmf, again beginning with a verification that it is indeed a pmf. Consider the sum

$$\sum_{k=0}^{\infty} p(k) = \sum_{k=0}^{\infty} \frac{\lambda^k e^{-\lambda}}{k!} = e^{-\lambda} \sum_{k=0}^{\infty} \frac{\lambda^k}{k!}.$$

Here the trick is to recognize the sum as the Taylor series expansion for an exponential, that is,

$$e^{\lambda} = \sum_{k=0}^{\infty} \frac{\lambda^k}{k!},$$

whence

$$\sum_{k=0}^{\infty} p(k) = e^{-\lambda} e^{\lambda} = 1,$$

proving the claim.

To evaluate the mean of the Poisson pmf, begin with

$$\sum_{k=0}^{\infty} k p(k) = \sum_{k=1}^{\infty} k \frac{\lambda^k e^{-\lambda}}{k!} = e^{-\lambda} \sum_{k=1}^{\infty} \frac{\lambda^k}{(k-1)!}.$$

Change variables $l = k - 1$ and pull a λ out of the sum to write

$$\sum_{k=0}^{\infty} k p(k) = \lambda e^{-\lambda} \sum_{l=0}^{\infty} \frac{\lambda^l}{l!}.$$

Recognizing the sum as e^{λ}, this yields

$$m = \lambda. \tag{2.51}$$

The second moment is found similarly, but with more bookkeeping. Analogous to the mean computation,

$$m^{(2)} = \sum_{k=1}^{\infty} k^2 \frac{\lambda^k e^{-\lambda}}{k!} = \sum_{k=2}^{\infty} k(k-1)\frac{\lambda^k e^{-\lambda}}{k!} + m$$
$$= \sum_{k=2}^{\infty} \frac{\lambda^k e^{-\lambda}}{(k-2)!} + m.$$

Change variables $l = k - 2$ and pull λ^2 out of the sum to obtain

$$m^{(2)} = \lambda^2 \sum_{l=0}^{\infty} \frac{\lambda^l e^{-\lambda}}{l!} + m = \lambda^2 + \lambda \tag{2.52}$$

so that from (2.41) the variance is

$$\sigma^2 = \lambda. \tag{2.53}$$

Multidimensional pmf's

Although the foregoing ideas were developed for scalar sample spaces such as $\mathcal{Z}_+$, they also apply to vector sample spaces. For example, if A is a discrete space, then so is the vector space $A^k = \{$all vectors $\mathbf{x} = (x_0, \ldots x_{k-1})$ with $x_i \in A$, $i = 0, 1, \ldots, k-1\}$. A common example of a pmf on vectors is the product pmf considered next.

[2.14] *The product pmf.*
Let $\{p_i;\ i = 0, 1, \ldots, k-1\}$, be a collection of one-dimensional pmf's; that is, for each $i = 0, 1, \ldots, k-1$ $p_i(r);\ r \in A$ satisfies (2.30) and (2.31). Define the product k-dimensional pmf p on A^k by

$$p(\mathbf{x}) = p(x_0, x_1, \ldots, x_{k-1}) = \prod_{i=0}^{k-1} p_i(x_i).$$

As a more specific example, suppose that all of the marginal pmf's are the same and are given by a Bernoulli pmf:

$$p(x) = p^x (1-p)^{1-x};\ x = 0, 1.$$

Then the corresponding product pmf for a k-dimensional vector becomes

$$p(x_0, x_1, \ldots, x_{k-1}) = \prod_{i=0}^{k-1} p^{x_i}(1-p)^{1-x_i}$$
$$= p^{w(x_0, x_1, \ldots, x_{k-1})}(1-p)^{k - w(x_0, x_1, \ldots, x_{k-1})},$$

where $w(x_0, x_1, \ldots, x_{k-1})$ is the number of ones occurring in the binary k-tuple $x_0, x_1, \ldots, x_{k-1}$, the *Hamming weight* of the vector.

2.5 Continuous probability spaces

Continuous spaces are handled in a manner analogous to discrete spaces, but with some fundamental differences. The primary difference is that usually probabilities are computed by integrating a density function instead of summing a mass function. The good news is that most formulas look the same with integrals replacing sums. The bad news is that there are some underlying theoretical issues that require consideration. The problem is that integrals are themselves limits, and limits do not always exist in the sense of converging to a finite number. Because of this, some care will be needed to clarify when the resulting probabilities are well defined.

[2.15] Let $(\Omega, \mathcal{F}) = (\Re, \mathcal{B}(\Re))$, the real line together with its Borel field. Suppose that we have a real-valued function f on the real line that satisfies the following properties

$$f(r) \geq 0, \quad \text{all } r \in \Omega. \tag{2.54}$$

$$\int_\Omega f(r)dr = 1, \tag{2.55}$$

that is, the function $f(r)$ has a well-defined integral over the real line. Define the set function P by

$$P(F) = \int_F f(r)\, dr = \int 1_F(r) f(r)\, dr, \quad F \in \mathcal{B}(\Re). \tag{2.56}$$

We note that a probability space defined as a probability measure on a Borel field is an example of a *Borel space*.

Again as in the discrete case, this integral is a special case of a more general weighted integral. Suppose that g is a real-valued function defined on Ω, i.e., $g : \Omega \to \Re$ assigns a real number $g(r)$ to every $r \in \Omega$. Recall that such a function is called a *random variable*. Given a pdf f, define the *expectation* of g (with respect to f) as

$$E(g) = \int g(r) f(r)\, dr. \tag{2.57}$$

With this definition we can rewrite (2.56) as

$$P(F) = E(1_F), \tag{2.58}$$

which has *exactly* the same form as in the discrete case. Thus probabilities can be considered as expectations of indicator functions in both the discrete case where the probability measure is described by a pmf and in the continuous case where the probability measure is described by a pdf.

As in the discrete case, there are several particularly important examples of expectations if the sample space is a subset of the real line, e.g., $\Re$ or $[0,1)$. The definitions are exact integral analogs of those for the discrete cases: the *mean* or *first moment*

$$m = \int r f(r)\, dr, \tag{2.59}$$

the kth moment

$$m^{(k)} = \int r^k f(r)\, dr, \tag{2.60}$$

including the second moment

$$m^{(2)} = \int r^2 f(r)\, dr, \tag{2.61}$$

the *centralized* moments formed by subtracting the mean before taking the power:

$$\int (r-m)^k f(r)\, dr, \tag{2.62}$$

including the *variance*

$$\sigma^2 = \int (r-m)^2 f(r)\, dr. \tag{2.63}$$

When more general complex-valued random variables are considered, often the kth absolute moment is used instead:

$$m_a^{(k)} = \int |r|^k f(r)\, dr. \tag{2.64}$$

As in the discrete case, the variance and the second moment are easily related as

$$\sigma^2 = m^{(2)} - m^2. \tag{2.65}$$

An important technical detail not yet considered is whether or not the set function defined as an integral over a pdf is actually a probability measure. In particular, are the probabilities of all events well defined and do they satisfy the axioms of probability? Intuitively this should be the case since (2.54) to (2.56) are the integral analogs of the summations of (2.30) to (2.32) and we have argued that summing pmf's provides a well-defined probability measure. In fact, this is a mathematically delicate issue which leads to the reasons behind the requirements for sigma-fields and Borel fields. Before exploring these issues in more depth in the next section, the easy portion of the answer should be recalled. We have already argued in the introduction to this chapter that if we define a set function $P(F)$ as the integral of a pdf over the set F, then if the integral exists for the sets in question, the set function must be nonnegative, normalized, and additive, that is, it must satisfy the first three axioms of probability. This is well and good, but it leaves some key points unanswered. First, is the candidate probability measure defined for all Borel sets? Equivalently, are we guaranteed that the integral will make sense for all sets (events) of interest? Second, is the candidate probability measure also countably additive or, equivalently, continuous from above or below? The answer to both questions is unfortunately *no* if one considers the integral to be a Riemann integral, the integral most engineers learn as undergraduates. The integral is not certain to exist for all Borel sets, even if the pdf is a simple uniform pdf. Riemann integrals in general do not have nice limiting properties, so the necessary continuity properties do not hold in general for Riemann integrals. These delicate issues are considered next in an optional subsection and further in Appendix B. The bottom line can be easily summarized as follows.

- Equation (2.56) defines a probability measure on the Borel space of the real line and its Borel sets provided that the integral is interpreted as a Lebesgue integral. In all practical cases of interest, the Lebesgue integral is either equal to the Riemann integral, usually more familiar to engineers, or to a limit of Riemann integrals on a converging sequence of sets.

⋆ *Probabilities as integrals*

The first issue is fundamental: does the integral of (2.56) make sense; i.e., is it well defined for all events of interest? Suppose first that we take the common engineering approach and use Riemann integration – the form of integration used in elementary calculus. Then the above integrals are defined at least for events F that are intervals. This implies from the linearity properties of Riemann integration that the integrals are also well defined for events F that are finite unions of intervals. It is not difficult, however, to construct sets F for which the indicator function 1_F is so nasty that the function $f(r)1_F(r)$ does not have a Riemann integral. For example, suppose that $f(r)$ is 1 for $r \in [0,1]$ and 0 otherwise. Then the Riemann integral $\int 1_F(r)f(r)\,dr$ is not defined for the set F of all irrational numbers, yet intuition should suggest that the set has probability 1. This intuition reflects the fact that if all points are somehow equally probable, then since the unit interval contains an uncountable infinity of irrational numbers and only a countable infinity of rational numbers, then the probability of the former set should be one and that of the latter 0. This intuition is not reflected in the integral definition, which is not defined for either set by the Riemann approach. Thus the definition of (2.56) has a basic problem: the integral in the formula giving the probability measure of a set might not be well defined.

A natural approach to escaping this dilemma would be to use the Riemann integral when possible, i.e., to define the probabilities of events that are finite unions of intervals, and then to obtain the probabilities of more complicated events by expressing them as a limit of finite unions of intervals, if the limit makes sense. We might hope that this would give us a reasonable definition of a probability measure on a class of events much larger than the class of all finite unions of intervals. Intuitively, it should give us a probability measure of all sets that can be expressed as increasing or decreasing limits of finite unions of intervals.

This larger class is, in fact, the Borel field, but the Riemann integral has the unfortunate property that in general we cannot interchange limits and integration; that is, the limit of a sequence of integrals of converging functions may not be itself an integral of a limiting function.

This problem is so important to the development of a rigorous probability theory that it merits additional emphasis: even though the familiar Riemann integrals of elementary calculus suffice for most engineering and computational purposes, they are too weak for building a useful theory, proving theorems, and evaluating the probabilities of some events which can be most easily expressed as limits of simple events. The problem is that the Riemann integral does not exist for sufficiently general functions and that limits and integration cannot be interchanged in general.

The solution is to use a different definition of integration – the Lebesgue integral. Here we need only concern ourselves with a few simple properties of the Lebesgue integral, which are summarized below. The interested reader is referred to Appendix B for a brief summary of basic definitions and properties of the Lebesgue integral which reinforce the following remarks.

The Riemann integral of a function $f(r)$ "carves up" or partitions the domain of the argument r and effectively considers weighted sums of the values of the function $f(r)$ as the partition becomes ever finer. Conversely, the Lebesgue integral "carves up" the values of the function itself and effectively defines an integral as a limit of simple integrals of quantized versions of the function. This simple change of definition results in two fundamentally important properties of Lebesgue integrals that are not possessed by Riemann integrals:

1. The integral is defined for all Borel sets.
2. Subject to suitable technical conditions (such as requiring that the integrands have bounded absolute value), one can interchange the order of limits and integration; e.g., if $F_n \uparrow F$, then

$$\begin{aligned} P(F) &= \int 1_F(r) f(r)\, dr = \int \lim_{n\to\infty} 1_{F_n}(r) f(r)\, dr \\ &= \lim_{n\to\infty} \int 1_{F_n}(r) f(r)\, dr = \lim_{n\to\infty} P(F_n) \end{aligned}$$

   that is, (2.28) holds, and hence the set function is continuous from below.

We have already seen that if the integral exists, then (2.56) ensures that the first three axioms hold. Thus the existence of the Lebesgue integral on all Borel sets coupled with continuity and the first three axioms ensures that a set function defined in this way is indeed a probability measure. We observe in passing that even if we confined interest to events for which the Riemann integral made sense, it would not follow that the resulting probability measure would be countably additive: as with continuity, these asymptotic properties hold for Lebesgue integration but not for Riemann integration.

How do we reconcile the use of a Lebesgue integral given the assumed prerequisite of traditional engineering calculus courses based on the Riemann integral? Here a standard result of real analysis comes to our aid: if the ordinary Riemann integral exists over a finite interval, then so does the Lebesgue integral, and the two are the same. If the Riemann integral does not exist, then we can try to find the probability as a limit of probabilities of simple events for which the Riemann integrals do exist, e.g., as the limit of probabilities of finite unions of intervals. It is possible for the improper Riemann integral to exist, i.e., a Riemann integral over infinite limits, and yet the Lebesgue integral not to be well defined. For example, $\int_0^\infty (\sin(x)/x)\, dx = \pi/2$ exists when evaluated as a Riemann improper integral but it does not exist as a Lebesgue integral. Riemann calculus will usually suffice for computation (at least if $f(r)$ is Riemann integrable over a finite interval) provided we realize that we may have to take limits of Riemann integrals for complicated events. Observe, for example, that in the case mentioned where $f(r)$ is 1 on $[0, 1]$, the probability of a single point $1/2$

can now be found easily as a limit of Riemann integrals:

$$P\left(\left\{\frac{1}{2}\right\}\right) = \lim_{\epsilon \to 0} \int_{(1/2-\epsilon,\, 1/2+\epsilon)} dr = \lim_{\epsilon \to 0} 2\epsilon = 0,$$

as expected.

In summary, our engineering compromise is this: we must realize that for the theory to be valid and for (2.56) indeed to give a probability measure on subsets of the real line, the integral must be interpreted as a Lebesgue integral and Riemann integrals may not exist. For computation, however, one will almost always be able to find probabilities either by Riemann integration or by taking limits of Riemann integrals over simple events. This distinction between Riemann integrals for computation and Lebesgue integrals for theory is analogous to the distinction between rational numbers and real numbers. Computational and engineering tasks use only arithmetic of finite precision in practice. However, in developing the theory irrational numbers such as $\sqrt{2}$ and π are essential. Imagine how hard it would be to develop a theory without using irrational numbers, and how unwise it would be to do so just because the eventual computations do not use them. So it is with Lebesgue integrals.

Probability density functions

The function f used in (2.54) to (2.56) is called a *probability density function* or *pdf* since it is a nonnegative function that is integrated to find a total mass of probability, just as a mass density function in physics is integrated to find a total mass. Like a pmf, a pdf is defined only for *points* in Ω and not for sets. Unlike a pmf, a pdf is not in itself the probability of anything; for example, a pdf can take on values greater than one, whereas a pmf cannot. Under a pdf, points frequently have probability zero, even though the pdf is nonzero. We can, however, interpret a pdf as being proportional to a probability in the following sense. For a pmf we had $p(x) = P(\{x\})$. Suppose now that the sample space is the real line and that a pdf f is defined. Let $F = [x, x + \Delta x)$, where Δx is extremely small. Then if f is sufficiently smooth, the mean value theorem of calculus implies that

$$P([x, x + \Delta x)) = \int_{x}^{x+\Delta x} f(\alpha)\, d\alpha \approx f(x)\Delta x, \tag{2.66}$$

Thus if a pdf $f(x)$ is multiplied by a differential Δx, it can be interpreted as approximately the probability of being within Δx of x.

Both probability functions, the pmf and the pdf, can be used to define and compute a probability measure: the pmf is summed over all points in the event, and the pdf is integrated over all points in the event. If the sample space is a subset of the real line, both can be used to compute expectations such as moments.

Some of the most common pdf's are listed below. As will be seen, these are indeed valid pdf's; that is, they satisfy (2.54) and (2.55). The pdf's are assumed to be 0 outside the specified domain and are given in terms of real-valued parameters $b, a, \lambda > 0$, m, and $\sigma > 0$.

The uniform pdf. Given $b > a$, $f(r) = 1/(b-a)$ for $r \in [a, b]$.

The exponential pdf. $f(r) = \lambda e^{-\lambda r}$; $r \geq 0$.

The doubly exponential (or Laplacian) pdf. $f(r) = \frac{\lambda}{2} e^{-\lambda |r|}$; $r \in \Re$.

The Gaussian (or Normal) pdf. $f(r) = (2\pi\sigma^2)^{-1/2} \exp(-(r-m)^2/2\sigma^2)$; $r \in \Re$. Since the density is completely described by two parameters, the mean m and variance $\sigma^2 > 0$, it is common to denote it by $\mathcal{N}(m, \sigma^2)$.

Other univariate pdf's may be found in Appendix C.

Just as we used a pdf to construct a probability measure on the space $(\Re, \mathcal{B}(\Re))$, we can also use it to define a probability measure on any smaller space $(A, \mathcal{B}(A))$, where A is a subset of $\Re$.

As a technical detail we note that to ensure that the integrals all behave as expected we must also require that A itself be a Borel set of $\Re$ so that it is precluded from being too intractable a set. Such probability spaces can be considered to have a sample space of either $\Re$ or A, as convenient. In the former case events outside A will have zero probability.

Computational examples

This section is less detailed than its counterpart for discrete probability because generally engineers are more familiar with common integrals than with common sums. We confine the discussion to a few observations and to an example of a multidimensional probability computation.

The uniform pdf is trivially a valid pdf because it is nonnegative and its integral is simply the length of the interval on which it is nonzero, $b - a$, divided by the length. For simplicity consider the case where $a = 0$ and $b = 1$ so that $b - a = 1$. In this case the probability of any interval within $[0, 1)$ is simply the length of the interval. The mean is easily found to be

$$m = \int_0^1 r\,dr = \left.\frac{r^2}{2}\right|_0^1 = \frac{1}{2}, \tag{2.67}$$

the second moment is

$$m^{(2)} = \int_0^1 r^2\,dr = \left.\frac{r^3}{3}\right|_0^1 = \frac{1}{3}, \tag{2.68}$$

and the variance is

$$\sigma^2 = \frac{1}{3} - \left(\frac{1}{2}\right)^2 = \frac{1}{12}. \tag{2.69}$$

The validation of the pdf and the mean, second moment, and variance of the exponential pdf can be found from integral tables or by the integral analog to the corresponding computations for the geometric pmf, as described in Appendix B. In particular, it follows from (B.9) that

$$\int_0^\infty \lambda e^{-\lambda r}\,dr = 1, \tag{2.70}$$

from (B.10) that

$$m = \int_0^\infty r\lambda e^{-\lambda r}\,dr = \frac{1}{\lambda} \tag{2.71}$$

and

$$m^{(2)} = \int_0^\infty r^2\lambda e^{-\lambda r}\,dr = \frac{2}{\lambda^2}, \tag{2.72}$$

and hence from (2.65)

$$\sigma^2 = \frac{2}{\lambda^2} - \frac{1}{\lambda^2} = \frac{1}{\lambda^2}. \tag{2.73}$$

The moments can also be found by integration by parts.

The Laplacian pdf is simply a mixture of an exponential pdf and its reverse, so its properties follow from those of an exponential pdf. The details are left as an exercise.

The Gaussian pdf example is more involved. In Appendix B, it is shown (in the development leading up to (B.13)) that

$$\int_{-\infty}^\infty \frac{1}{\sqrt{2\pi\sigma^2}} e^{-(x-m)^2/2\sigma^2}\,dx = 1. \tag{2.74}$$

It is reasonably easy to find the mean by inspection. The function $g(x) =$

$(x-m)\exp(-(x-m)^2/2\sigma^2)$ is an odd function, i.e., it has the form $g(-x) = -g(x)$, and hence its integral is 0 if the integral exists at all.

This means that

$$\int_{-\infty}^{\infty} \frac{1}{\sqrt{2\pi\sigma^2}} x e^{-(x-m)^2/2\sigma^2}\, dx = m. \tag{2.75}$$

The second moment and variance are most easily handled by the transform methods to be developed in Chapter 4. Their evaluation will be deferred until then, but we observe that the parameter σ^2 which we have called the variance is in fact the variance, i.e.,

$$\int_{-\infty}^{\infty} \frac{1}{\sqrt{2\pi\sigma^2}} (x-m)^2 e^{-(x-m)^2/2\sigma^2}\, dx = \sigma^2. \tag{2.76}$$

Computing probabilities with the various pdf's varies in difficulty. For simple pdf's one can easily find the probabilities of simple sets such as intervals. For example, with a uniform pdf on $[a,b]$, then for any $a \le c < d \le b$ $\Pr([c,d]) = (d-c)/(b-a)$, the probability of an interval is proportional to the length of the integral. For the exponential pdf, the probability of an interval $[c,d]$, $0 \le c < d$, is given by

$$\Pr([c,d]) = \int_c^d \lambda e^{-\lambda x}\, dx = e^{-\lambda c} - e^{-\lambda d}. \tag{2.77}$$

The Gaussian pdf does not yield nice closed-form solutions for the probabilities of simple sets like intervals, but it is well tabulated. Unfortunately there are several variations in table construction. The most common forms are the Φ- function

$$\Phi(\alpha) = \frac{1}{\sqrt{2\pi}} \int_{-\infty}^{\alpha} e^{-u^2/2}\, du, \tag{2.78}$$

which is the probability of the simple event $(-\infty,\alpha] = \{x : x \le \alpha\}$ for a zero-mean unit-variance Gaussian pdf $\mathcal{N}(0,1)$. The Q function is the complementary function

$$Q(\alpha) = \frac{1}{\sqrt{2\pi}} \int_{\alpha}^{\infty} e^{-u^2/2}\, du = 1 - \Phi(\alpha). \tag{2.79}$$

The Q function is used primarily in communications systems analysis where probabilities of exceeding a threshold describe error events in detection systems. The error function is defined by

$$\operatorname{erf}(\alpha) = \frac{2}{\sqrt{\pi}} \int_0^{\alpha} e^{-u^2}\, du \tag{2.80}$$

and it is related to the Q and Φ functions by

$$Q(\alpha) = \frac{1}{2}\left(1 - \operatorname{erf}\left(\frac{\alpha}{\sqrt{2}}\right)\right) = 1 - \Phi(\alpha). \tag{2.81}$$

Thus, for example, the probability of the set $(-\infty, \alpha)$ for $\mathcal{N}(m, \sigma^2)$ is found by changing variables $u = (x - m)/\sigma$ to be

$$\begin{aligned} P(\{x : x \leq \alpha\}) &= \int_{-\infty}^{\alpha} \frac{1}{\sqrt{2\pi\sigma^2}} e^{-(x-m)^2/2\sigma^2}\, dx \\ &= \int_{-\infty}^{(\alpha-m)/\sigma} \frac{1}{\sqrt{2\pi}} e^{-u^2/2}\, du \\ &= \Phi\left(\frac{\alpha - m}{\sigma}\right) = 1 - Q\left(\frac{\alpha - m}{\sigma}\right). \end{aligned} \tag{2.82}$$

The probability of an interval $(a, b]$ is then given by

$$P((a, b]) = P((-\infty, b]) - P((-\infty, a]) = \Phi\left(\frac{b - m}{\sigma}\right) - \Phi\left(\frac{a - m}{\sigma}\right). \tag{2.83}$$

Observe that the symmetry of a Gaussian density implies that

$$1 - \Phi(a) = \Phi(-a). \tag{2.84}$$

As a multidimensional example of probability computation, suppose that the sample space is $\Re^2$, the space of all pairs of real numbers. The probability space consists of this sample space, the corresponding Borel field, and a probability measure described by a pdf

$$f(x, y) = \begin{cases} \lambda\mu e^{-\lambda x - \mu y}; & x \in [0, \infty),\ y \in [0, \infty) \\ 0 & \text{otherwise} \end{cases}.$$

What is the probability of the event $F = \{(x, y) : x < y\}$? As an interpretation, the sample points (x, y) might correspond to the arrival times of two distinct types of particle at a sensor following its activation, say type A and type B for x and y, respectively. Then the event F is the event that a particle of type A arrives at the sensor before one of type B. Computation of the probability is then accomplished as

$$\begin{aligned} P(F) &= \int\int_{(x,y):(x,y)\in F} f(x, y)\, dx\, dy \\ &= \int\int_{(x,y):x\geq 0, y\geq 0, x<y} \lambda\mu e^{-\lambda x - \mu y}\, dx\, dy. \end{aligned}$$

This integral is a two-dimensional integral of its argument over the indicated region. Correctly describing the limits of integration is often the hardest part of computing probabilities. Note in particular the inclusion of the facts that both x and y are nonnegative (since otherwise the pdf is 0). The $x < y$ region for nonnegative x and y is most easily envisioned as the region of the first quadrant lying above the line $x = y$, if x and y correspond to the horizontal and vertical axes, respectively. Completing the calculus:

$$\begin{aligned} P(F) &= \lambda\mu \int_0^\infty dy \left(\int_0^y dx e^{-\lambda x} e^{-\mu y} \right) \\ &= \lambda\mu \int_0^\infty dy e^{-\mu y} \left(\int_0^y dx e^{-\lambda x} \right) \\ &= \lambda\mu \int_0^\infty dy e^{-\mu y} \frac{1}{\lambda}(1 - e^{-\lambda y}) \\ &= \mu \left(\int_0^\infty dy e^{-\mu y} - \int_0^\infty dy e^{-(\mu+\lambda) y} \right) \\ &= 1 - \frac{\mu}{\mu + \lambda} = \frac{\lambda}{\mu + \lambda}. \end{aligned}$$

Mass functions as densities

As in systems theory, discrete problems can be considered as continuous problems with the aid of the Dirac delta or unit impulse $\delta(t)$, a generalized function or singularity function (also, misleadingly, called a *distribution*) with the property that for any smooth function $\{g(r);\ r \in \Re\}$ and any $a \in \Re$

$$\int g(r)\delta(r - a)\, dr = g(a). \tag{2.85}$$

Given a pmf p defined on a subset of the real line $\Omega \subset \Re$, we can define a pdf f by

$$f(r) = \sum p(\omega)\delta(r - \omega). \tag{2.86}$$

This is indeed a pdf since $f(r) \geq 0$ and

$$\begin{aligned} \int f(r)\, dr &= \int \left(\sum p(\omega)\delta(r - \omega) \right) dr \\ &= \sum p(\omega) \int \delta(r - \omega)\, dr \\ &= \sum p(\omega) = 1. \end{aligned}$$

In a similar fashion, probabilities are computed as

$$\begin{aligned}\int 1_F(r)f(r)\,dr &= \int 1_F(r)\left(\sum p(\omega)\delta(r-\omega)\right)dr \\ &= \sum p(\omega)\int 1_F(r)\delta(r-\omega)\,dr \\ &= \sum p(\omega)1_F(\omega) = P(F).\end{aligned}$$

Given that discrete probability can be handled using the tools of continuous probability in this fashion, it is natural to inquire wheather pdf's should be used in both the discrete and continuous case. The main reason for not doing so is simplicity. Pmf's and sums are usually simpler to handle and evaluate than pdf's and integrals. Questions of existence and limits rarely arise, and the notation is simpler. In addition, the use of the theory of generalized functions is assumed in order to treat integrals involving Dirac deltas as if they were ordinary integrals, so additional mathematical machinery is required. As a result, this approach is rarely used in genuinely discrete problems. On the other hand, if one is dealing with a hybrid problem that has both discrete and continuous components, then this approach may make sense because it allows the use of a single probability function, a pdf, throughout.

Multidimensional pdf's

By considering multidimensional integrals we can also extend the construction of probabilities by integrals to finite-dimensional product spaces, e.g., $\Re^k$.

Given the measurable space $(\Re^k, \mathcal{B}(\Re)^k)$, say we have a real-valued function f on R^k with the properties that

$$f(\mathbf{x}) \geq 0\ ;\ \ \text{all } \mathbf{x} = (x_0, x_1, \ldots, x_{k-1}) \in \Re^k, \tag{2.87}$$

$$\int_{\Re^k} f(\mathbf{x})\,d\mathbf{x} = 1. \tag{2.88}$$

Then define a set function P by

$$P(F) = \int_F f(\mathbf{x})\,d\mathbf{x},\ \text{all } F \in \mathcal{B}(\Re)^k, \tag{2.89}$$

where the vector integral is shorthand for the k-dimensional integral, that is,

$$P(F) = \int_{(x_0, x_1, \ldots, x_{k-1}) \in F} f(x_0, x_1, \ldots, x_{k-1})\, dx_0\, dx_1 \ldots dx_{k-1}.$$

Note that (2.87) to (2.89) are exact vector equivalents of (2.54) to (2.56). As with multidimensional pmf's, a pdf is not itself the probability of anything. As in the scalar case, however, the mean value theorem of calculus can be used to interpret the pdf as being proportional to the probability of being in a very small region around a point, i.e., that

$$P(\{(\alpha_0, \alpha_1, \ldots, \alpha_{k-1}) : x_i \leq \alpha_i < x_i + \Delta_i;\ i = 0, 1, \ldots, n-1\}) \approx f(x_0, x_1, \ldots, x_{k-1}) \Delta_0 \Delta_1 \cdots \Delta_{n-1}. \quad (2.90)$$

Is P defined by (2.89) a probability measure? The answer is a qualified *yes* with exactly the same qualifications as in the one-dimensional case.

As in the one-dimensional sample space, a function f with the above properties is called a probability density function or pdf. To be more concise we will occasionally refer to a pdf on k-dimensional space as a k-dimensional pdf.

There are two common and important examples of k-dimensional pdf's. These are defined next. In both examples the dimension k of the sample space is fixed and the pdf's induce a probability measure on $(\Re^k, \mathcal{B}(\Re)^k)$ by (2.89).

[2.16] *The product pdf.*

Let f_i; $i = 0, 1, \ldots, k-1$, be a collection of one-dimensional pdf's; that is, $f_i(r)$; $r \in \Re$ satisfies (2.54) and (2.55) for each $i = 0, 1, \ldots, k-1$. Define the product k-dimensional pdf f by

$$f(\mathbf{x}) = f(x_0, x_1, \ldots, x_{k-1}) = \prod_{i=0}^{k-1} f_i(x_i).$$

The product pdf in k-dimensional space is simply the product of k pdf's on one-dimensional space. The one-dimensional pdf's are called the *marginal* pdf's, and the multidimensional pdf is sometimes called a *joint* pdf. It is easy to verify that the product pdf integrates to 1.

The case of greatest importance is when all of the marginal pdf's are identical, that is, when $f_i(r) = f_0(r)$ for all i. Note that any of the previously defined pdf's on $\Re$ yield a corresponding multidimensional pdf by this construction. In a similar manner we can construct pmf's on discrete product spaces as a product of marginal pmf's.

[2.17] *The multidimensional Gaussian pdf.*
Let $\mathbf{m} = (m_0, m_1, \ldots, m_{k-1})^t$ denote a column vector (the superscript t stands for "transpose"). Let Λ denote a k by k square matrix with entries $\{\lambda_{i,j};\ i = 0, 1, \ldots, k-1;\ j = 0, 1, \ldots, k-1\}$. Assume that Λ is symmetric; that is, that $\Lambda^t = \Lambda$ or, equivalently, that $\lambda_{i,j} = \lambda_{j,i}$, all i, j. Assume also that Λ is *positive definite;* that is, for any nonzero vector $\mathbf{y} \in \Re^k$ the quadratic form $\mathbf{y}^t \Lambda \mathbf{y}$ is positive;

$$\mathbf{y}^t \Lambda \mathbf{y} = \sum_{i=0}^{k-1} \sum_{j=0}^{k-1} y_i \lambda_{i,j} y_j > 0.$$

A multidimensional pdf is said to be Gaussian if it has the following form for some vector $\mathbf{m}$ and matrix Λ satisfying the above conditions:

$$f(\mathbf{x}) = (2\pi)^{-k/2} (\det \Lambda)^{-1/2} e^{-(\mathbf{x}-\mathbf{m})^t \Lambda^{-1} (\mathbf{x}-\mathbf{m})/2} \ ; \ \mathbf{x} \in \Re^k.$$

where $\det \Lambda$ is the determinant of the matrix Λ.

Since the matrix Λ is positive definite, the inverse of Λ exists and hence the pdf is well defined. It is also necessary for Λ to be positive definite if the integral of the pdf is to be finite. The Gaussian pdf may appear complicated, but it will later be seen to be one of the simplest to deal with. We shall later develop the significance of the vector $\mathbf{m}$ and matrix Λ. Note that if Λ is a diagonal matrix, Example [2.17] reduces to a special case of Example [2.16].

The reader must either accept on faith that the multidimensional Gaussian pdf integrates to 1 or seek out a derivation.

The Gaussian pdf can be extended to complex vectors if the constraints on Λ are modified to require that $\Lambda^* = \Lambda$, where the asterix denotes conjugate transpose, and where for any vector $\mathbf{y}$ not identically 0 it is required that $\mathbf{y}^* \Lambda \mathbf{y} > 0$.

[2.18] *Mixtures.*
Suppose that P_i, $i = 1, 2, \ldots, \infty$ is a collection of probability measures on a common measurable space $(\Omega, \mathcal{F})$, and let a_i, $i = 1, 2, \ldots$ be non-negative numbers that sum to 1. Then the set function determined by

$$P(F) = \sum_{i=1}^{\infty} a_i P_i(F)$$

is also a probability measure on $(\Omega, \mathcal{F})$. This relation is usually abbreviated to

$$P = \sum_{i=1}^{\infty} a_i P_i .$$

The first two axioms are obviously satisfied by P, and countable additivity follows from the properties of sums. (Finite additivity is easily demonstrated for the case of a finite number of nonzero a_i.) A probability measure formed in this way is called a *mixture*. Observe that this construction can be used to form a probability measure with both discrete and continuous aspects. For example, let Ω be the real line and $\mathcal{F}$ the Borel field; suppose that f is a pdf and p is a pmf; then for any $\lambda \in (0, 1)$ the measure P defined by

$$P(F) = \lambda \sum_{x \in F} p(x) + (1 - \lambda) \int_{x \in F} f(x)\, dx$$

combines a discrete portion described by p and a continuous portion described by f. Expectations can be computed in a similar way. Given a function g,

$$E(g) = \lambda \sum_{x \in \Omega} g(x) p(x) + (1 - \lambda) \int_{x \in \Omega} g(x) f(x)\, dx .$$

Note that this construction works for both scalar and vector spaces. This combination of discrete and continuous attributes is one of the main applications of mixtures. Another is in modeling a random process where there is some uncertainty about the parameters of the experiment. For example, consider a probability space for the following experiment: first a fair coin is flipped and a 0 or 1 (tail or head) observed. If the coin toss results in a 1, then a fair die described by a uniform pmf p_1 is rolled, and the outcome is the result of the experiment. If the coin toss results in a 0, then a biased die described by a nonuniform pmf p_2 is rolled, and the outcome is the result of the experiment. The pmf of the overall experiment is then the mixture $p_1/2 + p_2/2$. The mixture model captures our ignorance of which die we will be rolling.

2.6 Independence

Given a probability space $(\Omega, \mathcal{F}, P)$, two events F and G are defined to be *independent* if $P(F \cap G) = P(F)P(G)$. A collection of events $\{F_i;\ i = 0, 1, \ldots, k-1\}$ is said to be *independent* or *mutually independent* if for any distinct subcollection $\{F_{l_i};\ i = 0, 1, \ldots, m-1\}$, $l_m < k$, we have that

$$P\left(\bigcap_{i=0}^{m-1} F_{l_i}\right) = \prod_{i=0}^{m-1} P(F_{l_i}).$$

In words: the probability of the intersection of any subcollection of the given events equals the product of the probabilities of the separate events. Unfortunately it is not enough simply to require that $P\left(\bigcap_{i=0}^{k-1} F_i\right) = \prod_{i=0}^{k-1} P(F_i)$ as this does not imply a similar result for all possible subcollections of events, which is what will be needed. For example, consider the following case where $P(F \cap G \cap H) = P(F)P(G)P(H)$ for three events F, G, and H, yet it is not true that $P(F \cap G) = P(F)P(G)$

$$\begin{aligned}
P(F) = P(G) = P(H) &= \frac{1}{3} \\
P(F \cap G \cap H) &= \frac{1}{27} = P(F)P(G)P(H) \\
P(F \cap G) = P(G \cap H) = P(F \cap H) &= \frac{1}{27} \neq P(F)P(G).
\end{aligned}$$

The example places zero probability on the overlap $F \cap G$ except where it also overlaps H, i.e., $P(F \cap G \cap H^c) = 0$. Thus in this case $P(F \cap G \cap H) = P(F)P(G)P(H) = 1/27$, but $P(F \cap G) = 1/27 \neq P(F)P(G) = 1/9$.

The concept of independence in the probabilistic sense we have defined relates easily to the intuitive idea of independence of physical events. For example, if a fair die is rolled twice, one would expect the second roll to be unrelated to the first roll because there is no physical connection between the individual outcomes. Independence in the probabilistic sense is reflected in this experiment. The probability of any given outcome for either of the individual rolls is $1/6$. The probability of any given pair of outcomes is $(1/6)^2 = 1/36$: that is, the addition of a second outcome diminishes the overall probability by exactly the probability of the individual event, $1/6$. Note that the probabilities are not *added* – the probability of two successive outcomes cannot reasonably be greater than the probability of either of the outcomes alone. Do *not*, however, confuse the concept of independence with the concept of disjoint or mutually exclusive events. If you roll the die once, the event the roll is a one is not independent of the event the roll is a six. Given one event, the other cannot happen – they are neither physically nor probabilistically independent. They are mutually exclusive events.

2.7 Elementary conditional probability

Intuitively, independence of two events means that the occurrence of one event should not affect the occurrence of the other. For example, the knowledge of the outcome of the first roll of a die should not change the probabilities for the outcome of the second roll of the die if the die has no memory. To be more precise, the notion of conditional probability is required. Consider the following motivation. Suppose that $(\Omega, \mathcal{F}, P)$ is a probability space and that an observer is told that an event G has already occurred. The observer thus has a-posteriori knowledge of the experiment. The observer is then asked to calculate the probability of another event F given this information. We will denote this probability of F given G by $P(F|G)$. Thus instead of the ıa-priori or *unconditional* probability $P(F)$, the observer must compute the *a-posteriori* or *conditional* probability $P(F|G)$, read as "the probability that event F occurs given that the event G occurred." For a fixed G the observer should be able to find $P(F|G)$ for all events F; thus the observer is in fact being asked to describe a new probability measure, say P_G, on $(\Omega, \mathcal{F})$. How should this be defined? Intuition will lead to a useful definition and this definition will provide a useful interpretation of independence.

First, since the observer has been told that G has occurred and hence $\omega \in G$, clearly the new probability measure P_G must assign zero probability to the set of all ω outside G; that is, we should have

$$P(G^c|G) = 0 \tag{2.91}$$

or, equivalently,

$$P(G|G) = 1. \tag{2.92}$$

Equation (2.91) plus the axioms of probability in turn imply that

$$P(F|G) = P(F \cap (G \cup G^c)|G) = P(F \cap G|G). \tag{2.93}$$

Second, there is no reason to suspect that the *relative* probabilities within G should change because of the knowledge that G occurred. For example, if an event $F \subset G$ is twice as probable as an event $H \subset G$ with respect to P, then the same should be true with respect to P_G. For arbitrary events F and H, the events $F \cap G$ and $H \cap G$ are both in G, and hence this preservation of relative probability implies that

$$\frac{P(F \cap G|G)}{P(H \cap G|G)} = \frac{P(F \cap G)}{P(H \cap G)}.$$

But if we take $H = \Omega$ in this formula and use (2.92)–(2.93), we have that

$$P(F|G) = P(F \cap G|G) = \frac{P(F \cap G)}{P(G)}, \tag{2.94}$$

which is in fact the formula we now use to *define* the conditional probability of the event F given the event G. The conditional probability can be interpreted as "cutting down" the original probability space to a probability space with the smaller sample space G and with probabilities equal to the renormalized probabilities of the intersection of events with the given event G on the original space.

This definition meets the intuitive requirements of the derivation, but does it make sense and does it fulfill the original goal of providing an interpretation for independence? It does make sense provided $P(G) > 0$, that is, the conditioning event does not have zero probability. This is in fact the distinguishing requirement that makes the above definition work for what is known as *elementary* conditional probability. Nonelementary conditional probability will provide a more general definition that will work for conditioning events having zero probability, such as the event that a fair spin of a pointer results in a reading of exactly $1/\pi$. Further, if P is a probability measure, then it is easy to see that P_G defined by $P_G(F) = P(F|G)$ for $F \in \mathcal{F}$ is also a probability measure on the same space (remember G stays *fixed*), i.e., P_G is a normalized and countably additive function of events. As to independence, suppose that F and G are independent events and that $P(G) > 0$, then

$$P(F|G) = \frac{P(F \cap G)}{P(G)} = P(F);$$

the probability of F is not affected by the knowledge that G has occurred. This is exactly what one would expect from the intuitive notion of the independence of two events. Note, however, that it would not be as useful to define independence of two events by requiring $P(F) = P(F|G)$ since it would be less general than the product definition; it requires that one of the events have a nonzero probability.

Conditional probability provides a means of constructing new probability spaces from old ones by using conditional pmf's and elementary conditional pdf's.

Examples

[2.19] Suppose that $(\Omega, \mathcal{F}, P)$ is a probability space described by a pmf p and that A is an event with nonzero probability. Then the pmf p_A

defined by

$$p_A(\omega) = \begin{cases} p(\omega)/P(A) = P(\{\omega\}|A), & \omega \in A \\ 0 & \omega \notin A \end{cases}$$

implies a probability space $(\Omega, \mathcal{F}, P_A)$, where

$$P_A(F) = \sum_{\omega \in F} p_A(\omega) = P(F|A).$$

We call p_A a *conditional pmf*. More specifically, it is the conditional pmf given the event A. In some cases it may be more convenient to define the conditional pmf on the sample space A and hence the conditional probability measure on the original event space.

As an example, suppose that p is a geometric pmf and that $A = \{\omega : \omega \geq K\} = \{K, K+1, \ldots\}$. In this case the conditional pmf given that the outcome is greater than or equal to K is

$$\begin{aligned} p_A(k) &= \frac{(1-p)^{k-1}p}{\sum_{l=K}^{\infty}(1-p)^{l-1}p} = \frac{(1-p)^{k-1}p}{(1-p)^{K-1}} \\ &= (1-p)^{k-K}p;\ k = K+1, K+2, \ldots, \end{aligned} \tag{2.95}$$

which can be recognized as a geometric pmf that begins at $k = K + 1$.

[2.20] Suppose that $(\Omega, \mathcal{F}, P)$ is a probability space described by a pdf f and that A is an event with nonzero probability. Then the f_A defined by

$$f_A(\omega) = \begin{cases} f(\omega)/P(A) & \omega \in A \\ 0 & \omega \in A \end{cases}$$

is a pdf on A and describes a probability measure

$$P_A(F) = \int_{\omega \in F} f_A(\omega)\, d\omega = P(F|A). \tag{2.96}$$

f_A is called an *elementary conditional pdf* (given the event A). The word "elementary" reflects the fact that the conditioning event has nonzero probability. We will later see how conditional probability can be usefully extended to conditioning on events of zero probability and why they are important.

As a simple example, consider the continuous analog of the previous conditional geometric pmf example. Given an exponential pdf and $A = \{r : r \geq$

$c\}$, define

$$f_A(x) = \frac{\lambda e^{-\lambda x}}{\int_c^\infty \lambda e^{-\lambda y}\,dy} = \frac{\lambda e^{-\lambda x}}{e^{-\lambda c}}$$
$$= \lambda e^{-\lambda(x-c)};\ x \geq c, \tag{2.97}$$

which can be recognized as an exponential pdf that starts at c. The exponential pdf and geometric pmf share the unusual property that conditioning on the output being larger than some number does not change the basic form of the pdf or pmf, only its starting point. This has the discouraging implication that if, for example, the time for the next arrival of a bus is described by an exponential pdf, then knowing you have already waited for an hour does not change your pdf to the next arrival from what it was when you arrived.

2.8 Problems

1. Suppose that you have a set function P defined for all subsets $F \subset \Omega$ of a sample space Ω and suppose that you know that this set function satisfies (2.7)–(2.9). Show that for arbitrary (not necessarily disjoint) events,
$$P(F \cup G) = P(F) + P(G) - P(F \cap G).$$
2. Given a probability space $(\Omega, \mathcal{F}, P)$, and letting F, G, and H be events such that $P(F \cap G|H) = 1$, which of the following statements are true? Why or why not?
(a) $P(F \cap G) = 1$
(b) $P(F \cap G \cap H) = P(H)$
(c) $P(F^c|H) = 0$
(d) $H = \Omega$
3. Describe the sigma-field of subsets of $\Re$ generated by the points or singleton sets. Does this sigma-field contain intervals of the form (a, b) for $b > a$?
4. Given a finite subset A of the real line $\Re$, prove that the power set of A and $\mathcal{B}(A)$ are the same. Repeat for a countably infinite subset of $\Re$.
5. Given that the discrete sample space Ω has n elements, show that the power set of Ω consists of 2^n elements.
6. $\star$Let $\Omega = \Re$, the real line, and consider the collection $\mathcal{F}$ of subsets of $\Re$ defined as all sets of the form
$$\bigcup_{i=0}^{k}(a_i, b_i] \cup \bigcup_{j=0}^{m}(c_j, d_j]^c$$
for all possible choices of nonnegative integers k and m and all possible choices of real numbers $a_i < b_i$, $c_i < d_i$. If k or m is 0, then the respective unions are defined to be empty so that the empty set itself has the form given. In other

words, $\mathcal{F}$ contains all possible *finite* unions of half-open intervals of this form and complements of such half-open intervals. Every set of this form is in $\mathcal{F}$ and every set in $\mathcal{F}$ has this form. Prove that $\mathcal{F}$ is a field of subsets of Ω. Does $\mathcal{F}$ contain the points? For example, is the singleton set $\{0\}$ in $\mathcal{F}$? Is $\mathcal{F}$ a sigma-field?

7. Let $\Omega = [0, \infty)$ be a sample space and let $\mathcal{F}$ be the sigma-field of subsets of Ω generated by all sets of the form $(n, n+1)$ for $n = 1, 2, \ldots$
   (a) Are the following subsets of Ω in $\mathcal{F}$? (i) $[0, \infty)$, (ii) $\mathcal{Z}_+ = \{0, 1, 2, \ldots\}$, (iii) $[0, k] \cup [k+1, \infty)$ for any positive integer k, (iv) $\{k\}$ for any positive integer k, (v) $[0, k]$ for any positive integer k, (vi) $(1/3,\ 2)$.
   (b) Define the following set function on subsets of Ω :
   $$P(F) = c \sum_{i \in \mathcal{Z}_+ : i+1/2 \in F} 3^{-i}.$$
   (If there is no i for which $i + 1/2 \in F$, then the sum is taken as zero.) Is P a probability measure on $(\Omega, \mathcal{F})$ for an appropriate choice of c? If so, what is c?
   (c) Repeat part (b) with $\mathcal{B}$, the Borel field, replacing $\mathcal{F}$ as the event space.
   (d) Repeat part (b) with the power set of $[0, \infty)$ replacing $\mathcal{F}$ as the event space.
   (e) Find $P(F)$ for the sets F considered in part (a).
8. Show that an equivalent axiom to Axiom 2.3 of probability is the following:
   $$\text{If } F \text{ and } G \text{ are disjoint, then } P(F \cup G) = P(F) + P(G),$$
   that is, we really need only specify finite additivity for the special case of $n = 2$.
9. Consider the measurable space $([0,1], \mathcal{B}([0,1]))$. Define a set function P on this space as follows:
   $$P(F) = \begin{cases} 1/2 & \text{if } 0 \in F \text{ or } 1 \in F \text{ but not both} \\ 1 & \text{if } 0 \in F \text{ and } 1 \in F \\ 0 & \text{otherwise} \end{cases}.$$
   Is P a probability measure?
10. Let $\mathcal{S}$ be a sphere in $\Re^3 : \mathcal{S} = \{(x, y, z) : x^2 + y^2 + z^2 \le r^2\}$, where r is a fixed radius. In the sphere are fixed N molecules of gas, each molecule being considered as an infinitesimal volume (that is, it occupies only a point in space). Define for *any* subset F of $\mathcal{S}$ the function
    $$\#(F) = \{\text{the number of molecules in } F\}.$$
    Show that $P(F) = \#(F)/N$ is a probability measure on the measurable space consisting of $\mathcal{S}$ and its power set.
11. $\star$ Suppose that you are given a probability space $(\Omega, \mathcal{F}, P)$ and that a collection

$\mathcal{F}_P$ of subsets of Ω is defined by

$$\mathcal{F}_P = \{F \cup N; \ \text{all } F \in \mathcal{F}, \ \text{all } N \subset G \text{ for which } G \in \mathcal{F} \text{ and } P(G) = 0\}. \tag{2.98}$$

In words: $\mathcal{F}_P$ contains every event in $\mathcal{F}$ along with every subset N which is a subset of zero probability event $G \in \mathcal{F}$, whether or not N is itself an event (a member of $\mathcal{F}$). Thus $\mathcal{F}_P$ is formed by adding any sets not already in $\mathcal{F}_P$ which happen to be subsets of zero probability events. We can define a set function $\overline{P}$ for the measurable space $(\Omega, \mathcal{F}_P)$ by

$$\overline{P}(F \cup N) = P(F) \text{ if } F \in \mathcal{F} \text{ and } N \subset G \in \mathcal{F}, \text{ where } P(G) = 0. \tag{2.99}$$

Show that $(\Omega, \mathcal{F}_P, \overline{P})$ is a probability space, i.e., show that $\mathcal{F}_P$ is an event space and that $\overline{P}$ is a probability measure. A probability space with the property that all subsets of zero probability events are also events is said to be *complete* and the probability space $(\Omega, \mathcal{F}_P, \overline{P})$ is called the *completion* of the probability space $(\Omega, \mathcal{F}, P)$.

In Problems 2.12 to 2.18 let $(\Omega, \mathcal{F}, P)$ be a probability space and assume that all given sets are events.

12. If $G \subset F$, prove that $P(F - G) = P(F) - P(G)$. Use this fact to prove that if $G \subset F$, then $P(G) \leq P(F)$.
13. Let $\{F_i\}$ be a countable partition of a set G. Prove that for any event H,

$$\sum_i P(H \cap F_i) = P(H \cap G).$$

14. If $\{F_i, \, i = 1, 2, \ldots\}$ forms a partition of Ω and $\{G_i; \ i = 1, 2, \ldots\}$ forms a partition of Ω, prove that for any H,

$$P(H) = \sum_{i=1}^{\infty} \sum_{j=1}^{\infty} P(H \cap F_i \cap G_j).$$

15. Prove that $|P(F) - P(G)| \leq P(F \Delta G)$.
    In words: if the probability of the symmetric difference of two events is small, then the two events must have approximately the same probability.
16. Prove that $P(F \cup G) \leq P(F) + P(G)$. Prove more generally that for any sequence (i.e., countable collection) of events F_i,

$$P\left(\bigcup_{i=1}^{\infty} F_i\right) \leq \sum_{i=1}^{\infty} P(F_i).$$

    This inequality is called the *union bound* or the *Bonferroni inequality*. (*Hint:* use Problem A.2 or Problem 2.1.)
17. Prove that for any events F, G, and H,

$$P(F \Delta G) \leq P(F \Delta H) + P(H \Delta G).$$

The astute observer may recognize this as a form of the triangle inequality; one can consider $P(F\Delta G)$ as a *distance* or *metric* on events.

18. Prove that if $P(F) \geq 1-\delta$ and $P(G) \geq 1-\delta$, then also $P(F \cap G) \geq 1-2\delta$. In other words, if two events have probability nearly one, then their intersection has probability nearly one.
19. $\star$*The Cantor set.* Consider the probability space $(\Omega, \mathcal{B}(\Omega), P)$ where P is described by a uniform pdf on $\Omega = [0,1)$. Let $F_1 = (1/3, 2/3)$, the middle third of the sample space. Form the set $G_1 = \Omega - F_1$ by removing the middle third of the unit interval. Next define F_2 as the union of the middle thirds of all of the intervals in G_1, i.e., $F_2 = (1/9, 2/9)\bigcup(7/9, 8/9)$. Define G_2 as what remains when remove F_2 from G_1, that is,

$$G_2 = G_1 - F_2 = [0,1] - (F_1 \bigcup F_2).$$

Continue in this manner. At stage n F_n is the union of the middle thirds of all of the intervals in $G_{n-1} = [0,1] - \bigcup_{k=1}^{n-1} F_n$. The Cantor set is defined as the limit of the G_n, that is,

$$C = \bigcap_{n=1}^{\infty} G_n = [0,1] - \bigcup_{n=1}^{\infty} F_n. \tag{2.100}$$

(a) Prove that $C \in \mathcal{B}(\Omega)$, i.e., that it is an event.
(b) Prove that

$$P(F_n) = \frac{1}{3}\left(\frac{2}{3}\right)^{n-1} ; \; n = 1, 2, \ldots \tag{2.101}$$

(c) Prove that $P(C) = 0$, i.e., that the Cantor set has zero probability.
One thing that makes this problem interesting is that unlike most simple examples of nonempty events with zero probability, the Cantor set has an uncountable infinity of points and is not a discrete set of points. This can be shown by first showing that a point $x \in C$ if and only if the point can be expressed as a ternary number $x = \sum_{n=1}^{\infty} a_n 3^{-n}$ where all the a_n are either 0 or 2. Thus the number of points in the Cantor set is the same as the number of real numbers that can be expressed in this fashion, which is the same as the number of real numbers that can be expressed in a binary expansion (since each a_n can have only two values), which is the same as the number of points in the unit interval, which is uncountably infinite.
20. Six people sit at a circular table and pass around and roll a single fair die (equally probable to have any face 1 through 6 showing) beginning with person no. 1. The game continues until the first 6 is rolled, the person who rolled it wins the game. What is the probability that player no. 2 wins?
21. Show that given Equations (2.22) through (2.24), (2.28) or (2.29) implies (2.25). Thus (2.25), (2.28), and (2.29) provide equivalent candidates for the fourth axiom of probability.

22. Suppose that P is a probability measure on the real line and define the sets $F_n = (0, 1/n)$ for all positive integer n. Evaluate $\lim_{n\to\infty} P(F_n)$.
23. Answer true or false for each of the following statements. Answers must be justified.
    (a) The following is a valid probability measure on the sample space $\Omega = \{1, 2, 3, 4, 5, 6\}$ with event space $\mathcal{F}$ = all subsets of Ω:
$$P(F) = \frac{1}{21} \sum_{i \in F} i; \text{ all } F \in \mathcal{F}.$$
    (b) The following is a valid probability measure on the sample space $\Omega = \{1, 2, 3, 4, 5, 6\}$ with event space $\mathcal{F}$ = all subsets of Ω:
$$P(F) = \begin{cases} 1 & \text{if } 2 \in F \text{ or } 6 \in F \\ 0 & \text{otherwise} \end{cases}.$$
    (c) If $P(G \cup F) = P(F) + P(G)$, then F and G are independent.
    (d) $P(F|G) \geq P(G)$ for all events F and G.
    (e) Mutually exclusive (disjoint) events with nonzero probability cannot be independent.
    (f) For any finite collection of events F_i; $i = 1, 2, \ldots, N$
$$P(\cup_{i=1}^{N} F_i) \leq \sum_{i=1}^{N} P(F_i).$$
24. Prove or provide a counterexample for the relation $P(F|G) + P(F|G^c) = P(F)$.
25. Find the mean, second moment, and variance of a uniform pdf on an interval $[a, b)$.
26. Given a sample space $\Omega = \{0, 1, 2, \ldots\}$ define
$$p(k) = \frac{\gamma}{2^k}; \; k = 0, 1, 2, \ldots$$
    (a) What must γ be in order for $p(k)$ to be a pmf?
    (b) Find the probabilities $P(\{0, 2, 4, 6, \ldots\})$, $P(\{1, 3, 5, 7, \ldots\})$, and $P(\{0, 1, 2, 3, 4, \ldots, 20\})$.
    (c) Suppose that K is a fixed integer. Find $P(\{0, K, 2K, 3K, \ldots\})$.
    (d) Find the mean, second moment, and variance of this pmf.
27. Suppose that $p(k)$ is a geometric pmf. Define $q(k) = (p(k) + p(-k))/2$. Show that this is a pmf and find its mean and variance. Find the probability of the sets $\{k : |k| \geq K\}$ and $\{k : k \text{ is a multiple of } 3\}$. Find the probability of the sets $\{k : k \text{ is odd }\}$
28. Define a pmf $p(k) = C\lambda^{|k|}/|k|!$ for $k \in \mathcal{Z}$, $\lambda > 0$. Evaluate the constant C and find the mean and variance of this pmf.
29. A probability space consists of a sample space Ω = all pairs of positive integers (that is, $\Omega = \{1, 2, 3, \ldots\}^2$) and a probability measure P described by the pmf p

defined by

$$p(k,m) = p^2(1-p)^{k+m-2}.$$

(a) Find $P(\{(k,m) : k \geq m\})$.

(b) Find the probability $P(\{(k,m) : k+m=r\})$ as a function of r for $r = 2,3,\ldots$ Show that the result is a pmf.

(c) Find the probability $P(\{(k,m) : k \text{ is an odd number}\})$.

(d) Define the event $F = \{(k,m) : k \geq m\}$. Find the conditional pmf $p_F(k,m) = P(\{k,m\}|F)$. Is this a product pmf?

30. The probability that Riddley Walker goes for a run in the morning before work is 2/5. If he runs then the probability that he catches the train to work is 2/5. If he does not run then the probability that he catches the train to work is 1/2. If he does not catch the train, then he catches the bus. The model holds from Monday to Friday.
    (a) What is the probability that Riddley gets the train any morning?
    (b) You are told that Riddley made the train – what is the probability that he ran?
    (c) What is the probability that Riddley catches the train exactly twice this week (out of the five working days)?
    (d) What is the expected number of times during the week that Riddley will catch the train? You can leave your answer in terms of a sum.
31. You roll a six-sided die until either a 2 shows or an odd number shows. What is the probability of rolling a 2 before rolling an odd number?
32. Rita and Ravi are starting a company. Together they must raise at least $100,000. Each raises money with a uniform distribution between $0 and $100,000. (Assume that money is continuous and that this is a uniform pdf – it is easier that way.) If either of them individually raises more than $75,000 they have to fill out extra Internal Revenue Service forms. What is the probability that they raise enough money but neither has to fill out extra forms?
33. The probability that a man has a particular disease is 1/20. John is tested for the disease but the test is not totally accurate. The probability that a person with the disease tests negative is 1/50 while the probability that a person who does not have the disease tests positive is 1/10. John's test returns positive.
    (a) Find the probability that John has the disease.
    (b) You are now told that this disease is hereditary. The probability that a son suffers from the disease if his father does is 4/5, the probability that a son is infected with the disease even though his father is not is 1/95. What is the probability that Max has the disease given that his son Peter has the disease? (Note: you may assume that the disease affects only males so you can ignore the dependence on Peter's mother's health.)
    (c) Michael is also tested but he worries about the accuracy of the test so he

takes the test 10 times. One of the ten tests turns out positive, the other nine negative. What is the probability that Michael has the disease?

34. Define the uniform probability density function on $[0, 1)$ in the usual way as

$$f(r) = \begin{cases} 1 & 0 \le r < 1 \\ 0 & \text{otherwise} \end{cases}.$$

(a) Define the set $F = \{0.25, 0.75\}$, a set with only two points. What is the value of

$$\int_F f(r)\, dr?$$

The Riemann integral is well defined for a finite collection of points and this should be easy. What is $\int_{F^c} f(r)\, dr$?

(b) Now define the set F as the collection of all rational numbers in $[0, 1)$, that is, all numbers that can be expressed as k/n for some integers $0 \le k < n$. What is the integral $\int_F f(r)\, dr$? Is it defined? Thinking intuitively, what should it be? Suppose instead you consider the set F^c, the set of all irrational numbers in $[0, 1)$. What is $\int_{F^c} f(r)\, dr$?

35. Given the uniform pdf on $[0, 1], f(x) = 1;\ x \in [0, 1]$, find an expression for $P((a, b))$ for all real $b > a$. Define the *cumulative distribution function* or *cdf* F as the probability of the event $\{x : x \le r\}$ as a function of $r \in \Re$:

$$F(r) = P((-\infty, r]) = \int_{-\infty}^{r} f(x)\, dx. \tag{2.102}$$

Find the cdf for the uniform pdf. Find the probability of the event

$$\begin{aligned} G &= \left\{\omega : \omega \in \left[\frac{1}{2^k}, \frac{1}{2^k} + \frac{1}{2^{k+1}}\right) \text{ for some even } k\right\} \\ &= \bigcup_{k \text{ even}} \left[\frac{1}{2^k}, \frac{1}{2^k} + \frac{1}{2^{k+1}}\right). \end{aligned}$$

36. $\star$ Let Ω be a unit square $\{(x, y) : (x, y) \in \Re^2, -1/2 \le x \le 1/2, -1/2 \le y \le 1/2\}$ and let $\mathcal{F}$ be the corresponding product Borel field. Is the circle $\{(x, y) : (x^2 + y^2)^{1/2} \le 1/2\}$ in $\mathcal{F}$? (Give a plausibility argument.) If so, find the probability of this event if one assumes a uniform density function on the unit square.

37. Given a pdf f, find the cumulative distribution function or cdf F defined as in (2.102) for the exponential, Laplacian, and Gaussian pdf's. In the Gaussian case, express the cdf in terms of the Φ function. Prove that if $a \ge b$, then $F(a) \ge F(b)$. What is $\frac{dF(r)}{dr}$?

38. Let $\Omega = \Re^2$ and suppose we have a pdf $f(x, y)$ such that

$$f(x, y) = \begin{cases} C & \text{if } x \ge 0,\ y \ge 0,\ x + y \le 1 \\ 0 & \text{otherwise} \end{cases}.$$

Find the probability $P(\{(x,y) : 2x > y\})$. Find the probability $P(\{(x,y) : x \leq \alpha\})$ for all real α. Is f a product pdf?

39. Prove that the product k-dimensional pdf integrates to 1 over $\Re$.
40. Given the one-dimensional exponential pdf, find $P(\{x : x > r\})$ and the cumulative distribution function $P(\{x : x \leq r\})$ for $r \in \Re$.
41. Given the k-dimensional product doubly exponential pdf, find the probabilities of the following events in $\Re^k$: $\{\mathbf{x} : x_0 \geq 0\}$, $\{\mathbf{x} : x_i > 0$, all $i = 0, 1, \ldots, k-1\}$, $\{\mathbf{x} : x_0 > x_1\}$.
42. Let $(\Omega, \mathcal{F}) = (\Re, \mathcal{B}(\Re))$. Let P_1 be the probability measure on this space induced by a geometric pmf with parameter p and let P_2 be the probability measure induced on this space by an exponential pdf with parameter λ. Form the mixture measure $P = P_1/2 + P_2/2$. Find $P(\{\omega : \omega > r\})$ for all $r \in [0, \infty)$.
43. Let $\Omega = \Re^2$ and suppose we have a pdf $f(x,y)$ such that

$$f(x,y) = Ce^{-(1/2\sigma^2)\,x^2} e^{-\lambda y} \;;\; x \in (-\infty, \infty),\; y \in [0, \infty).$$

Find the constant C. Is f a product pdf? Find the probability $\Pr(\{(x,y) : \sqrt{|x|} \leq \alpha\})$ for all possible values of a parameter α. Find the probability $\Pr(\{(x,y) : x^2 \leq y\})$.

44. Define $g(x)$ by

$$g(x) = \begin{cases} \lambda e^{-\lambda x} & x \in [0, \infty) \\ 0 & \text{otherwise} \end{cases}.$$

Let $\Omega = \Re^2$ and suppose we have a pdf $f(x,y)$ such that

$$f(x,y) = Cg(x)g(y-x).$$

Find the constant C. Find an expression for the probability $P(\{(x,y) : y \leq \alpha\})$ as a function of the parameter α. Is f a product pdf?

45. Let $\Omega = \Re^2$ and suppose we have a pdf such that

$$f(x,y) = \begin{cases} C|x| & -1 \leq x \leq 1;\; -1 \leq y \leq x \\ 0 & \text{otherwise} \end{cases}.$$

Find the constant C. Is f a product pdf?

46. Suppose that a probability space has sample space $\mathcal{R}^n$, the n-dimensional Euclidean space. (This is a product space.) Suppose that a multidimensional pdf f is defined on this space by

$$f(\mathbf{x}) = \begin{cases} C; & \max_i |x_i| \leq 1/2 \\ 0; & \text{otherwise} \end{cases};$$

that is, $f(\mathbf{x}) = C$ when $-1/2 \leq x_i \leq 1/2$ for $i = 0, 1, \ldots, n-1$ and is 0 otherwise.
(a) What is C?
(b) Is f a product pdf?

(c) What is $P(\{\mathbf{x} : \min_i x_i \geq 0\})$, that is, the probability that the smallest coordinate value is nonnegative?

Suppose next that we have a pdf g defined by

$$g(\mathbf{x}) = \begin{cases} K; & ||\mathbf{x}|| \leq 1 \\ 0; & \text{otherwise} \end{cases},$$

where

$$||\mathbf{x}|| = \sqrt{\sum_{i=0}^{n-1} x_i^2}$$

is the Euclidean norm of the vector $\mathbf{x}$. Thus g equals K inside an n-dimensional sphere of radius 1 centered at the origin.

(d) What is the constant K? (You may need to go to a book of integral tables to find this.)

(e) Is this density a product pdf?

47. Let $(\Omega, \mathcal{F}, P)$ be a probability space and consider events F, G, and H for which $P(F) > P(G) > P(H) > 0$. Events F and G form a partition of Ω, and events F and H are independent. Can events G and H be disjoint?
48. (Courtesy of Professor T. Cover) Suppose that the evidence of an event F increases the likelihood of a criminal's guilt; that is, if G is the event that the criminal is guilty, then $P(G|F) \geq P(G)$. The prosecutor discovers that the event F did *not* occur. What do you now know about the criminal's guilt? Prove your answer.

3
Random variables, vectors, and processes

3.1 Introduction

This chapter provides theoretical foundations and examples of of random variables, vectors, and processes. All three concepts are variations on a single theme and may be included in the general term of *random object*. We will deal specifically with random variables first because they are the simplest conceptually – they can be considered to be special cases of the other two concepts.

3.1.1 Random variables

The name *random variable* suggests a variable that takes on values randomly. In a loose, intuitive way this is the right interpretation – e.g., an observer who is measuring the amount of noise on a communication link sees a random variable in this sense. We require, however, a more precise mathematical definition for analytical purposes. Mathematically a random variable is neither random nor a variable – it is just a function mapping one sample space into another space. The first space is the sample space portion of a probability space, and the second space is a subset of the real line (some authors would call this a "real-valued" random variable). The careful mathematical definition will place a constraint on the function to ensure that the theory makes sense, but for the moment we informally define a random variable as a function.

A random variable is perhaps best thought of as a measurement on a probability space; that is, for each sample point ω the random variable produces some value, denoted functionally as $f(\omega)$. One can view ω as the result of some experiment and $f(\omega)$ as the result of a measurement made on the experiment, as in the example of the simple binary quantizer introduced in the introduction to Chapter 2. The experiment outcome ω is from an abstract space, e.g., real numbers, integers, ASCII characters, waveforms, sequences, Chinese characters, etc. The resulting value of the measurement or random variable $f(\omega)$, however, must be "concrete" in the sense of being a real number, e.g., a meter reading. The randomness is all in the original probability space and not in the random variable; that is, once the ω is selected in a "random" way, the output value or *sample value* of the random variable is determined.

Alternatively, the original point ω can be viewed as an "input signal" and the random variable f can be viewed as "signal processing," i.e., the input signal ω is converted into an "output signal" $f(\omega)$ by the random variable. This viewpoint becomes both precise and relevant when we choose our original sample space to be a signal space and we generalize random variables by random vectors and processes.

Before proceeding to the formal definition of random variables, vectors, and processes, we motivate several of the basic ideas by simple examples, beginning with random variables constructed on the fair wheel experiment of the introduction to Chapter 2.

A Coin Flip

We have already encountered an example of a random variable in the introduction to Chapter 2, where we defined a random variable q on the spinning wheel experiment which produced an output with the same pmf as a uniform coin flip. We begin by summarizing the idea with some slight notational changes and then consider the implications in additional detail.

Begin with a probability space $(\Omega, \mathcal{F}, P)$ where $\Omega = \Re$ and the probability P is defined by (2.2) using the uniform pdf on $[0, 1)$ of (2.4) Define the function $Y : \Re \to \{0, 1\}$ by

$$Y(r) = \begin{cases} 0 & \text{if } r \leq 0.5 \\ 1 & \text{otherwise} \end{cases}. \tag{3.1}$$

When Tyche performs the experiment of spinning the pointer, we do not actually observe the pointer, but only the resulting binary value of Y, which can be thought of as signal processing or as a measurement on the original experiment. Subject to a technical constraint to be introduced later, any function defined on the sample space of an experiment is called a *random variable*. The "randomness" of a random variable is "inherited" from the underlying experiment and in theory the probability measure describing its outputs should be derivable from the initial probability space and the structure of the function. To avoid confusion with the probability measure P of the original experiment, we refer to the probability measure associated with outcomes of Y as P_Y. P_Y is called the *distribution* of the random variable Y. The probability $P_Y(F)$ can be defined in a natural way as the probability computed using P of all the original samples that are mapped by Y into the subset F:

$$P_Y(F) = P(\{r : Y(r) \in F\}). \tag{3.2}$$

In this simple discrete example P_Y is naturally defined for any subset F of $\Omega_Y = \{0, 1\}$, but in preparation for more complicated examples we assume that P_Y is to be defined for all suitably defined events, that is, for $F \in \mathcal{B}_Y$, where $\mathcal{B}_Y$ is an event space consisting of subsets of Ω_Y. The probability measure for the output sample space can be computed from the probability measure for the input using the formula (3.2), which will shortly be generalized. This idea of deriving new probabilistic descriptions for the outputs of some operation on an experiment which produces inputs to the operation is fundamental to the theories of probability, random processes, and signal processing.

In our simple example, (3.2) implies that

$$\begin{aligned} P_Y(\{0\}) &= P(\{r : Y(r) = 0\}) = P(\{r : 0 \le r \le 0.5\}) \\ &= P([0, 0.5]) = 0.5 \\ P_Y(\{1\}) &= P((0.5, 1.0]) = 0.5 \\ P_Y(\Omega_Y) &= P_Y(\{0, 1\}) = P(\Re) = 1 \\ P_Y(\emptyset) &= P(\emptyset) = 0, \end{aligned}$$

so that every output event can be assigned a probability by P_Y by computing the probability of the corresponding input event under the input probability measure P.

Equation (3.2) can be written in a convenient compact manner by means of the definition of the *inverse image* of a set F under a mapping $Y : \Omega \to \Omega_Y$:

$$Y^{-1}(F) = \{r : Y(r) \in F\}. \tag{3.3}$$

With this notation (3.2) becomes

$$P_Y(F) = P(Y^{-1}(F)); \; F \subset \Omega_Y; \tag{3.4}$$

that is, the inverse image of a given set (output) under a mapping is the collection of all points in the original space (input points) which map into the given (output) set. This result is sometimes called the fundamental derived distribution formula or the *inverse image formula*. It will be seen in a variety of forms throughout the book. When dealing with random variables it is common to interpret the probability $P_Y(F)$ as "the probability that the random variable Y takes on a value in F" or "the probability that the event $Y \in F$ occurs." These English statements are often abbreviated to the form $\Pr(Y \in F)$.

The probability measure P_Y can be computed by summing a pmf, which we denote p_Y. In particular, if we define

$$p_Y(y) = P_Y(\{y\}); \; y \in \Omega_Y, \tag{3.5}$$

then additivity implies that

$$P_Y(F) = \sum_{y \in F} p_Y(y); \; F \in \mathcal{B}_Y. \tag{3.6}$$

Thus the pmf describing a random variable can be computed as a special case of the inverse image formula (3.5), and then used to compute the probability of any event.

The indirect method provides a description of the fair coin flip in terms of a random variable. The idea of a random variable can also be applied to the direct description of a probability space. As in the introduction to Chapter 2, we describe directly a single coin flip by choosing $\Omega = \{0, 1\}$ and assign a probability measure P on this space as in (2.12). Now define a random variable $V : \{0, 1\} \to \{0, 1\}$ on this space by

$$V(r) = r. \tag{3.7}$$

Here V is trivial: it is just the *identity mapping*. The measurement just puts out the outcome of the original experiment and the inverse image formula trivially yields

$$\begin{aligned} P_V(F) &= P(F) \\ p_V(v) &= p(v). \end{aligned}$$

Note that this construction works on any probability space having the real line or a Borel subset thereof as a sample space. Thus for each of the named pmf's and pdf's there is an associated random variable.

If we have two random variables V and Y (which may be defined on completely separate experiments as in the present case), we say that they are *equivalent* or *identically distributed* if $P_V(F) = P_Y(F)$ for all events F, that is, the two probability measures agree exactly on all events. It is easy to show with the inverse image formula that V is equivalent to Y and hence that

$$p_Y(y) = p_V(y) = 0.5; \; y = 0, 1. \tag{3.8}$$

Thus we have two equivalent random variables, either of which can be used to model the single coin flip. Note that we do not say the random variables are *equal* since they need not be. For example, you could spin a pointer and find Y and I could flip my own coin to find V. The probabilities are the same, but the outcomes might differ.

3.1.2 Random vectors

The issue of the possible equality of two random variables raises an interesting point. If you are told that Y and V are two separate random variables with pmf's p_Y and p_V, then the question of whether they are equivalent can be answered from these pmf's alone. If you wish to determine whether or not the two random variables are in fact *equal*, however, then they must be considered together or jointly. In the case where we have a random variable Y with outcomes in $\{0, 1\}$ and a random variable V with outcomes in $\{0, 1\}$, we could consider the two together as a single random *vector* $\{Y, V\}$ with outcomes in the Cartesian product space $\Omega_{YV} = \{0, 1\}^2 \triangleq \{(0, 0), (0, 1), (1, 0), (1, 1)\}$ with some pmf $p_{Y,V}$ describing the combined behavior

$$p_{Y,V}(y, v) = \Pr(Y = y, V = v) \tag{3.9}$$

so that

$$\Pr((Y, V) \in F) = \sum_{y,v:(y,v)\in F} p_{Y,V}(y, v); \; F \in \mathcal{B}_{YV},$$

where in this simple discrete problem we take the event space $\mathcal{B}_{YV}$ to be the power set of Ω_{YV}. Now the question of equality makes sense as we can evaluate the probability that the two are equal:

$$\Pr(Y = V) = \sum_{y,v:y=v} p_{Y,V}(y, v).$$

If this probability is one, then we know that the two random variables are in fact equal with probability one. In any particular example "equal with probability one" does not mean identically equal since they can be different on Ω with probability zero.

A random two-dimensional random vector (Y, V) is simply two random variables described on a common probability space. Knowledge of the individual pmf's p_Y and p_V alone is not sufficient in general to determine $p_{Y,V}$. More information is needed. Either the joint pmf must be given to us or we must be told the definitions of the two random variables (two components of the two-dimensional binary vector) so that the joint pmf can be derived. For example, if we are told that the two random variables Y and V of our example are in fact equal, then $\Pr(Y = V) = 1$ and $p_{Y,V}(y, v) = 0.5$ for $y = v$, and 0 for $y \neq v$. This experiment can be thought of as flipping two coins that are soldered together on the edge so that the result is two heads or two tails.

To see an example of a radically different behavior, consider the random variable $W : [0, 1) \to \{0, 1\}$ by

$$W(r) = \begin{cases} 0 & r \in [0.0, 0.25) \bigcup [0.5, 0.75) \\ 1 & \text{otherwise} \end{cases}. \tag{3.10}$$

It is easy to see that W is equivalent to the random variables Y and V of this section, but W and Y are not equal even though they are equivalent and defined on a common experiment. We can easily derive the joint pmf for W and Y since the inverse image formula extends immediately to random vectors. Now the events involve the outputs of two random variables so some care is needed to keep the notation from getting out of hand. As in the random variable case, any probability measure on a discrete space can be expressed as a sum over a pmf on points, that is,

$$P_{Y,W}(F) = \sum_{y,w:(y,w)\in F} p_{Y,W}(y, w), \tag{3.11}$$

where $F \subset \{0, 1\}^2$, and where

$$\begin{aligned} p_{Y,W}(y, w) &= P_{Y,W}(\{y, w\}) \\ &= \Pr(Y = y, W = w);\ y \in \{0, 1\}, w \in \{0, 1\}. \end{aligned} \tag{3.12}$$

As previously observed, pmf's describing the joint behavior of several random variables are called *joint* pmf's and the corresponding distribution is called a joint distribution. Thus finding the entire distribution only requires finding the pmf, which can be done via the inverse image formula. For example, if $(y, w) = (0, 0)$, then

$$\begin{aligned} p_{Y,W}(0, 0) &= P(\{r : Y(r) = 0, W(r) = 0\}) \\ &= P([0, 0.5) \bigcap ([0.0, 0.25) \bigcup [0.5, 0.75))) \\ &= P([0, 0.25)) = 0.25. \end{aligned}$$

Similarly it can be shown that

$$p_{Y,W}(0,1) = p_{Y,W}(1,0) = p_{Y,W}(1,1) = 0.25.$$

Joint and marginal pmf's can both be calculated from the underlying distribution, but the marginals can also be found directly from the joint pmf's without reference to the underlying distribution. For example, $p_Y(y_0)$ can be expressed as $P_{Y,W}(F)$ by choosing $F = \{(y,w) : y = y_0\}$. Then the pmf formula for $P_{Y,W}$ can be used to write

$$\begin{aligned} p_Y(y_0) = P_{Y,W}(F) &= \sum_{y,w:(y,w)\in F} p_{Y,W}(y,w) \\ &= \sum_{w\in\Omega_W} p_{Y,W}(y_0,w). \end{aligned} \tag{3.13}$$

Similarly

$$p_W(w_0) = \sum_{y\in\Omega_Y} p_{Y,W}(y,w_0). \tag{3.14}$$

This is an example of the consistency of probability – using different pmf's derived from a common experiment to calculate the probability of a single event must produce the same result; the marginals must agree with the joints. Consistency means that we can find marginals by "summing out" joints without knowing the underlying experiment on which the random variables are defined.

This completes the derived distribution of the two random variables Y and W (or the single random vector (Y,W)) defined on the original uniform pdf experiment. For this particular example the joint pmf and the marginal pmf's have the interesting property

$$p_{Y,W}(y,w) = p_Y(y)p_W(w), \tag{3.15}$$

that is, the joint distribution is a product distribution. A product distribution better models our intuitive feeling of experiments such as flipping two fair coins and letting the outputs Y and W be 1 or 0 according to the coins landing heads or tails.

In both of these examples cases the joint pmf had to be *consistent* with the individual pmf's p_Y and p_V (i.e., the *marginal pmf's*) in the sense of giving the same probabilities to events where both joint and marginal probabilities make sense. In particular,

$$
\begin{aligned}
p_Y(y) = \Pr(Y = y) &= \Pr(Y = y, V \in \{0,1\}) \\
&= \sum_{v=0}^{1} p_{Y,V}(y, v),
\end{aligned}
$$

an example of a *consistency* property.

The two examples just considered of a random vector (Y, V) with the property $\Pr(Y = V) = 1$ and the random vector (Y, W) with the property $p_{Y,W}(y, w) = p_Y(y)p_W(w)$ represent extreme cases of two-dimensional random vectors. In the first case $Y = V$ and hence being told, say, that $V = v$ also tells us that necessarily $Y = v$. Thus V depends on Y in a particularly strong manner and the two random variables can be considered to be extremely *dependent*. The product distribution, on the other hand, can be interpreted as implying that knowing the outcome of one of the random variables tells us absolutely nothing about the other, as is the case when flipping two fair coins. Two discrete random variables Y and W will be defined to be *independent* if they have a product pmf, that is, if $p_{Y,W}(y, w) = p_Y(y)p_W(w)$. Independence of random variables will be shortly related to the idea of independence of events as introduced in Chapter 2, but for the moment simply observe that it can be interpreted as meaning that knowing the outcome of one random variable does not affect the probability distribution of the other. This is a very special case of general joint pmf's. It may be surprising that two random variables defined on a common probability space can be independent of one another, but this was ensured by the specific construction of the two random variables Y and W.

Note that we have also defined a three-dimensional random vector (Y, V, W) because we have defined three random variables on a common experiment. Hence you should be able to find the joint pmf p_{YVW} using the same ideas.

Note also that in addition to the indirect derivations of a specific examples of a two-dimensional random variable, a direct development is possible. For example, let $\{0,1\}^2$ be a sample space with all of its four points having equal probability. Any point r in the sample space can be expressed as $r = (r_0, r_1)$, where $r_i \in \{0,1\}$ for $i = 0, 1$. Define the random variables $V : \{0,1\}^2 \to \{0,1\}$ and $U : \{0,1\}^2 \to \{0,1\}$ by $V(r_0, r_1) = r_0$ and $U(r_0, r_1) = r_1$. You should convince yourself that

$$
p_{Y,W}(y, w) = p_{V,U}(y, w); \; y = 0, 1; \; w = 0, 1
$$

and that $p_Y(y) = p_W(y) = p_V(y) = p_U(y)$, $y = 0, 1$. Thus the random vectors (Y, W) and (V, U) are equivalent.

In a similar manner pdf's can be used to describe continuous random vectors, but we shall postpone this step until a later section and instead move to the idea of random processes.

3.1.3 Random processes

It is straightforward conceptually to go from one random variable to k random variables constituting a k-dimensional random vector. It is perhaps a greater leap to extend the idea to a random process. The idea is at least easy to state, but it will take more work to provide examples and the mathematical details will be more complicated. A *random process* is a sequence of random variables $\{X_n;\ n = 0, 1, \ldots\}$ defined on a common experiment. It can be thought of as an infinite-dimensional random vector. To be more accurate, this is an example of a discrete time, one-sided random process. It is called "discrete time" because the index n which corresponds to time takes on discrete values (here the nonnegative integers) and it is called "one-sided" because only nonnegative times are allowed. A discrete time random process is also called a *time series* in the statistics literature and is often denoted as $\{X(n)\ n = 0, 1, \ldots\}$ Sometimes it is denoted by $\{X[n]\}$ in the digital signal processing literature. Two questions might occur to the reader: how does one construct an infinite family of random variables on a single experiment? How can one provide a direct development of a random process as accomplished for random variables and vectors? The direct development might appear hopeless since infinite-dimensional vectors are involved, but it is not.

The first question is reasonably easy to handle by example. Consider the usual uniform pdf experiment. Rename the random variables Y and W as X_0 and X_1, respectively. Consider the following definition of an infinite family of random variables $X_n : [0, 1) \to \{0, 1\}$ for $n = 0, 1, \ldots$. Every $r \in [0, 1)$ can be expanded as a binary expansion of the form

$$r = \sum_{n=0}^{\infty} b_n(r) 2^{-n-1}. \tag{3.16}$$

This simply replaces the usual decimal representation by a binary representation. For example, $1/4$ is 0.25 in decimal and 0.01 or $0.010000\ldots$ in binary, $1/2$ is 0.5 in decimal and yields the binary sequence $0.1000\ldots$, $1/4$ is 0.25 in decimal and yields the binary sequence $0.0100\ldots$, $3/4$ is 0.75 in decimal and $0.11000\ldots$, and $1/3$ is $0.3333\ldots$ in decimal and $0.010101\ldots$ in binary.

Define the random process by $X_n(r) = b_n(r)$, that is, the nth term in the binary expansion of r. When $n = 0, 1$ this reduces to the specific X_0 and X_1 already considered.

The inverse image formula can be used to compute probabilities, although the calculations can get messy. Given the simple two-dimensional example, however, the pmf's for random vectors $X^n = (X_0, X_1, \ldots, X_{n-1})$ can be evaluated as

$$p_{X^n}(x^n) = \Pr(X^n = x^n) = 2^{-n};\ x^n \in \{0,1\}^n, \tag{3.17}$$

where $\{0,1\}^n$ is the collection of all 2^n binary n-tuples. In other words, the first n binary digits in a binary expansion for a uniformly distributed random variable are all equally probable. Note that in this special case the joint pmf's are again related to the marginal pmf's in a product fashion, that is,

$$p_{X^n}(x^n) = \prod_{i=0}^{n-1} p_{X_i}(x_i), \tag{3.18}$$

in which case the random variables $X_0, X_1, \ldots, X_{n-1}$ are said to be *mutually independent* or, more simply, *independent*. If a random process is such that any finite collection of the random variables produced by the process are independent and the marginal pmf's are all the same (as in the case under consideration), the process is said to be *independent identically distributed* or *iid* for short. An iid process is also called a *Bernoulli process*, although the name is sometimes reserved for a binary iid process.

Something fundamentally important has happened here. If we have a random process, then the probability distribution for any random vectors formed by collecting outputs of the random process can be found (at least in theory) from the inverse image formula. The calculations may be a mess, but at least in some cases such as this one they can be done. Furthermore these pmf's are consistent in the sense noted before. In particular, if we use (3.13)–(3.14) to compute the already computed pmf's for X_0 and X_1 we get the same thing we did before: they are each equiprobable binary random variables. If we compute the joint pmf for X_0 and X_1 using (3.17) we also get the same joint pmf we got before. This observation probably seems trivial at this point (and it should be natural that the mathematics does not give any contradictions), but it emphasizes a property that is critically important when trying to describe a random process in a more direct fashion.

Suppose now that a more direct model of a random process is desired without a complicated construction on an original experiment. Here the problem is not as simple as in the random variable or random vector case where all that was needed was a consistent assignment of probabilities and an identity mapping. The solution is known as the Kolmogorov extension theorem, named after A. N. Kolmogorov, the primary developer of modern probability theory. The theorem will be stated formally later in this chapter, but its complicated proof will be left to other texts. The basic idea, however, can be stated in a few words. If one can specify a consistent family of pmf's $p_{X^n}(x^n)$ for all n (we have done this for $n = 1$ and 2), then there exists a random process described by those pmf's. Thus, for example, there will exist a random process described by the family of pmf's $p_{X^n}(x^n) = 2^{-n}$ for $x^n \in \{0, 1\}^n$ for all positive integers n if and only if the family is consistent. We have already argued that the family is indeed consistent, which means that even without the indirect construction previously followed we can argue that there is a well-defined random process described by these pmf's. In particular, one can think of a "grand experiment" where Nature selects a one-sided binary sequence according to some mysterious probability measure on sequences that we have difficulty envisioning. Nature then reveals the chosen sequence to us one coordinate at a time, producing the process $X_0, X_1, X_2, \ldots$ The distributions of any finite collection of these random variables are known from the given pmf's p_{X^n}. Putting this in yet another way, describing or specifying the finite-dimensional distributions of a process is enough to completely describe the process (provided of course the given family of distributions is consistent).

In this example the abstract probability measure on semi-infinite binary sequences is not all that mysterious From our construction the sequence space can be considered to be essentially the same as the unit interval (each point in the unit interval corresponding to a binary sequence) and the probability measure is described by a uniform pdf on this interval.

The second method of describing a random process is by far the most common in practice. One usually describes a process by its finite sample behavior and not by a construction on an abstract experiment. The Kolmogorov extension theorem ensures that this works. Consistency is easy to demonstrate for iid processes, but unfortunately it becomes more difficult to verify in more general cases (and even more difficult to define and demonstrate for continuous time examples).

Having toured the basic ideas to be explored in this chapter, we now delve into the details required to make the ideas precise and general.

3.2 Random variables

We now develop the precise definition of a random variable. As you might guess, a technical condition for random variables is required because of certain subtle pathological problems that have to do with the ability to determine probabilities for the random variable. To arrive at the precise definition, we start with the informal definition of a random variable that we have already given and then show the inevitable difficulty that results without the technical condition. We have informally defined a random variable as being a function on a sample space. Suppose we have a probability space $(\Omega, \mathcal{F}, P)$. Let $f : \Omega \to \Re$ be a function mapping the same space into the real line so that f is a candidate random variable. Since the selection of the original sample point ω is random, that is, governed by a probability measure, the output of our measurement of random variable $f(\omega)$ should also be random. That is, we should be able to find the probability of an "output event' such as the event "the outcome of the random variable f was between a and b," that is, the event $F \subset \Re$ given by $F = (a, b)$. Observe that there are two different kinds of events being considered here:

1. output events or members of the event space of the range or range space of the random variable, that is, events consisting of subsets of possible output values of the random variable; and
2. input events or Ω events, events in the original sample space of the original probability space.

Can we find the probability of this output event? That is, can we make mathematical sense out of the quantity "the probability that f assumes a value in an event $F \subset \Re$"? On reflection it seems clear that we can. The probability that f assumes a value in some set of values must be the probability of all values in the original sample space that result in a value of f in the given set. We will make this concept more precise shortly. To save writing we will abbreviate such English statements to the form $\Pr(f \in F)$, or $\Pr(F)$, that is, when the notation $\Pr(F)$ is encountered it should be interpreted as shorthand for the English statement "the probability of an event F" or "the probability that the event F will occur" and not as a precise mathematical quantity.

Recall from Chapter 2 that for a subset F of the real line $\Re$ to be an event, it must be in a sigma-field or event space of subsets of $\Re$. Recall also that we adopted the Borel field $\mathcal{B}(\Re)$ as our basic event space for the real line. Hence it makes sense to require that our output event F be a Borel set.

We can now state the question as follows: Given a probability space $(\Omega, \mathcal{F}, P)$ and a function $f : \Omega \to \Re$, is there a reasonable and useful precise definition for the probability $\Pr(f \in F)$ for any $F \in \mathcal{B}(\Re)$, the Borel field or event space of the real line? Since the probability measure P sits on the original measurable space $(\Omega, \mathcal{F})$ and since f assumes a value in F if and only if $\omega \in \Omega$ is chosen so that $f(\omega) \in F$, the desired probability is obviously $\Pr(f \in F) = P(\{\omega : f(\omega) \in F\}) = P(f^{-1}(F))$. In other words, the probability that a random variable f takes on a value in a Borel set F is the probability (defined in the original probability space) of the set of all (original) sample points ω that yield a value $f(\omega) \in F$. This, in turn, is the probability of the inverse image of the Borel set F under the random variable f. This idea of computing the probability of an output event of a random variable using the original probability measure of the corresponding inverse image of the output event under the random variable is depicted in Figure 3.1.

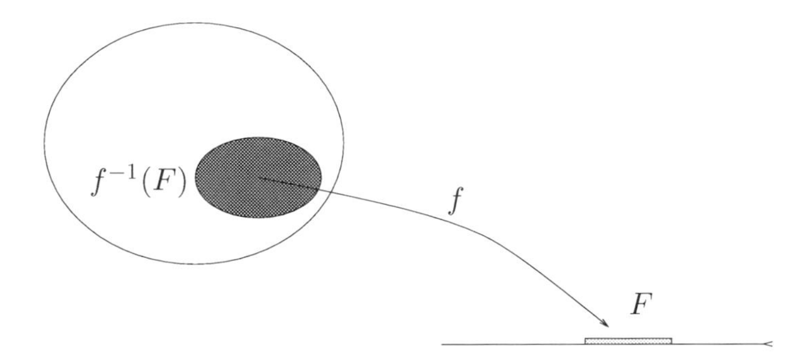

Figure 3.1 The inverse image method: $\Pr(f \in F) = P(\{\omega : f(\omega) \in F\}) = P(f^{-1}(F))$

This natural definition of the probability of an output event of a random variable indeed makes sense if and only if the probability $P(f^{-1}(F))$ makes sense, that is, if the subset $f^{-1}(F)$ of Ω corresponding to the output event F is itself an event, in this case an input event or member of the event space $\mathcal{F}$ of the original sample space. This, then, is the required technical condition: a function f mapping the sample space of a probability space $(\Omega, \mathcal{F}, P)$ into the real line $\Re$ is a random variable if and only if the inverse images of all Borel sets in $\Re$ are members of $\mathcal{F}$, that is, if all of the Ω sets corresponding to output events (members of $\mathcal{B}(\Re)$) are input events (members of $\mathcal{F}$). Unlike some of the other pathological conditions that we have met, it is easy to display some trivial examples where the technical condition is not met (as we will see in Example [3.11]). We now formalize the definition.

Given a probability space $(\Omega, \mathcal{F}, P)$ a (real-valued) *random variable* is a function $f : \Omega \to \Re$ with the property that if $F \in \mathcal{B}(\Re)$, then also $f^{-1}(F) = \{\omega : f(\omega) \in F\} \in \mathcal{F}$.

Given a random variable f defined on a probability space $(\Omega, \mathcal{F}, P)$, the set function

$$\begin{aligned} P_f(F) &\triangleq P(f^{-1}(F)) = P(\{\omega : f(\omega) \in F\}) \\ &= \Pr(f \in F); \;\; F \in \mathcal{B}(\Re) \end{aligned} \tag{3.19}$$

is well defined since by definition $f^{-1}(F) \in \mathcal{F}$ for all $F \in \mathcal{B}(\Re)$. In the next section the properties of distributions will be explored.

In some cases one may wish to consider a random variable with a more limited range space than the real line, e.g., when the random variable is binary. (Recall from Appendix A that the range space of f is the image of Ω.) If so, $\Re$ can be replaced in the definition by the appropriate subset, say $A \subset \Re$. This is really just a question of semantics since the two definitions are equivalent. One or the other view may, however, be simpler to deal with for a particular problem.

A function meeting the condition in the definition we have given is said to be *measurable.* This is because such functions inherit a probability measure on their output events.

If a random variable has a distribution described by a pmf or a pdf with a specific name, then the name is often applied also to the random variable; e.g., a continuous random variable with a Gaussian pdf is called a Gaussian random variable.

Examples

In every case we are given a probability space $(\Omega, \mathcal{F}, P)$. For the moment, however, we will concentrate on the sample space Ω and the random variable that is defined functionally on that space. Note that the function must be defined for *every* value in the sample space if it is to be a valid function. On the other hand, the function does not have to assume every possible value in its range.

There is nothing particularly special about the names of the random variables. So far we have used the lower-case letter f. On occasion we will use other lower-case letters such as g and h. As we progress we will follow custom and more often use upper-case letters late in the alphabet, such as X, Y, Z, U, V, and W. Capital Greek letters like Θ and Ψ are also popular.

The reader should keep the signal processing interpretation in mind while considering these examples. Several very common types of signal processing are considered, including quantization, sampling, and filtering.

[3.1] Let $\Omega = \Re$, the real line, and define the random variable $X : \Omega \to \Omega$ by $X(\omega) = \omega^2$ for all $\omega \in \Omega$. Thus the random variable is the square of the sample point. Note that since the square of a real number is always nonnegative, we could replace the range Ω by the range space $[0, \infty)$ and consider X as a mapping $X : \Omega \to [0, \infty)$. Other random variables mapping Ω into itself are $Y(\omega) = |\omega|$, $Z(\Omega) = \sin(\omega)$, $U(\omega) = 3 \times \omega + 321.5$, and so on. We can also consider the identity mapping as a random variable; that is, we can define a random variable $W : \Omega \to \Omega$ by $W(\omega) = \omega$.

[3.2] Let $\Omega = \Re$ as in Example [3.1] and define the random variable $f : \omega \to \{-V, V\}$ by

$$f(r) = \begin{cases} +V & \text{if } r \geq 0 \\ -V & \text{otherwise} \end{cases}.$$

This example is a variation of the binary *quantizer* of a real input considered in the introduction to Chapter 2. With this specific choice of output levels it is also called a hard limiter.

So far we have used ω exclusively to denote the argument of the random variable. We can, however, use any letter to denote the dummy variable (or argument or independent variable) of the function, provided that we specify its domain; that is, we do not need to use ω all the time to specify elements of Ω: r, x, or any other dummy variable will do. We will, however, as a convention, always use *only lower-case letters* to denote dummy variables.

When referring to a function, we will use several methods of specification. Sometimes we will only give its name, say f; sometimes we will specify its domain and range, as in $f : \Omega \to A$; sometimes we will provide a specific dummy variable, as in $f(r)$; and sometimes we will provide the dummy variable and its domain, as in $f(r); r \in \Omega$. Finally, functions can be shown with a place for the dummy variable marked by a period to avoid anointing any particular dummy variable as being somehow special, as in $f(\cdot)$. These various notations are really just different means of denoting the same thing while emphasizing certain aspects of the functions. The only real danger of this notation is the same as that of calculus and trigonometry: if one encounters a function, say $\sin t$, does this mean the sine of a *particular* t (and hence a real number) or does it mean the entire waveform of $\sin t$ for *all* t? The distinction should be clear from the context, but the ambiguity can be removed, for example, by defining something like $\sin t_0$ to mean a particular value and $\{\sin t;\ t \in \Re\}$ or $\sin(\cdot)$ to mean the entire waveform.

[3.3] Let U be as in Example [3.1] and f as in Example [3.2]. Then the function $g : \Omega \to \Omega$ defined by $g(\omega) = f(U(\omega))$ is also a random variable. This relation is often abbreviated by dropping the explicit dependence on ω to write $g = f(U)$. More generally, any function of a function is another function, called a "composite" function. Thus *a function of a random variable is another random variable.* Similarly, one can consider a random variable formed by a complicated combination of other random variables – for example, $g(\omega) = (1/\omega)\sinh^{-1}[\pi \times e^{\cos(|\omega|^{3.4})}]$.

[3.4] Let $\Omega = \Re^k$, k-dimensional Euclidean space. Occasionally it is of interest to focus attention on the random variable which is defined as a particular coordinate of a vector $\omega = (x_0, x_1, \ldots, x_{k-1}) \in \Re^k$. Toward this end we can define for each $i = 0, 1, \ldots, k-1$ a *sampling function* (or *coordinate function*) $\Pi_i : \Re^k \to \Re$ as the following random variable:

$$\Pi_i(\omega) = \Pi_i((x_0, \ldots, x_{k-1})) = x_i.$$

The sampling functions are also called "projections" of the higher-dimensional space onto the lower. (This is the reason for the choice of Π, Greek P – not to be confused with the product symbol $\prod$ – to denote the functions.)

Similarly, we can define a sampling function for any product space, e.g., for sequence and waveform spaces.

⋆[3.5] Given a space A, an index set $\mathcal{T}$, and the product space $A^{\mathcal{T}}$, define as a random variable, for any fixed $t \in \mathcal{T}$, the sampling function Π_i : $A^{\mathcal{T}} \to A$ as follows. Since any $\omega \in A^{\mathcal{T}}$ is a vector or function of the form $\{x_s;\ s \in \mathcal{T}\}$, define for each t in $\mathcal{T}$ the mapping

$$\Pi_t(\omega) = \Pi_t(\{x_s;\ s \in \mathcal{T}\}) = x_t.$$

Thus, for example, if Ω is a one-sided binary sequence space

$$\prod_{i \in \mathcal{Z}_+} \{0, 1\}_i = \{0, 1\}^{\mathcal{Z}_+},$$

and hence every point has the form $\omega = (x_0, x_1, \ldots)$, then

$$\Pi_3((0, 1, 1, 0, 0, 0, 1, 0, 1, \ldots)) = 0.$$

As another example, if for all t in the index set $\Re_t$ is a replica of $\Re$ and Ω is the space

$$\Re^{\Re} = \prod_{t \in \Re} \Re_t$$

of all real-valued waveforms $\{x(t); t \in (-\infty, \infty)\}$, then for $\omega = \{\sin t; t \in \Re\}$, the value of the sampling function at the particular time $t = 2\pi$ is

$$\Pi_{2\pi}(\{\sin t; t \in \Re\}) = \sin 2\pi = 0.$$

[3.6] Suppose that we have a one-sided binary sequence space $\{0,1\}^{\mathcal{Z}_+}$. For any $n \in \{1, 2, \ldots\}$, define the random variable Y_n by $Y_n(\omega) = Y_n((x_0, x_1, x_2, \ldots)) =$ the index (time) of occurrence of the nth 1 in ω. For example, $Y_2((0, 0, 0, 1, 0, 1, 1, 0, 1, \ldots)) = 5$ because the second sample to be 1 is x_5.

[3.7] Say we have a one-sided sequence space $\Omega = \prod_{i \in \mathcal{Z}_+} \Re_i$, where $\Re_i$ is a replica of the real line for each i in the index set. Since every ω in this space has the form $\{x_0, x_1, \ldots\} = \{x_i; i \in \mathcal{Z}_+\}$, we can define for each positive integer n the random variable, depending on n,

$$S_n(\omega) = S_n(\{x_i; i \in \mathcal{Z}_+\}) = n^{-1} \sum_{i=0}^{n-1} x_i$$

the arithmetic average or "mean" of the first n coordinates of the infinite sequence.

For example, if $\omega = \{1, 1, 1, 1, 1, 1, 1, \ldots\}$, then $S_n = 1$. This average is also called a *Césaro mean* or *sample average* or *time average* since the index being summed over often corresponds to time; for example, we are adding the outputs at times 0 through $n - 1$ in the preceding equation. Such arithmetic means will later be seen to play a fundamental role in describing the long-term average behavior of random processes. The arithmetic mean can also be written using coordinate functions as

$$S_n(\omega) = n^{-1} \sum_{i=0}^{n-1} \Pi_i(\omega), \tag{3.20}$$

which we abbreviate to

$$S_n = n^{-1} \sum_{i=0}^{n-1} \Pi_i \tag{3.21}$$

by suppressing the dummy variable or argument ω. Equation (3.21) is shorthand for (3.20) and says the same thing: the arithmetic average of the first

n terms of a sequence is the sum of the first n coordinates or samples of the sequence.

[3.8] As a generalization of the sample average consider weighted averages of sequences. Such weighted averages occur in the convolutions of linear system theory. Let Ω be the space $\prod_{i\in\mathcal{Z}}\Re_i$, where $\Re_i$ are all copies of the real line. Suppose that $\{h_k;\ k=0,1,2,\ldots\}$ is a fixed sequence of real numbers that can be used to form a weighted average of the coordinates of $\omega\in\Omega$. Each ω in this space has the form $\omega=(\ldots,x_{-1},x_0,x_1,\ldots)=\{x_i;\ i\in\mathcal{Z}\}$ and hence a weighted average can be defined for each integer n:

$$Y_n(\omega)=\sum_{k=0}^{\infty}h_k x_{n-k}.$$

Thus the random variable Y_n is formed as a linear combination of the coordinates of the sequence constituting the point ω in the double-sided sequence space. This is a discrete time convolution of an input sequence with a linear weighting. In linear system theory the weighting is called a *unit pulse response* (or *Kronecker delta response* or δ*-response*), and it is the discrete time equivalent of an impulse response. Note that we could also use the sampling function notation to write Y_n, as a weighted sum of the sample random variables.

[3.9] In a similar fashion, complicated random variables can be defined on waveform spaces. For example, let $\Omega=\prod_{t\in\mathcal{R}}\Re_t$, the space of all real-valued functions of time such as voltage–time waveforms. For each T, define a time average

$$Y_T(\omega)=Y_T(\{x(t);\ t\in\Re\})=T^{-1}\int_0^T x(t)\,dt,$$

or given the impulse response $h(t)$ of a causal, linear time-invariant system, we define a weighted average

$$W_T(\omega)=\int_0^{\infty}h(t)x(T-t)\,dt.$$

Are these also random variables? They are certainly functions defined on the underlying sample space, but as one might suspect, the sample space of all real-valued waveforms is quite large and contains some bizarre waveforms. For example, the waveforms can be sufficiently pathological to preclude the existence of the integrals cited (see Chapter 2 for a discussion of this point).

These examples are sufficiently complicated to force us to look a bit closer at a proper definition of a random variable and to develop a technical condition that constrains the generality of our definition but ensures that the definition will lead to a useful theory. It should be pointed out, however, that this difficulty is no accident and is not easily solved: waveforms are truly more complicated than sequences because of the wider range of possible waveforms owing to the uncountability of the time variable. Continuous time random processes are more difficult to deal with rigorously than are discrete time processes. One can write equations such as the integrals and then find that the integrals do not make sense even in the general Lebesgue sense. Often fairly advanced mathematics are required to patch up the problems properly. For purposes of simplicity we usually concentrate on sequences (and hence on discrete time) rather than waveforms, and we gloss over the technical problems when we consider continuous time examples. In Chapter 5 we will return to add some rigor to the continuous case using the idea of mean square convergence to define the integrals.

One must know the event space being considered in order to determine whether or not a function is a random variable. Although we will virtually always assume the usual event spaces (that is, the power set for discrete spaces, the Borel field for the real line or subsets of the real line, and the corresponding product event spaces for product sample spaces), it is useful to consider some other examples to help clarify the basic definition.

[3.10] First consider $(\Omega, \mathcal{F}, P)$ where Ω is itself a discrete subset of the real line $\Re$, e.g., $\{0, 1\}$ or $\mathcal{Z}_+$. If, as usual, we take $\mathcal{F}$ to be the power set, then *any* function $f : \Omega \to \Re$ is a random variable. This follows since the inverse image of any Borel set in $\Re$ must be a subset of Ω and hence must be in the collection of all subsets of Ω.

Thus with the usual event space for a discrete sample space – the power set – *any* function defined on the probability space is a random variable. This is why all of the structure of event spaces and random variables is not seen in elementary texts that consider only discrete spaces: there is no need.

It should be noted that for any Ω, discrete or not, if $\mathcal{F}$ is the power set, then *all* functions defined on Ω are random variables. This fact is useful, however, only for discrete sample spaces. The power set is not a useful event space in the continuous case (as we cannot endow it with useful probability measures).

If, however, $\mathcal{F}$ is *not* the power set, some functions defined on Ω are *not* random variables, as the following simple example shows:

[3.11] Let Ω be arbitrary, but let $\mathcal{F}$ be the trivial sigma-field $\{\Omega, \emptyset\}$. On this space it is easy to construct functions that are not random variables (and hence are nonmeasurable functions). For example, let $\Omega = \{0, 1\}$ and define $f(\omega) = \omega$, the identity function. Then $f^{-1}(\{0\}) = \{0\}$ is not in $\mathcal{F}$, and hence this simple function is not a random variable. In fact, it is obvious that any function that assigns different values to 0 and 1 is not a random variable. Note, however, that some functions are random variables.

The problem illustrated by this example is that the input event space is not big enough or "fine" enough to contain all input sets corresponding to output events. This apparently trivial example suggests an important technique for dealing with advanced random process theory, especially for continuous time random processes: if the event space is not large enough to include the inverse image of all Borel sets, then enlarge the event space to include all such events, namely by using the power set as in Example [3.10]. Alternatively, we might try to force $\mathcal{F}$ to contain *all* sets of the form $f^{-1}(F)$, $F \in \mathcal{B}(\Re)$; that is, make $\mathcal{F}$ the sigma-field generated by such sets. Further treatment of this subject is beyond the scope of the book. However, it is worth remembering that if a sigma-field is not big enough to make a function a random variable, it can often be enlarged to be big enough. This is not idle twiddling; such a procedure is required for important applications, e.g., to make integrals over time defined on a waveform space into random variables.

On a more hopeful tack, if the probability space $(\Omega, \mathcal{F}, P)$ is chosen with $\Omega = \Re$ and $\mathcal{F} = \mathcal{B}(\Re)$, then all functions f normally encountered in the real world are in fact random variables. For example, continuous functions, polynomials, step functions, trigonometric functions, limits of measurable functions, maxima and minima of measurable functions, and so on are random variables. It is, in fact, extremely difficult to construct functions on Borel spaces that are *not* random variables. The same statement holds for functions on sequence spaces. The difficulty is comparable to constructing a set on the real line that is not a Borel set and is beyond the scope of this book.

So far we have considered abstract philosophical aspects in the definition of random variables. We are now ready to develop the probabilistic properties of the defined random variables.

3.3 Distributions of random variables

3.3.1 Distributions

Suppose we have a probability space $(\Omega, \mathcal{F}, P)$ with a random variable, X, defined on the space. The random variable X takes values on its range space which is some subset A of $\Re$ (possibly $A = \Re$). The range space A of a random variable is often called the *alphabet* of the random variable. As we have seen, since X is a random variable, we know that all subsets of Ω of the form $X^{-1}(F) = \{\omega : X(\omega) \in F\}$, with $F \in \mathcal{B}(A)$, must be members of $\mathcal{F}$ by definition. Thus the set function P_X defined by

$$P_X(F) = P(X^{-1}(F)) = P(\{\omega : X(\omega) \in F\}); \quad F \in \mathcal{B}(A) \tag{3.22}$$

is well defined and assigns probabilities to *output* events involving the random variable in terms of the original probability of *input* events in the original experiment. The three written forms in Equation (3.22) are all read as $\Pr(X \in F)$ or "the probability that the random variable X takes on a value in F." Furthermore, since inverse images preserve all set-theoretic operations (see Problem A.12), P_X satisfies the axioms of probability as a probability measure on $(A, \mathcal{B}(A))$ – it is nonnegative, $P_X(A) = 1$, and it is countably additive. Thus P_X is a probability measure on the measurable space $(A, \mathcal{B}(A))$. Therefore, given a probability space and a random variable X, we have constructed a new probability space $(A, \mathcal{B}(A), P_X)$ where the events describe outcomes of the random variable. The probability measure P_X is called the *distribution* of X (as opposed to a "cumulative distribution function" of X to be introduced later).

If two random variables have the same distribution, then they are said to be *equivalent* since they have the same probabilistic description, whether or not they are defined on the same underlying space or have the same functional form (see Problem 3.30).

A substantial part of the application of probability theory to practical problems is devoted to determining the distributions of random variables, that is, performing the "calculus of probability." One begins with a probability space. A random variable is defined on that space. The distribution of the random variable is then derived, and this results in a new probability space. This topic is called variously "derived distributions" or "transformations of random variables" and is often developed in the literature as a sequence of apparently unrelated subjects. When the points in the original sample space can be interpreted as "signals," then such problems can be viewed as "signal processing" and derived distribution problems are fundamental to the analysis of statistical signal processing systems. We shall emphasize that all such examples are just applications of the basic inverse image formula (3.22) and form a unified whole. In fact, this formula, with its vector analog, is one of the most important applications of probability theory. Its specialization to discrete input spaces using sums and to continuous input spaces using integrals will be seen and used often throughout this book.

It is useful to bear in mind both the mathematical and the intuitive concepts of a random variable when studying them. Mathematically, a random variable, say X, is a "nice" (that is, measurable) real-valued function defined on the sample space of a probability space $(\Omega, \mathcal{F}, P)$. Intuitively, a random variable is something that takes on values at random. The randomness is described by a distribution P_X, that is, by a probability measure on an event space of the real line. When doing computations involving random variables, it is usually simpler to concentrate on the probability space $(A, \mathcal{B}(A), P_X)$, where A is the range space of X, than on the original probability space $(\Omega, \mathcal{F}, P)$. Many experiments can yield equivalent random variables, and the space $(A, \mathcal{B}(A), P_X)$ can be considered as a canonical description of the random variable that is often more useful for computation. The original space is important, however, for two reasons. First, all distribution properties of random variables are inherited from the original space. Therefore much of the theory of random variables is just the theory of probability spaces specialized to the case of real sample spaces. If we understand probability spaces in general, then we understand random variables in particular. Second, and more important, we will often have many interrelated random variables defined on a common probability space. Because of the interrelationships, we cannot consider the random variables independently with separate probability spaces and distributions. We must refer to the original space in order to study the dependencies among the various random variables (or to consider the random variables jointly as a random vector).

Since a distribution is a special case of a probability measure, in many cases it can be induced or described by a probability function, i.e., a pmf or a pdf. If a range space of the random variable is discrete or, more generally, if there is a discrete subset of the range space A such that $P_X(A) = 1$, then there is a pmf, say p_X, corresponding to the distribution P_X. The two are related via the formulas

$$p_X(x) = P_X(\{x\}), \text{ all } x \in A, \tag{3.23}$$

where A is the range space or alphabet of the random variable, and

$$P_X(F) = \sum_{x \in F} p_X(x); \quad F \in \mathcal{B}(A). \tag{3.24}$$

In (3.23) both quantities are read as $\Pr(X = x)$.

The pmf and the distribution imply each other from (3.23) and (3.24), and hence either formula specifies the random variable.

If the range space of the random variable is continuous and if a pdf f_X exists, then we can write the integral analog to (3.24):

$$P_X(F) = \int_F f_X(x)\,dx; \quad F \in \mathcal{B}(A). \tag{3.25}$$

There is no direct analog of (3.23) for a pdf since a pdf is not a probability. An approximate analog of (3.23) follows from the mean value theorem of calculus. Suppose that $F = [x, x + \Delta x)$, where Δx is extremely small. Then if f_X is sufficiently smooth, the mean value theorem implies that

$$P_X([x, x + \Delta x)) = \int_x^{x+\Delta x} f_X(\alpha)\,d\alpha \approx f_X(x)\Delta x, \tag{3.26}$$

so that if we multiply a pdf $f_X(x)$ by a differential Δx, it can be interpreted as (approximately) the probability of being within Δx of x. It is desirable, however, to have an exact pair of results like (3.23) and (3.24) that show how to go both ways, that is, to get the probability function from the distribution as well as vice versa. From considerations of elementary calculus it seems that we should somehow differentiate both sides of (3.25) to yield the pdf in terms of the distribution. This is not immediately possible, however, because F is a set and not a real variable. Instead to find a pdf from a distribution, we use the intermediary of a *cumulative distribution function* or *cdf*. We pause to give the formal definition.

Given a random variable X with distribution P_X, the *cumulative distribution function* or *cdf* F_X is defined by

$$F_X(\alpha) = P_X((-\infty, \alpha]) = P_X(\{x : x \leq \alpha\}); \quad \alpha \in \Re.$$

The cdf is seen to represent the *cumulative* probability of all values of the random variable in the infinite interval from minus infinity up to and including the real number argument of the cdf. The various forms can be summarized as $F_X(\alpha) = \Pr(X \leq \alpha)$. If the random variable X is defined on the probability space $(\Omega, \mathcal{F}, P)$, then by definition

$$F_X(\alpha) = P(X^{-1}((-\infty, \alpha])) = P(\{\omega : X(\omega) \leq \alpha\}).$$

If a distribution possesses a pdf, then the cdf and pdf are related through the distribution and (3.25) by

$$F_X(\alpha) = P(X^{-1}((-\infty, \alpha])) = \int_{-\infty}^{\alpha} f_X(x)\,dx; \; \alpha \in \Re. \tag{3.27}$$

The motivation for the definition of the cdf in terms of our previous discussion is now obvious. Since integration and differentiation are mutually inverse operations, the pdf is determined from the cdf (and hence the distribution) by

$$f_X(\alpha) = \frac{dF_X(\alpha)}{d\alpha}; \quad \alpha \in \Re. \tag{3.28}$$

where, as is customary, the right-hand side is shorthand for

$$\left. \frac{dF_X(x)}{dx} \right|_{x=\alpha},$$

the derivative evaluated at α. Alternatively, (3.28) also follows from the fundamental theorem of calculus and the observation that

$$P_X((a,b]) = \int_a^b f_X(x)\, dx = F_X(b) - F_X(a). \tag{3.29}$$

Thus (3.27) and (3.28) together show how to find a pdf from a distribution and hence to provide the continuous analog of (3.23). Equation (3.23) is useful, however only if the derivative, and hence the pdf, exists. Observe that the cdf is always well defined (because the semi-infinite interval is a Borel set and therefore an event), regardless of whether or not the pdf exists. This is true in both the continuous and the discrete alphabet cases. For example, if X is a discrete alphabet random variable with alphabet $\mathcal{Z}$ and pmf p_X, then the cdf is

$$F_X(x) = \sum_{k=-\infty}^{x} p_X(k), \tag{3.30}$$

the analogous sum to the integral of (3.27). Furthermore, for this example, the pmf can be determined from the cdf (as well as the distribution) as

$$p_X(x) = F_X(x) - F_X(x-1), \tag{3.31}$$

a difference analogous to the derivative of (3.28).

It is desirable to use a single notation for the discrete and continuous cases whenever possible. This is accomplished for expressing the distribution in terms of the probability functions by using a *Stieltjes integral*, which is defined as

$$P_X(F) = \int_F dF_X(x) = \int 1_F(x)\, dF_X(x)$$
$$\triangleq \begin{cases} \sum_{x \in F} p_X(x) & \text{if } X \text{ is discrete} \\ \int_F f_X(x)\, dx & \text{if } X \text{ has a pdf} \end{cases}. \tag{3.32}$$

Thus (3.32) is a combination of both (3.24) and (3.25).

3.3.2 Mixture distributions

More generally, we may have a random variable that has both discrete and continuous aspects and hence is not describable by either a pmf alone or a pdf alone. For example, we might have a probability space $(\Re, \mathcal{B}(\Re), P)$, where P is described by a Gaussian pdf $f(\omega)$; $\omega \in \Re$. The sample point $\omega \in \Re$ is input to a soft limiter with output $X(\omega)$ – a device with input/output characteristic X defined by

$$X(\omega) = \begin{cases} -1 & \omega \leq -1 \\ \omega & \omega \in (-1, 1) \\ 1 & 1 \leq \omega \end{cases}. \tag{3.33}$$

As long as $|\omega| \leq 1$, $X(\omega) = \omega$. But for values outside this range, the output is set equal to -1 or $+1$. Thus all of the probability density outside the limiting range "piles up" on the ends so that $\Pr(X(\omega) = 1) = \int_{\omega \geq 1} f(\omega) d\omega$ is not zero. As a result X will have a mixture distribution, described by a pdf in $(-1, 1)$ and by a pmf at the points ± 1.

Random variables of this type can be described by a distribution that is the weighted sum of two other distributions – a discrete distribution and a continuous distribution. The weighted sum is an example of a mixture distribution, that is, a mixture of probability measures as in Example [2.18]. Specifically, let P_1 be a discrete distribution with corresponding pmf p, and let P_2 be a continuous distribution described by a pdf f. For any positive weights c_1, c_2 with $c_1 + c_2 = 1$, the following mixture distribution P_X is defined for all $F \in \mathcal{B}(\Re)$:

$$P_X(F) = c_1 P_1(F) + c_2 P_2(F) = c_1 \sum_{k \in F} p(k) + c_2 \int_F f(x)\, dx$$
$$= c_1 \sum 1_F(k) p(k) + c_2 \int 1_F(r) f(x)\, dx. \tag{3.34}$$

Continuing the example, the output of the limiter of (3.33) has a pmf and a pdf. The pmf places probability 1/2 on ± 1, while the pdf is Gaussian-shaped for magnitudes less than unity (i.e., it is a truncated Gaussian pdf normalized so that the pdf integrates to one over the range $(-1, 1)$). The constant c_2 is the integral of the input Gaussian pdf over $(-1, 1)$ and $c_1 = 1 - c_2$. Observe that the cdf for a random variable with a mixture distribution is

$$\begin{aligned} F_X(\alpha) &= c_1 \sum_{k:k \leq \alpha} p(k) + c_2 \int_{\infty}^{\alpha} f(x)\, dx \\ &= c_1 F_1(\alpha) + c_2 F_2(\alpha), \end{aligned} \tag{3.35}$$

where F_1 and F_2 are the cdf's corresponding to P_1 and P_2 respectively.

The combined notation for discrete and continuous alphabets using the Stieltjes integral notation of (3.32) also can be used as follows. Given a random variable with a mixture distribution of the form (3.34), then

$$P_X(F) = \int_F dF_X(x) = \int 1_F(x)\, dF_X(x); \quad F \in \mathcal{B}(\Re). \tag{3.36}$$

where

$$\int 1_F(x)\, dF_X(x) \triangleq c_1 \sum 1_F(x) p(x) + c_2 \int 1_F(x) f(x)\, dx. \tag{3.37}$$

Observe that (3.36) and (3.37) include (3.32) as a special case where either c_1 or c_2 is 0. Equations (3.36) and (3.37) provide a general means for finding the distribution of a random variable X given its cdf, provided the distribution has the form of (3.35).

All random variables can be described by a cdf. But, more subtly, do all random variables have a cdf of the form (3.35)? The answer is almost yes. Certainly all of the random variables encountered in this book and in engineering practice have this form. It can be shown, however, that the most general cdf has the form of a mixture of three cdf's: a continuous and differentiable piece induced by a pdf, a discrete piece induced by a pmf, and a third pathological piece. The third piece is an odd beast wherein the cdf is something called a singular function – the cdf is continuous (it has no jumps as it does in the discrete case), and is differentiable almost everywhere (here "almost everywhere" means that the cdf is differentiable at all points except some set F for which $\int_F dx = 0$), but this derivative is 0 almost everywhere and hence it cannot be integrated to find a probability! Thus for this third piece, one cannot use pmf's or pdf's to compute probabilities. The construction of such a cdf is beyond the scope of this text, but we can point out for the curious that the typical example involves placing probability measures on the Cantor set that was considered in Problem 2.19. At any rate, as such examples almost never arise in practice, we shall ignore them and henceforth consider only random variables for which (3.36) and (3.37) hold.

Although the general mixture distribution random variable has both discrete and continuous pieces, for pedagogical purposes it is usually simplest to treat the two pieces separately – i.e., to consider random variables that have either a pdf or a pmf. Hence we will rarely consider mixture distribution random variables and will almost always focus on those that are described either by a pmf or by a pdf and not both.

To summarize our discussion, we will define a random variable to be a *discrete, continuous,* or *mixture* random variable depending on whether it is described probabilistically by a pmf, pdf, or mixture as in (3.36) and (3.37) with $c_1, c_2 > 0$.

We note in passing that some texts endeavor to use a uniform approach to mixture distributions by permitting pdf's to possess Dirac delta or impulse functions. The purpose of this approach is to permit the use of the continuous ideas in discrete cases, as in our limiter output example. If the cdf is differentiated, then a legitimate pdf results (without the need for a pmf) if a delta function is allowed at the two discontinuities of the cdf. As a general practice we prefer the Stieltjes notation, however, because of the added notational clumsiness resulting from using pdf's to handle inherently discrete problems. For example, compare the notation for the geometric pmf with the corresponding pdf that is written using Dirac delta functions.

3.3.3 Derived distributions

[3.12] Let $(\Omega, \mathcal{F}, P)$ be a discrete probability space with Ω a discrete subset of the real line and $\mathcal{F}$ the power set. Let p be the pmf corresponding to P, that is,

$$p(\omega) = P(\{\omega\}), \text{ all } \omega \in \Omega.$$

(*Note:* there is a very subtle possibility for confusion here. One could consider $p(\omega)$ to be a random variable because it satisfies the definition for a random variable. We do not use it in this sense, however; we use it as a pmf for evaluating probabilities in the context given. In addition, no confusion should result because we rarely use lower-case letters for random variables.) Let X be a random variable defined on this space. Since the domain of X is discrete, its range space, A, is also discrete (refer to the definition of a function to understand this point). Thus the probability measure P_X must also correspond to a pmf, say p_X; that is, (3.23) and (3.24) must hold. Then we can derive either the distribution P_X or the simpler pmf p_X in order to complete a probabilistic description of X. Using (3.22) yields

$$p_X(x) = P_X(\{x\}) = P(X^{-1}(\{x\})) = \sum_{\omega:X(\omega)=x} p(\omega). \tag{3.38}$$

Equation (3.38) provides a formula for computing the pmf and hence the distribution of any random variable defined on a discrete probability space. As a specific example, consider a discrete probability space $(\Omega, \mathcal{F}, P)$ with $\Omega = \mathcal{Z}_+$, $\mathcal{F}$ the power set of Ω, and P the probability measure induced by the geometric pmf. Define a random variable Y on this space by

$$Y(\omega) = \begin{cases} 1 & \text{if } \omega \text{ even} \\ 0 & \text{if } \omega \text{ odd} \end{cases}.$$

where we consider 0 (which has probability zero under the geometric pmf) to be even. Thus we have a random variable $Y : \mathcal{Z}_+ \to \{0, 1\}$. Using the formula (3.38) for the pmf for $Y(\omega) = 1$ results in

$$\begin{aligned} p_Y(1) &= \sum_{\omega:\,\omega \text{ even}} (1-p)^{k-1}p = \sum_{k=2,4,\ldots} (1-p)^{k-1}p \\ &= \frac{p}{(1-p)} \sum_{k=1}^{\infty} ((1-p)^2)^k = p(1-p)\sum_{k=0}^{\infty}((1-p)^2)^k \\ &= p\frac{(1-p)}{1-(1-p)^2} = \frac{1-p}{2-p}, \end{aligned}$$

where we have used the standard geometric series summation formula (in a thinly disguised variation of an example of Section 2.2.4). We can calculate the remaining point in the pmf from the axioms of probability: $p_Y(0) = 1 - p_Y(1)$. Thus we have found a nonobvious derived distribution by computing a pmf via (3.38), a special case of (3.22). Of course, given the pmf, we could now calculate the distribution from (3.24) for all four sets in the power set of $\{0, 1\}$.

[3.13] Say we have a probability space $(\Re, \mathcal{B}(\Re), P)$ where P is described by a pdf g; that is, g is a nonnegative function of the real line with total integral 1 and

$$P(F) = \int_{r\in F} g(r)\,dr; \quad F \in \mathcal{B}(\Re).$$

Suppose that we have a random variable $X : \Re \to \Re$. We can use (3.22) or (3.25) to write a general formula for the distribution of X:

$$P_X(F) = P(X^{-1}(F)) = \int_{r:\, X(r)\in F} g(r)\, dr.$$

Ideally, however, we would like to have a simpler description of X. In particular, if X is a "reasonable function" it should have either a discrete range space (e.g., a quantizer) or a continuous range space (or possibly both, as in the general mixture case). If the range space is discrete, then X can be described by a pmf, and the preceding formula (with the requisite change of dummy variable) becomes

$$p_X(x) = P_X(\{x\}) = \int_{r:\, X(r)=x} g(r)\, dr.$$

If, however, the range space is continuous, then there should exist a pdf for X, say f_X, such that (3.25) holds. How do we find this pdf? As previously discussed, to find a pdf from a distribution, we first find the cdf F_X. Then we differentiate the cdf with respect to its argument to obtain the pdf. As a nontrivial example, suppose that we have a probability space $(\Re, \mathcal{B}(\Re), P)$ with P the probability measure induced by the Gaussian pdf. Define a random variable $W : \Re \to \Re$ by $W(r) = r^2$; $r \in \Re$. Following the described procedure, we first attempt to find the cdf F_W for W:

$$\begin{aligned} F_W(w) &= \Pr(W \le w) = P(\{\omega : W(\omega) = \omega^2 \le w\}) \\ &= P([-w^{1/2}, w^{1/2}]);\ \text{if } w \ge 0. \end{aligned}$$

The cdf is clearly 0 if $w < 0$. Since P is described by a pdf, say g (the specific Gaussian form is not yet important), then

$$F_W(w) = \int_{-w^{1/2}}^{w^{1/2}} g(r)\, dr.$$

If one should now try to plug in the specific form for the Gaussian density, one would quickly discover that no closed form solution exists. Happily, however, the integral does not have to be evaluated explicitly – we need only its derivative. Therefore we can use the following handy formula from elementary calculus for differentiating the integral:

$$\frac{d}{dw}\int_{a(w)}^{b(w)} g(r)\, dr = g(b(w))\frac{db(w)}{dw} - g(a(w))\frac{da(w)}{dw}. \tag{3.39}$$

Application of the formula yields

$$f_W(w) = g(w^{1/2})\left(\frac{w^{-1/2}}{2}\right) - g(-w^{1/2})\left(-\frac{w^{-1/2}}{2}\right). \tag{3.40}$$

The final answer is found by plugging in the Gaussian form of g. For simplicity we do this only for the special case where $m = 0$. Then g is symmetric; that is, $g(w) = g(-w)$, so that

$$f_W(w) = w^{-1/2} g(w^{1/2}); \;\; w \in [0, \infty),$$

and finally

$$f_W(w) = \frac{w^{-1/2}}{\sqrt{2\pi\sigma^2}} e^{-w/2\sigma^2}; \;\; w \in [0, \infty).$$

This pdf is called a *chi-squared pdf* with one degree of freedom. Note that the functional form of the pdf is valid only for the given domain. By implication the pdf is zero outside the given domain – in this example, negative values of W cannot occur. One should always specify the domain of the dummy variable of a pdf; otherwise the description is incomplete.

In practice one is likely to encounter the following trick for deriving densities for certain simple one-dimensional problems. The approach can be used whenever the random variable is a monotonic (increasing or decreasing) function of its argument. Suppose first that we have a random variable $Y = g(X)$, where g is a differentiable monotonically increasing function. Since g is monotonic, it is invertible and we can write $X = g^{-1}(Y)$, that is, $x = g^{-1}(y)$ is the value of x for which $g(x) = y$. Then

$$\begin{aligned} F_Y(y) &= \Pr(g(X) \leq y) = \Pr(X \leq g^{-1}(y)) \\ &= F_X(g^{-1}(y)) = \int_{-\infty}^{g^{-1}(y)} f_X(x)\, dx. \end{aligned}$$

From (3.39) the density can be found as

$$f_Y(y) = \frac{d}{dy} F_Y(y) = f_X(g^{-1}(y)) \frac{dg^{-1}(y)}{dy}.$$

A similar result can be derived for a monotonically decreasing g except that a minus sign results. The final formula is that if $Y = g(X)$ and g is monotone, then

$$f_Y(y) = f_X(g^{-1}(y)) \left|\frac{dg^{-1}(y)}{dy}\right|. \tag{3.41}$$

This result is a one-dimensional special case of the so-called *Jacobian* approach to derived distributions. The result could be used to solve the previous problem by separately considering negative and nonnegative values of the input r since r^2 is a monotonic increasing function for nonnegative r and monotonic decreasing for negative r. As in this example, the direct approach from the inverse image formula is often simpler than using the Jacobian "shortcut," unless one is dealing with a monotonic function.

It can be seen that although the details may vary from application to application, all derived distribution problems are solved by the general formula (3.22). In some cases the solution will result in a pmf; in others the solution will result in a pdf.

To review the general philosophy, one uses the inverse image formula to compute the probability of an output event. This is accomplished by finding the probability with respect to the original probability measure of all input events that result in the given output event. In the discrete case one concentrates on output events of the form $X = x$ and thereby finds a pmf. In the continuous case, one concentrates on output events of the form $X \leq x$ and thereby finds a cdf. The pdf is then found by differentiating.

[3.14] As a final example of derived distributions, suppose that we are given a probability space $(\Omega, \mathcal{B}(\Omega), P)$ with $\Omega \subset \Re$. Define the identity mapping $X : \Omega \to \Omega$ by $X(\omega) = \omega$. The identity mapping on the real line with the Borel field is always a random variable because the measurability requirement is automatically satisfied. Obviously the distribution P_X is identical to the original probability measure P. Thus all probability spaces with real sample spaces provide examples of random variables through the identity mapping. A random variable described in this form instead of as a general function (not the identity mapping) on an underlying probability space is called a "directly given" random variable.

3.4 Random vectors and random processes

Thus far we have emphasized random variables, scalar functions on a sample space that assume real values. In some cases we may wish to model processes or measurements with complex values. Complex outputs can be considered as two-dimensional real vectors with the components being the real and imaginary parts or, equivalently, the magnitude and phase. This special case can be equally well described as a single complex-valued random variable or as a two-dimensional random vector.

More generally, we may have k-dimensional real vector outputs. A random variable is a real-valued function on a sample space (with a technical condition), that is, a function mapping a sample space into the real line $\Re$. The obvious random vector definition is a vector-valued function definition. Under this definition, a random vector is a vector of random variables, a function mapping the sample space into $\Re^k$ instead of $\Re$. Yet even more generally, we may have vectors that are not finite-dimensional, e.g., sequences and waveforms whose values at each time are random variables. This is essentially the definition of a *random process.* Fundamentally speaking, both random vectors and random processes are simply collections of random variables defined on a common probability space.

Given a probability space $(\Omega, \mathcal{F}, P)$, a finite collection of random variables $\{X_i;\ i = 0, 1, \ldots, k-1\}$ is called a *random vector.* We will often denote a random vector in boldface as $\mathbf{X}$. Thus a random vector is a vector-valued function $\mathbf{X} : \Omega \to \Re^k$ defined by $\mathbf{X} = (X_0, X_1, \ldots, X_{k-1})$ with each of the components being a random variable. It is also common to use an ordinary non-boldface X and let context indicate whether X has dimension 1 or not. Another common notation for the k-dimensional random vector is X^k. Each of these forms is convenient in different settings, but we begin with the boldface notation in order to distinguish the random vectors from scalar random variables. As we progress, however, the non-boldface notation will be used with increasing frequency to match current style in the literature. The boldface notation is still found, but it is far less common then it used to be. When vectors are used in linear algebra manipulations with matrices and other vectors, we will assume that they are column vectors so that strictly speaking the vector should be denoted $\mathbf{X} = (X_0, X_1, \ldots, X_{k-1})^t$, where t denotes transpose.

A slightly different notation will ease the generalization to random processes. A random vector $\mathbf{X} = (X_0, X_1, \ldots, X_{k-1})$ can be defined as an *indexed* family of random variables $\{X_i;\ i \in \mathcal{T}\}$ where $\mathcal{T}$ is the index set $\mathcal{Z}_k = \{0, 1, \ldots, k-1\}$. The index set in some examples will correspond to time; e.g., X_i is a measurement on an experiment at time i for k different times. We get a random process by using the same basic definition with an infinite index set, which almost always corresponds to time. A *random process* or *stochastic process* is an indexed family of random variables $\{X_i;\ t \in \mathcal{T}\}$ or, equivalently, $\{X(t);\ t \in \mathcal{T}\}$, defined on a common probability space $(\Omega, \mathcal{F}, P)$. The process is said to be *discrete time* if $\mathcal{T}$ is discrete, e.g., $\mathcal{Z}_+$ or $\mathcal{Z}$, and *continuous time* if the index set is continuous, e.g., $\Re$ or $[0, \infty)$. A discrete time random process is often called a *time series.* It is said to be *discrete alphabet* or *discrete amplitude* if all finite-length random vectors of random variables drawn from the random process are discrete random vectors. The process is said to be *continuous alphabet* or *continuous amplitude* if all finite-length random vectors of random variables drawn from the random process are continuous random vectors. The process is said to have a *mixed alphabet* if all finite-length random vectors of random variables drawn from the random process are mixture random vectors.

Thus a random process is a collection of random variables indexed by time, usually into the indefinite future and sometimes into the infinite past as well. For each value of time t, X_t or $X(t)$ is a random variable. Both notations are used, but X_t or X_n is more common for discrete time processes whereas $X(t)$ is more common for continuous time processes. It is useful to recall that random variables are functions on an underlying sample space Ω and hence implicitly depend on $\omega \in \Omega$. Thus a random process (and a random vector) is actually a function of two arguments, written explicitly as $X(t, \omega)$; $t \in \mathcal{T}, \omega \in \Omega$ (or $X_t(\omega)$ – we use the first notation for the moment). Observe that for a fixed value of time, $X(t, \omega)$ is a random variable whose value depends probabilistically on ω. On the other hand, if we fix ω and allow t to vary deterministically, we have either a sequence ($\mathcal{T}$ discrete) or a waveform ($\mathcal{T}$ continuous). If we fix both t and ω, we have a number. Overall we can consider a random process as a two-space mapping $X : \Omega \times \mathcal{T} \to \Re$ or as a one-space mapping $X : \Omega \to \Re^{\mathcal{T}}$ from sample space into a space of sequences or waveforms.

There is a common notational ambiguity and hence confusion when dealing with random processes. It is the same problem we encountered with functions in the context of random variables at the beginning of the chapter. The notation $X(t)$ or X_t usually means a sample of the random process at a specified time t, i.e., a random variable, just as $\sin t$ means the sine of a specified value t. Often in the literature, however, the notation is used as an abbreviation for $\{X(t);\ t \in \mathcal{T}\}$ or $\{X_t;\ t \in \mathcal{T}\}$, that is, for the entire random process or family of random variables. The abbreviation is the same as the common use of $\sin t$ to mean $\{\sin t; t \in (-\infty, \infty)\}$, that is, the entire waveform and not just a single value. In summary, the common (and sometimes unfortunate) ambiguity is in whether the dummy variable t means a *specific* value or is implicitly allowed to vary over its entire domain. Of course, as noted at the beginning of the chapter, the problem could be avoided by reserving a different notation to specify a fixed time value, say t_0, but this is usually not done, to avoid a proliferation of notation. In this book we will attempt to avoid the potential confusion by using the abbreviations $\{X(t)\}$ and $\{X_t\}$ for the random processes when the index set is clear from context and reserving the notation $X(t)$ and X_t to mean the tth random variable of the process, that is, the sample of the random process at time t. The reader should beware in reading other sources, however, because this sloppiness will undoubtedly be encountered at some point in the literature; when this happens one can only hope that the context will make the meaning clear.

There is also an ambiguity regarding the alphabet of the random process. If $X(t)$ takes values in A_t, then strictly speaking the alphabet of the random process is $\prod_{t\in\mathcal{T}} A_t$, the space of all possible waveforms or sequences with coordinate taking values in A_t. If all of the A_t are the same, say $A_t = A$, this process alphabet is $A^{\mathcal{T}}$. In this case, however, the alphabet of the process is commonly said to be simply A, the set of values from which all of the coordinate random variables are drawn. We will frequently use this convention.

3.5 Distributions of random vectors

Since a random vector takes values in a space $\Re^k$, one might expect that the events in this space, that is, the members of the event space $\mathcal{B}(\Re)^k$, should inherit a probability measure from the original probability space. This is in fact true as one would expect by analogy to scalar random variables. Also analogous to the case of a random variable, the probability measure is called a *distribution* and is defined as

$$\begin{aligned} &P_{\mathbf{X}}(F) \\ &= P(\mathbf{X}^{-1}(F)) = P(\{\omega : \mathbf{X}(\omega) \in F\}) \\ &= P(\{\omega : (X_0(\omega), X_1(\omega), \ldots, X_{k-1}(\omega)) \in F\}), F \in \mathcal{B}(\Re)^k, \end{aligned} \tag{3.42}$$

where the various forms are equivalent and all stand for $\Pr(\mathbf{X} \in F)$. Equation (3.42) is the vector generalization of the inverse image equation (3.22) for random variables. Hence (3.42) is the fundamental formula for deriving vector distributions, that is, probability distributions describing random vector events. Keep in mind that the random vectors might be composed of a collection of samples from a random process.

By definition the distribution given by (3.22) is valid for each component random variable. This does not immediately imply, however, that the distribution given by (3.42) for events on all components together is valid. As in the case of a random variable, the distribution will be valid if the output events $F \in \mathcal{B}(\Re)^k$ have inverse images under $\mathbf{X}$ that are input events, that is, if $\mathbf{X}^{-1}(F) \in \mathcal{F}$ for every $F \in \mathcal{B}(\Re)^k$. The following subsection treats this subtle issue in further detail, but the only crucial point for our purposes is the following. Given that we consider real-valued vectors $\mathbf{X} = (X_0, X_1, \ldots, X_{k-1})$, knowing that each coordinate X_i is a random variable (i.e., $X_i^{-1}(F)$ for each real event F) guarantees that $\mathbf{X}^{-1}(F) \in \mathcal{F}$ for every $F \in \mathcal{B}(\Re)^k$ and hence the basic derived distribution formula is valid for random vectors.

3.5.1 ⋆Multidimensional events

From the discussion following Example [2.11] we can at least resolve the issue for certain types of output events, namely, events that are rectangles. Rectangles are special events in that the values assumed by any component in the event are not constrained by any of the other components (compare a two-dimensional rectangle with a circle, as in Problem 2.36). Specifically $F \in \mathcal{B}(\Re)^k$ is a rectangle if it has the form

$$F = \{\mathbf{x} : s_i \in F_i;\ i = 0, 1, \ldots, k-1\} = \bigcap_{i=0}^{k-1} \{\mathbf{x} : s_i \in F_i\} = \prod_{i=0}^{k-1} F_i,$$

where all $F_i \in \mathcal{B}(\Re)$; $i = 0, 1, \ldots, k-1$ (refer to Figure 2.3(d) for a two-dimensional illustration of such a rectangle). Because inverse images preserve set operations (see Problem A.12), the inverse image of F can be specified as the intersection of the inverse images of the individual events:

$$\mathbf{X}^{-1}(F) = \{\omega : X_i(\omega) \in F_i;\ i = 0, 1, \ldots, k-1\} = \bigcap_{i=0}^{k-1} X_i^{-1}(F_i).$$

Since each of the X_i is a random variable, the inverse images of the individual events $X_i^{-1}(F_i)$ must all be in $\mathcal{F}$. Since $\mathcal{F}$ is an event space, the intersection of events must also be an event, and hence $\mathbf{X}^{-1}(F)$ is indeed an event.

Thus we conclude that the distribution is well defined for rectangles. As to more general output events, we simply observe that a result from measure theory ensures that if (1) inverse images of rectangles are events and (2) rectangles are used to generate the output event space, then the inverse images of all output events are events. These two conditions are satisfied by our definition. Thus the distribution of the random vector $\mathbf{X}$ is well defined. Although a detailed proof of the measure theoretic result will not be given, the essential concept can be given: any event in $\mathcal{F}$ can be approximated arbitrarily closely by finite unions of rectangles (e.g., a circle can be approximated by lots of very small squares). The union of the rectangles is an event. Finally, the limit of the events as the approximation gets better must also be an event.

3.5.2 Multidimensional probability functions

Given a probability space $(\Omega, \mathcal{F}, P)$ and a random vector $\mathbf{X} : \Omega \to \Re^k$, we have seen that there is a probability measure $P_{\mathbf{X}}$ that the random vector inherits from the original probability space. With the new probability measure we define a new probability space $(\Re^k, \mathcal{B}(\Re)^k, P_{\mathbf{X}})$. As in the scalar case, the distribution can be described by probability functions, that is, cdf's and either pmf's or pdf's (or both). If the random vector has a discrete range space, then the distribution can be described by a multidimensional pmf $p_{\mathbf{X}}(\mathbf{x}) = P_{\mathbf{X}}(\{\mathbf{x}\}) = \Pr(\mathbf{X} = \mathbf{x})$ as

$$p_{\mathbf{X}}(F) = \sum_{\mathbf{x} \in F} p_{\mathbf{X}}(\mathbf{x}) = \sum_{(x_0, x_1, \ldots, x_{k-1}) \in F} p_{X_0, X_1, \ldots, X_{k-1}}(x_0, x_1, \ldots, x_{k-1}),$$

where the last form points out the economy of the vector notation of the previous line. If the random vector $\mathbf{X}$ has a continuous range space, then in a similar fashion its distribution can be described by a multidimensional pdf $f_{\mathbf{X}}$ with $P_{\mathbf{X}}(F) = \int_F f_{\mathbf{X}}(\mathbf{x})\, d\mathbf{x}$. In order to derive the pdf from the distribution, as in the scalar case, we use a cdf.

Given a k-dimensional random vector $\mathbf{X}$, define its *cumulative distribution function* $F_{\mathbf{X}}$ by

$$\begin{aligned} F_{\mathbf{X}}(\alpha) &= F_{X_0, X_1, \ldots, X_{k-1}}(\alpha_0, \alpha_1, \ldots, \alpha_{k-1}) \\ &= P_{\mathbf{X}}(\{\mathbf{x} : x_i \le \alpha_i;\ i = 0, 1, \ldots, k-1\}). \end{aligned}$$

In English, $F_{\mathbf{X}}(\mathbf{x}) = \Pr(X_i \le x_i;\ i = 0, 1, \ldots, k-1)$. Note that the cdf for any value of its argument is the probability of a special kind of rectangle. For example, if we have a two-dimensional random vector (X, Y), then the cdf $F_{X,Y}(\alpha, \beta) = \Pr(X \le \alpha, Y \le \beta)$ is the probability of the semi-infinite rectangle $\{(x, y) : x \le \alpha,\ y \le \beta\}$.

Observe that we can also write this probability in several other ways, e.g.,

$$\begin{aligned} F_{\mathbf{X}}(\mathbf{x}) &= P_{\mathbf{X}}\left(\prod_{i=0}^{k-1} (-\infty, x_i]\right) \\ &= P(\{\omega : X_i(\omega) \le x_i;\ i = 0, 1, \ldots, k-1\}) \\ &= P\left(\bigcap_{i=0}^{k-1} X_i^{-1}((-\infty, x_i])\right). \end{aligned}$$

Since integration and differentiation are inverses of each other, it follows that

$$f_{X_0,X_1,\ldots,X_{k-1}}(x_0, x_1, \ldots, x_{k-1})$$
$$= \frac{\partial^k}{\partial x_0 \partial x_1 \ldots \partial x_{k-1}} F_{X_0,X_1,\ldots,X_{k-1}}(x_0, x_1, \ldots, x_{k-1}).$$

As with random variables, random vectors can, in general, have discrete and continuous parts with a corresponding mixture distribution. We will concentrate on random vectors that are described completely by either pmf's or pdf's. Also as with random variables, we can always unify notation using a multidimensional Stieltjes integral to write

$$P_{\mathbf{X}}(F) = \int_F dF_{\mathbf{X}}(\mathbf{x}) \, ; \quad F \in \mathcal{B}(\Re)^k,$$

where the integral is defined as the usual integral if $\mathbf{X}$ is described by a pdf, as a sum if $\mathbf{X}$ is described by a pmf, and by a weighted average if $\mathbf{X}$ has both a discrete and a continuous part. Random vectors are said to be continuous, discrete, or mixture random vectors in accordance with the above analogy to random variables.

3.5.3 Consistency of joint and marginal distributions

By definition a random vector $\mathbf{X} = (X_0, X_1, \ldots, X_{k-1})$ is a collection of random variables defined on a common probability space $(\Omega, \mathcal{F}, P)$. Alternatively, $\mathbf{X}$ can be considered to be a random vector that takes on values randomly as described by a probability distribution $P_{\mathbf{X}}$, without explicit reference to the underlying probability space. Either the original probability measure P or the induced distribution $P_{\mathbf{X}}$ can be used to compute probabilities of events involving the random vector. In turn, $P_{\mathbf{X}}$ may be induced by a pmf $p_{\mathbf{X}}$ or a pdf $f_{\mathbf{X}}$. From any of these probabilistic descriptions we can find a probabilistic description for any of the component random variables or any collection of thereof. That is, $P_{\mathbf{X}}$ is evaluated on rectangles of the form $\{\mathbf{x} = (x_0, \ldots, x_{k-1}) : x_i \in G\}$ for any $G \in \mathcal{B}(\Re)$ as

$$P_{X_i}(G) = P_{\mathbf{X}}(\{\mathbf{x} : x_i \in G\}), \quad G \in \mathcal{B}(\Re). \tag{3.43}$$

For example, given a value of i in $\{0, 1, \ldots, k-1\}$, the distribution of the random variable X_i is found by evaluating the distribution $P_{\mathbf{X}}$ for the random vector on one-dimensional rectangles where only the component X_i is constrained to lie in some set – the rest of the components can take on any value. Of course the probability can also be evaluated using the underlying probability measure P via the usual formula

$$P_{X_i}(G) = P(X_i^{-1}(G)).$$

Alternatively, we can consider this a derived distribution problem on the vector probability space $(\Re^k, \mathcal{B}(\Re)^k, P_{\mathbf{X}})$ using a sampling function $\Pi_i : \Re^k \to \Re$ as in Example [3.4]. Specifically, let $\Pi_i(\mathbf{X}) = X_i$. Using (3.22) we write

$$P_{\Pi_i}(G) = P_{\mathbf{X}}(\Pi_i^{-1}(G)) = P_{\mathbf{X}}(\{\mathbf{x} : x_i \in G\}). \qquad (3.44)$$

The two formulas (3.43) and (3.44) demonstrate that Π_i and X_i are equivalent random variables, and indeed they correspond to the same physical events – the outputs of the ith coordinate of the random vector $\mathbf{X}$. They are related through the formula $\Pi_i(\mathbf{X}(\omega)) = X_i(\omega)$. Intuitively, the two random variables provide different models of the same thing. As usual, which is "better" depends on which is the simpler model to handle for a given problem.

Another fundamental observation implicit in these ruminations is that there are many ways to compute the probability of a given event such as "the ith coordinate of the random vector $\mathbf{X}$ takes on a value in an event F," and all these methods must yield the same answer because they all can be referred back to a common definition in terms of the underlying probability measure P. This is called *consistency*; the various probability measures (P, P_{X_i}, and $P_{\mathbf{X}}$) are all *consistent* in that they assign the same number to any given physical event for which they all are defined. In particular, if we have a random process $\{X_t;\ t \in \mathcal{T}\}$, then there is an infinite number of ways we could form a random vector $(X_{t_0}, X_{t_1}, \ldots, X_{t_{k-1}})$ by choosing a finite number k and sample times $t_0, t_1, \ldots, t_{k-1}$ and each of these would result in a corresponding k-dimensional probability distribution $P_{X_{t_0}, X_{t_1}, \ldots, X_{t_{k-1}}}$. The calculus derived from the axioms of probability implies that all of these distributions must be consistent in the same sense, i.e., all must yield the same answer when used to compute the probability of a given event.

The distribution P_{X_i} of a single component X_i of a random vector $\mathbf{X}$ is referred to as a *marginal distribution*, while the distribution $P_{\mathbf{X}}$ of the random vector is called a *joint distribution*. As we have seen, joint and marginal distributions are related by consistency with respect to the original probability measure, i.e.,

$$\begin{aligned} P_{X_i}(G) &= P_{\mathbf{X}}(\{\mathbf{x} : x_i \in G\}) \\ &= P(\{\omega : X_i(\omega) \in G\}) \\ &= \Pr(X_i \in G). \end{aligned} \qquad (3.45)$$

For the cases where the distributions are induced by pmf's (marginal pmf's and joint pmf's) or pdf's (marginal pdf's or joint pdf's), the relation becomes, respectively,

$$p_{X_i}(\alpha) = \sum_{x_0,x_1,\ldots,x_{i-1},x_{i+1},\ldots,x_{k-1}} p_{X_0,X_1,\ldots,X_{k-1}}(x_0,x_1,\ldots,x_{i-1},\alpha,x_{i+1},\ldots,x_{k-1})$$

or

$$f_{X_i}(\alpha) = \int_{x_0,\ldots,x_{i-1},x_{i+1},\ldots,x_{k-1}} dx_0 \ldots dx_{i-1} dx_{i+1} \ldots dx_{k-1} \, f_{X_0,\ldots,X_{k-1}}(x_0,\ldots,x_{i-1},\alpha,x_{i+1},\ldots,x_{k-1}).$$

That is, we sum or integrate over all of the dummy variables corresponding to the unwanted random variables in the vector to obtain the pmf or pdf for the random variable X_i. The two formulas look identical except that one formula sums for discrete random variables and the other integrates for continuous ones. We repeat the fact that both formulas are simple consequences of (3.45).

One can also use (3.43) to derive the cdf of X_i by setting $G = (-\infty, \alpha]$. The cdf is

$$F_{X_i}(\alpha) = F_{\mathbf{X}}(\infty, \infty, \ldots, \infty, \alpha, \infty, \ldots, \infty),$$

where the α appears in the ith position. This equation states that $\Pr(X_i \leq \alpha) = \Pr(X_i \leq \alpha \text{ and } X_j \leq \infty)$, all $j \neq i$. The expressions for pmf's and pdf's also can be derived from the expression for cdf's.

The details of notation with k random variables can cloud the meaning of the relations we are discussing. Therefore we rewrite them for the special case of $k = 2$ to emphasize the essential form. Suppose that (X, Y) is a random vector. Then the marginal distribution of X is obtained from the joint distribution of X and Y by leaving Y unconstrained, i.e., as in (3.43):

$$P_X(F) = P_{X,Y}(\{(x, y) : x \in F\}); \quad F \in \mathcal{B}(\Re).$$

Furthermore, the marginal cdf of X is

$$F_X(\alpha) = F_{X,Y}(\alpha, \infty).$$

If the range space of the vector (X, Y) is discrete, the marginal pmf of X is

$$p_X(x) = \sum_y p_{X,Y}(x, y).$$

If the range space of the vector (X, Y) is continuous and the cdf is differentiable so that $f_{X,Y}(x, y)$ exists, the marginal pdf of X is

$$f_X(x) = \int_{-\infty}^{\infty} f_{X,Y}(x, y)\, dy,$$

with similar expressions for the distribution and probability functions for the random variable Y.

In summary, given a probabilistic description of a random vector, we can always determine a probabilistic description for any of the component random variables of the random vector. This follows from the consistency of probability distributions derived from a common underlying probability space. It is important to keep in mind that *the opposite statement is not true.* As considered in the introduction to this chapter, given all the marginal distributions of the component random variables, we cannot find the joint distribution of the random vector formed from the components unless we further constrain the problem. This is true because the marginal distributions provide none of the information about the interrelationships of the components that is contained in the joint distribution.

In a similar manner we can deduce the distributions or probability functions of "subvectors" of a random vector, that is, if we have the distribution for $\mathbf{X} = (X_0, X_1, \ldots, X_{k-1})$ and if k is big enough, we can find the distribution for the random vector (X_1, X_2) or the random vector (X_5, X_{10}, X_{15}), and so on. Writing the general formulas in detail in tedious and adds little insight. The basic idea, however, is extremely important. One always starts with a probability space $(\Omega, \mathcal{F}, P)$ from which one can proceed in many ways to compute the probability of an event involving any combination of random variables defined on the space. No matter how one proceeds, the probability computed for a given event must be the same. In other words, all joint and marginal probability distributions for random variables on a common probability space must be consistent since they all follow from the common underlying probability measure. For example, after finding the distribution of a random vector $\mathbf{X}$, the marginal distribution for the specific component X_i can be found from the joint distribution. This marginal distribution must agree with that obtained for X_i directly from the probability space. As another possibility, one might first find a distribution for a subvector containing X_i, say the vector $\mathbf{Y} = (X_{i-1}, X_i, X_{i+1})$. This distribution can be used to find the marginal distribution for X_i. All answers must be the same since all can be expressed in the form $P(X^{-1}(F))$ using the original probability space must be consistent in the sense that they agree with one another on events.

Examples

We now give examples of the computation of marginal probability functions from joint probability functions.

[3.15] Say that we are given a pair of random variables X and Y such that the random vector (X, Y) has a pmf of the form

$$p_{X,Y}(x, y) = r(x)q(y),$$

where r and q are both valid pmf's. In other words, $p_{X,Y}$ is a product pmf. Then it is easily seen that

$$\begin{aligned} p_X(x) &= \sum_y p_{X,Y}(x, y) = \sum_y r(x)q(y) \\ &= r(x)\sum_y q(y) = r(x). \end{aligned}$$

Thus in the special case of a product distribution, knowing the marginal pmf's is enough to know the joint distribution.

[3.16] Consider flipping two fair coins connected by a piece of rubber that is fairly flexible. Unlike the example where the coins were soldered together, it is not certain that they will show the same face; it is, however, more probable. To quantify the pmf, say that the probability of the pair (0,0) is 0.4, the probability of the pair (1,1) is 0.4, and the probabilities of the pairs (0,1) and (1,0) are each 0.1. As with the soldered-coins case, this is clearly not a product distribution, but a simple computation shows that as in Example [3.15], p_X and p_Y both place probability 1/2 on 0, and 1. Thus this distribution, the soldered-coins distribution, and the product distribution of Example [3.15] *all* yield the same marginal pmf's. The point again is that the marginal probability functions are not enough to describe a vector experiment. We need the joint probability function to describe the interrelations or dependencies among the random variables.

[3.17] A gambler has a pair of very special dice: the sum of the two dice comes up as seven on every roll. Each die has six faces with values in $A = \{1, 2, 3, 4, 5, 6\}$. All combinations have equal probability; e.g., the probability of a one and a six has the same probability as a three and a four. Although the two dice are identical, we will distinguish between them by number for the purposes of assigning two random variables. The outcome of the roll of the first die is denoted X and the outcome of the roll of the second die is called Y so that (X, Y) is a random vector taking values in A^2, the space of all pairs of numbers drawn from A. The joint pmf of X and Y is

$$p_{X,Y}(x,y) = \frac{1}{6}, \quad x+y=7, \quad (x,y) \in A^2.$$

The pmf of X is determined by summing the pmf with respect to y. However, for any given $X \in A$, the value of Y is determined: for example, $Y = 7 - X$. Therefore the pmf of X is

$$p_X(x) = 1/6, \quad x \in A.$$

Note that this pmf is the same as one would derive for the roll of a single unbiased die! Note also that the pmf for Y is identical with that for X. Obviously, then, it is impossible to tell that the gambler is using unfair dice *as a pair* from looking at outcomes of the rolls of each die alone. The joint pmf cannot be deduced from the marginal pmf's alone.

[3.18] Let (X,Y) be a random vector with a pdf that is constant on the unit disk in the XY plane; i.e.,

$$f_{X,Y}(x,y) = C, x^2 + y^2 \leq 1.$$

The constant C is determined by the requirement that the pdf integrate to 1; i.e.,

$$\int_{x^2+y^2 \leq 1} C\, dx\, dy = 1.$$

Since this integral is just the area of a circle multiplied by C, we have immediately that $C = 1/\pi$. For the moment, however, we leave the joint pdf in terms of C and determine the pdf of X in terms of C by integrating with respect to y:

$$f_X(x) = \int_{-\sqrt{1-x^2}}^{+\sqrt{1-x^2}} C\, dy = 2C\sqrt{1-x^2}, \quad x^2 \leq 1.$$

Observe that we could now also find C by a second integration:

$$\int_{-1}^{+1} 2C\sqrt{1-x^2}\, dx = \pi C = 1,$$

or $C = \pi^{-1}$. Thus the pdf of X is

$$f_X(x) = 2\pi^{-1}\sqrt{1-x^2}, \quad x^2 \leq 1.$$

By symmetry Y has the same pdf. Note that the marginal pdf is *not constant*, even though the joint pdf is. Furthermore, it is obvious that it would be impossible to determine the joint density from the marginal pdf's alone.

[3.19] Consider the two-dimensional Gaussian pdf of Example [2.17] with $k = 2$, $\mathbf{m} = (0,0)^t$, and $\Lambda = \{\lambda(i,j) : \lambda(1,1) = \lambda(2,2) = 1, \lambda(1,2) = \lambda(2,1) = \rho\}$. Since the inverse matrix is

$$\begin{bmatrix} 1 & \rho \\ \rho & 1 \end{bmatrix}^{-1} = \frac{1}{1-\rho^2}\begin{bmatrix} 1 & -\rho \\ -\rho & 1 \end{bmatrix},$$

the joint pdf for the random vector (X,Y) is

$$f_{X,Y}(x,y) = \frac{\exp\left(-\frac{1}{2(1-\rho^2)}(x^2+y^2-2\rho xy)\right)}{2\pi\sqrt{1-\rho^2}}, \ (x,y) \in \Re^2.$$

Here, ρ is called the "correlation coefficient". between X and Y and must satisfy $\rho^2 < 1$ for Λ to be positive definite. To find the pdf of X we complete the square in the exponent so that

$$\begin{aligned} f_{X,Y}(x,y) &= \frac{\exp\left(-\frac{(y-\rho x)^2}{2(1-\rho^2)} - \frac{x^2}{2}\right)}{2\pi\sqrt{1-\rho^2}} \\ &= \frac{\exp\left(-\frac{(y-\rho x)^2}{2(1-\rho^2)}\right)}{\sqrt{2\pi(1-\rho^2)}} \frac{\exp\left(-\frac{x^2}{2}\right)}{\sqrt{2\pi}}. \end{aligned}$$

The pdf of X is determined by integrating with respect to y on $(-\infty,\infty)$. To perform this integration, refer to the form of the one-dimensional Gaussian pdf with $m = \rho x$ (note that x is fixed while the integration is with respect to y) and $\sigma^2 = 1-\rho^2$. The first factor in the preceding equation has this form. Because the one-dimensional pdf must integrate to one, the pdf of X that results from integrating y out from the two-dimensional pdf is also a one-dimensional Gaussian pdf; i.e.,

$$f_X(x) = (2\pi)^{-1/2}e^{-x^2/2}.$$

As in Examples [3.16], [3.17], and [3.18], Y has the same pdf as X. Note that by varying ρ there is a whole family of joint Gaussian pdf's with the same marginal Gaussian pdf's.

3.6 Independent random variables

In Chapter 2 it was seen that events are independent if the probability of a joint event can be written as a product of probabilities of individual events.

The notion of independent events provides a corresponding notion of independent random variables and, as will be seen, results in random variables being independent if their joint distributions are product distributions.

Two random variables X and Y defined on a probability space are *independent* if the events $X^{-1}(F)$ and $Y^{-1}(G)$ are independent for all F and G in $\mathcal{B}(\Re)$. A collection of random variables $\{X_i, i = 0, 1, \ldots, k-1\}$ is said to be *independent* or *mutually independent* if all collections of events of the form $\{X_i^{-1}(F_i); i = 0, 1, \ldots, k-1\}$ are mutually independent for any $F_i \in \mathcal{B}(\Re); i = 0, 1, \ldots, k-1$.

Thus two random variables are independent if and only if their output events correspond to independent input events. Translating this statement into distributions yields the following.

Random variables X and Y are independent if and only if

$$P_{X,Y}(F_1 \times F_2) = P_X(F_1)P_Y(F_2), \text{all } F_1, F_2 \in \mathcal{B}(\Re).$$

Recall that $F_1 \times F_2$ is an alternative notation for $\prod_{i=1}^2 F_i$ – we will frequently use the alternative notation when the number of product events is small. Note that a *product* and not an *intersection* is used here. The reader should be certain that this is understood. The intersection is appropriate if we refer back to the original ω events; that is, using the inverse image formula to write this statement in terms of the underlying probability space yields

$$P(X^{-1}(F_1) \cap Y^{-1}(F_2) = P(X^{-1}(F_1))P(Y^{-1}(F_2)).$$

Random variables $X_0, \ldots, X_{k-1}$ are independent or mutually independent if and only if

$$P_{X_0,\ldots,X_{k-1}}\left(\prod_{i=0}^{k-1} F_i\right) = \prod_{i=0}^{k-1} P_{X_i}(F_i);$$

for all $F_i \in \mathcal{B}(\Re); i = 0, 1, \ldots, k-1$.

The general form for distributions can be specialized to pmf's, pdf's, and cdf's as follows. Two discrete random variables X and Y are independent if and only if the joint pmf factors as

$$p_{X,Y}(x, y) = p_X(x)p_Y(y) \ \forall\, x, y.$$

A collection of discrete random variables $X_i; i = 0, 1, \ldots, k-1$ is mutually independent if and only if the joint pmf factors as

$$p_{X_0,\ldots,X_{k-1}}(x_0,\ldots,x_{k-1}) = \prod_{i=0}^{k-1} p_{X_i}(x_i);\ \forall x_i.$$

Similarly, if the random variables are continuous and described by pdf's, then two random variables are independent if and only if the joint pdf factors as

$$f_{X,Y}(x,y) = f_X(x) f_Y(y);\ \forall\ x,y \in \Re.$$

A collection of continuous random variables is independent if and only if the joint pdf factors as

$$f_{X_0,\ldots,X_{k-1}}(x_0,\ldots,x_{k-1}) = \prod_{i=0}^{k-1} f_{X_i}(x_i).$$

Two general random variables (discrete, continuous, or mixture) are independent if and only if the joint cdf factors as

$$F_{X,Y}(x,y) = F_X(x) F_Y(y)\ ;\ \forall\ x,y \in \Re.$$

A collection of general random variables is independent if and only if the joint cdf factors as

$$F_{X_0,\ldots,X_{k-1}}(x_0,\ldots,x_{k-1}) = \prod_{i=0}^{k-1} F_{X_i}(x_i);\quad (x_0,x_1,\ldots,x_{k-1}) \in \Re^k.$$

We have separately stated the two-dimensional case because of its simplicity and common occurrence. The student should be able to prove the equivalence of the general distribution form and the pmf form. If one does not consider technical problems regarding the interchange of limits of integration, then the equivalence of the general form and the pdf form can also be proved.

Independent identically distributed random vectors

A random vector is said to be *independent, identically distributed* or *iid* if the coordinate random variables are independent and identically distributed; that is, if

- the distribution is a product distribution, i.e., it has the form

$$P_{X_0,\ldots,X_{k-1}}\left(\prod_{i=0}^{k-1} F_i\right) = \prod_{i=0}^{k-1} P_{X_i}(F_i)$$

for all choices of $F_i \in \mathcal{B}(\Re), i = 0, 1, \ldots, k-1$, and
- if all the marginal distributions are the same (the random variables are all equivalent), i.e., if there is a distribution P_X such that $P_{X_i}(F) = P_X(F)$; all $F \in \mathcal{B}(\Re)$ for all i.

For example, a random vector will have a product distribution if it has a joint pdf or pmf that is a product pdf or pmf as described in Example [2.16]. The general property is easy to describe in terms of probability functions. The random vector is iid if it has a joint pdf with the form

$$f_{\mathbf{X}}(\mathbf{x}) = \prod_i f_X(x_i)$$

for some pdf f_X defined on $\Re$ or if it has a joint pmf with the form

$$p_{\mathbf{X}}(\mathbf{x}) = \prod_i p_X(x_i)$$

for some pmf p_X defined on some discrete subset of the real line. Both of these cases are included in the following statement: a random vector will be iid if and only if its cdf has the form

$$F_{\mathbf{X}}(\mathbf{x}) = \prod_i F_X(x_i)$$

for some cdf F_X.

Note that, in contrast with earlier examples, the specification "product distribution," along with the marginal pdf's or pmf's or cdf's, is sufficient to specify the joint distribution.

3.7 Conditional distributions

The idea of conditional probability can be used to provide a general representation of a joint distribution as a product, but a more complicated product than arises with an iid vector. As one would hope, the complicated form reduces to the simpler form when the vector is iid. The individual terms of the product have useful interpretations.

The use of conditional probabilities allows us to break up many problems in a convenient form and focus on the relations among random variables. Examples to be treated include statistical detection, statistical classification, and additive noise.

3.7.1 Discrete conditional distributions

We begin with the discrete alphabet case as elementary conditional probability suffices in this simple case. We can derive results that appear similar for the continuous case, but nonelementary conditional probability will be required to interpret the results correctly.

Begin with the simple case of a discrete random vector (X, Y) with alphabet $A_X \times A_Y$ described by a pmf $p_{X,Y}(x, y)$. Let p_X and p_Y denote the corresponding marginal pmf's. Define for each $x \in A_X$ for which $p_X(x) > 0$ the *conditional pmf* $p_{Y|X}(y|x)$; $y \in A_Y$ as the elementary conditional probability of $Y = y$ given $X = x$, that is,

$$\begin{aligned} p_{Y|X}(y|x) &= P(Y = y|X = x) \\ &= \frac{P(Y = y \text{ and } X = x)}{P(X = x)} \\ &= \frac{P(\{\omega : Y(\omega) = y\} \cap \{\omega : X(\omega) = x\})}{P(\{\omega : X(\omega) = x\})} \\ &= \frac{p_{X,Y}(x, y)}{p_X(x)}, \end{aligned} \tag{3.46}$$

where we have assumed that $p_X(x) > 0$ for all suitable x to avoid dividing by 0. Thus a conditional pmf is just a special case of an elementary conditional probability. For each x a conditional pmf is itself a pmf, since it is clearly nonnegative and sums to 1:

$$\begin{aligned} \sum_{y \in A_Y} p_{Y|X}(y|x) = \sum_{y \in A_Y} \frac{p_{X,Y}(x, y)}{p_X(x)} &= \frac{1}{p_X(x)} \sum_{y \in A_Y} p_{X,Y}(x, y) \\ &= \frac{1}{p_X(x)} p_X(x) = 1. \end{aligned}$$

We can compute conditional probabilities by summing conditional pmf's, i.e.,

$$P(Y \in F|X = x) = \sum_{y \in F} p_{Y|X}(y|x). \tag{3.47}$$

The joint probability can be expressed as a product as

$$p_{X,Y}(x, y) = p_{Y|X}(y|x)p_X(x). \tag{3.48}$$

Unlike the independent case, the terms of the product do not each depend on only a single independent variable. If X and Y are independent, then $p_{Y|X}(y|x) = p_Y(y)$ and the joint pmf reduces to the product of two marginals.

Given the conditional pmf $p_{Y|X}$ and the pmf p_X, the conditional pmf with the roles of the two random variables reversed can be computed by marginal pmf's by

$$p_{X|Y}(x|y) = \frac{p_{X,Y}(x,y)}{p_Y(y)} = \frac{p_{Y|X}(y|x)p_X(x)}{\sum_u p_{Y|X}(y|u)p_X(u)}, \tag{3.49}$$

a result often referred to as *Bayes' rule.*

The ideas of conditional pmf's immediately extend to random vectors. Suppose we have a random vector $(X_0, X_1, \ldots, X_{k-1})$ with a pmf $p_{X_0,X_1,\ldots,X_{k-1}}$, then (provided none of the denominators are 0) we can define for each $l = 1, 2, \ldots, k-1$ the conditional pmf's

$$p_{X_l|X_0,\ldots,X_{l-1}}(x_l|x_0, \ldots, x_{l-1}) = \frac{p_{X_0,\ldots,X_l}(x_0, \ldots, x_l)}{p_{X_0,\ldots,X_{l-1}}(x_0, \ldots, x_{l-1})}. \tag{3.50}$$

Then simple algebra leads to the *chain rule* for pmf's:

$$\begin{aligned}
&p_{X_0,X_1,\ldots,X_{n-1}}(x_0, x_1, \ldots, x_{n-1}) \\
&= \left(\frac{p_{X_0,X_1,\ldots,X_{n-1}}(x_0, x_1, \ldots, x_{n-1})}{p_{X_0,X_1,\ldots,X_{n-2}}(x_0, x_1, \ldots, x_{n-2})}\right) p_{X_0,X_1,\ldots,X_{n-2}}(x_0, x_1, \ldots, x_{n-2}) \\
&\vdots \\
&= p_{X_0}(x_0) \prod_{i=1}^{n-1} \frac{p_{X_0,X_1,\ldots,X_i}(x_0, x_1, \ldots, x_i)}{p_{X_0,X_1,\ldots,X_{i-1}}(x_0, x_1, \ldots, x_{i-1})} \\
&= p_{X_0}(x_0) \prod_{l=1}^{n-1} p_{X_l|X_0,\ldots,X_{l-1}}(x_l|x_0, \ldots, x_{l-1}),
\end{aligned} \tag{3.51}$$

a product of conditional probabilities. This provides a general form of the iid product form and reduces to the iid product form if indeed the random variables are mutually independent. This formula plays an important role in characterizing memory in random vectors and processes. It can be used to construct joint pmf's, and can be used to specify a random process.

3.7.2 Continuous conditional distributions

The situation with continuous random vectors is more complicated if rigor is required, but the mechanics are similar. Again begin with the simple case of two random variables X and Y with a joint distribution, now taken to be described by a pdf $f_{X,Y}$. We *define* the conditional pdf as an exact analog to that for pmf's:

$$f_{Y|X}(y|x) \triangleq \frac{f_{X,Y}(x,y)}{f_X(x)}. \tag{3.52}$$

This looks the same as the pmf, but it is not the same because pmf's are probabilities and pdf's are not. A conditional pmf is an elementary conditional probability. A conditional pdf is not. It is also not the same as the conditional pdf of Example [2.19] as in that case the conditioning event had nonzero probability. The conditional pdf $f_{Y|X}$ can, however, be related to a probability in the same way that an ordinary pdf (and the conditional pdf of Example [2.19]) can. An ordinary pdf is a density of probability, it is integrated to compute a probability. In the same way, a conditional pdf can be interpreted as a density of conditional probability, something you integrate to get a conditional probability. Now, however, the conditioning event can have probability zero and this does not really fit into the previous development of elementary conditional probability. Note that a conditional pdf is indeed a pdf, a nonnegative function that integrates to one. This follows from

$$\begin{aligned}\int f_{Y|X}(y|x)\,dy &= \int \frac{f_{X,Y}(x,y)}{f_X(x)}\,dy \\ &= \frac{1}{f_X(x)}\int f_{X,Y}(x,y)\,dy \\ &= \frac{1}{f_X(x)} f_X(x) = 1,\end{aligned}$$

provided we require that $f_X(x) > 0$ over the region of integration.

To be more specific, given a conditional pdf $f_{Y|X}$, we will make a tentative definition of the (nonelementary) conditional probability that $Y \in F$ given $X = x$ is

$$P(Y \in F|X = x) = \int_F f_{Y|X}(y|x)\,dy. \tag{3.53}$$

Note the close resemblance to the elementary conditional probability formula in terms of conditional pmf's of (3.47). For all practical purposes (and hence for virtually all of this book), this constructive definition of nonelementary conditional probability will suffice. Unfortunately it does not provide sufficient rigor to lead to a useful advanced theory. Section 3.17 discusses the problems and the correct general definition in some depth, but it is not required for most applications.

Via almost identical manipulations to the pmf case in (3.49), conditional pdf's satisfy a Bayes' rule:

$$f_{X|Y}(x|y) = \frac{f_{X,Y}(x,y)}{f_Y(y)} = \frac{f_{Y|X}(y|x)f_X(x)}{\int f_{Y|X}(y|u)f_X(u)\,du}. \tag{3.54}$$

As a simple but informative example of a conditional pdf, consider the generalization of Example [3.19] to the case of a two-dimensional vector $U = (X, Y)$ with a Gaussian pdf having a mean vector $(m_X, m_Y)^t$ and a covariance matrix

$$\Lambda = \begin{bmatrix} \sigma_X^2 & \rho\sigma_X\sigma_Y \\ \rho\sigma_X\sigma_Y & \sigma_Y^2 \end{bmatrix}, \tag{3.55}$$

where ρ is the correlation coefficient of X and Y. Straightforward algebra yields

$$\det(\Lambda) = \sigma_X^2\sigma_Y^2(1-\rho^2) \tag{3.56}$$

$$\Lambda^{-1} = \frac{1}{(1-\rho^2)}\begin{bmatrix} 1/\sigma_X^2 & -\rho/(\sigma_X\sigma_Y) \\ -\rho/(\sigma_X\sigma_Y) & 1/\sigma_Y^2 \end{bmatrix} \tag{3.57}$$

so that the two-dimensional pdf becomes

$$\begin{aligned} &f_{XY}(x,y) \\ &= \frac{1}{2\pi\sqrt{\det\Lambda}} e^{-\frac{1}{2}(x-m_X, y-m_Y)\Lambda^{-1}(x-m_X, y-m_Y)^t} \\ &= \frac{1}{2\pi\sigma_X\sigma_Y\sqrt{1-\rho^2}} \exp\left(-\frac{1}{2(1-\rho^2)}\right. \\ &\quad \left.\times\left[\left(\frac{x-m_X}{\sigma_X}\right)^2 - 2\rho\frac{(x-m_X)(y-m_Y)}{\sigma_X\sigma_Y} + \left(\frac{y-m_Y}{\sigma_Y}\right)^2\right]\right). \end{aligned} \tag{3.58}$$

A little algebra to rearrange the expression yields

$$f_{XY}(x,y) = \frac{\exp\left(-\frac{1}{2}\left(\frac{x-m_X}{\sigma_X}\right)^2\right)}{\sqrt{2\pi\sigma_X^2}} \frac{\exp\left(-\frac{1}{2}\left(\frac{y-m_Y-(\rho\sigma_Y/\sigma_X)(x-m_X)}{\sqrt{1-\rho^2}\,\sigma_Y}\right)^2\right)}{\sqrt{2\pi\sigma_Y^2(1-\rho^2)}} \tag{3.59}$$

from which it follows immediately that the conditional pdf is

$$f_{Y|X}(y|x) = \frac{\exp\left(-\frac{1}{2}\left(\frac{y-m_Y-(\rho\sigma_Y/\sigma_X)(x-m_X)}{\sqrt{1-\rho^2}\,\sigma_Y}\right)^2\right)}{\sqrt{2\pi\sigma_Y^2(1-\rho^2)}}, \tag{3.60}$$

which is itself a Gaussian density with variance $\sigma^2_{Y|X} \stackrel{\Delta}{=} \sigma^2_Y(1-\rho^2)$ and mean $m_{Y|X} \stackrel{\Delta}{=} m_Y + \rho(\sigma_Y/\sigma_X)(x-m_X)$. Integrating y out of the joint pdf then shows that as in Example [3.19] the marginal pdf is also Gaussian:

$$f_X(x) = \frac{e^{-(x-m_X)^2/2\sigma_X^2}}{\sqrt{2\pi\sigma_X^2}}. \tag{3.61}$$

A similar argument shows that $f_Y(y)$ and $f_{X|Y}(x|y)$ are also Gaussian pdf's. Observe that if X and Y are jointly Gaussian, then they are also both individually and conditionally Gaussian!

A chain rule for pdf's follows in exactly the same way as that for pmf's. Assuming $f_{X_0,X_1,\ldots,X_i}(x_0,x_1,\ldots,x_i) > 0$,

$$\begin{aligned}
&f_{X_0,X_1,\ldots,X_{n-1}}(x_0,x_1,\ldots,x_{n-1}) \\
&= \frac{f_{X_0,X_1,\ldots,X_{n-1}}(x_0,x_1,\ldots,x_{n-1})}{f_{X_0,X_1,\ldots,X_{n-2}}(x_0,x_1,\ldots,x_{n-2})} f_{X_0,X_1,\ldots,X_{n-2}}(x_0,x_1,\ldots,x_{n-2}) \\
&\vdots \\
&= f_{X_0}(x_0) \prod_{i=1}^{n-1} \frac{f_{X_0,X_1,\ldots,X_i}(x_0,x_1,\ldots,x_i)}{f_{X_0,X_1,\ldots,X_{i-1}}(x_0,x_1,\ldots,x_{i-1})} \\
&= f_{X_0}(x_0) \prod_{l=1}^{n-1} f_{X_l|X_0,\ldots,X_{l-1}}(x_l|x_0,\ldots,x_{l-1}).
\end{aligned} \tag{3.62}$$

3.8 Statistical detection and classification

Consider a simple, but nonetheless very important, example of the application of conditional probability mass functions describing discrete random vectors. Suppose that X is a binary random variable described by a pmf p_X, with $p_X(1) = p$. It might be one bit in some data coming through a modem. You receive a random variable Y, which is equal to X with probability $1-\epsilon$. In terms of a conditional pmf this is

$$p_{Y|X}(y|x) = \begin{cases} \epsilon & x \neq y \\ 1-\epsilon & x = y \end{cases}. \tag{3.63}$$

This can be written in a simple form using the idea of modulo 2 (or mod 2) arithmetic which will often be useful when dealing with binary variables. Modulo 2 arithmetic or the "Galois field of 2 elements" arithmetic consists of an operation $\oplus$ called *modulo 2 addition* defined on the binary alphabet $\{0, 1\}$ as follows:

$$0 \oplus 1 = 1 \oplus 0 = 1 \tag{3.64}$$

$$0 \oplus 0 = 1 \oplus 1 = 0. \tag{3.65}$$

The operation $\oplus$ corresponds to an "exclusive or" in logic; that is, it produces a 1 if one or the other but not both of its arguments is 1. An equivalent definition for the conditional pmf is

$$p_{Y|X}(y|x) = \epsilon^{x \oplus y}(1-\epsilon)^{1-x \oplus y}. \tag{3.66}$$

For example, the channel over which the bit is being sent is *noisy* in that the receiver occasionally makes an error. Suppose that it is known that the probability of such an error is ϵ. The error might be very small on a good phone line, but it might be very large if an evil hacker is trying to corrupt your data. Given the observed Y, what is the best guess $\hat{X}(Y)$ of what is actually sent? In other words, what is the best *decision rule* or *detection rule* for guessing the value of X given the observed value of Y? A reasonable parameter for judging the quality of an arbitrary rule $\hat{X}$ is the resulting probability of error

$$P_e(\hat{X}) = \Pr(\hat{X}(Y) \neq X). \tag{3.67}$$

A decision rule is *optimal* if it yields the smallest possible probability of error over all possible decision rules. A little probability manipulation quickly yields the optimal decision rule. Instead of minimizing the error probability, we maximize the probability of being correct:

$$\begin{aligned}
\Pr(\hat{X} = X) = 1 - P_e(\hat{X}) &= \sum_{(x.y):\hat{X}(y)=x} p_{X.Y}(x, y) \\
&= \sum_{(x.y):\hat{X}(y)=x} p_{X|Y}(x|y) p_Y(y) \\
&= \sum_y p_Y(y) \left(\sum_{x:\hat{X}(y)=x} p_{X|Y}(x|y) \right) \\
&= \sum_y p_Y(y) p_{X|Y}(\hat{X}(y)|y).
\end{aligned}$$

To maximize this sum, we want to maximize the terms within the sum for each y. Clearly the maximum value of the conditional probability, $p_{X|Y}(\hat{X}(y)|y) = \max_u p_{X|Y}(u|y)$, will be achieved if we define the decision rule $\hat{X}(y)$ to be the value of u achieving the maximum of $p_{X|Y}(u|y)$ over u, that is, define $\hat{X}$ to be $\arg\max_u p_{X|Y}(u|y)$ (also denoted $\max_u^{-1} p_{X|Y}(u|y)$). In words: the optimal estimate of X given the observation Y in the sense of minimizing the probability of error is the most probable value of X given the observation. This is called the *maximum a posteriori* or *MAP* decision rule. In our binary example it reduces to choosing $\hat{x} = y$ if $\epsilon < 1/2$ and $\hat{x} = 1 - y$ if $\epsilon > 1/2$. If $\epsilon = 1/2$ you can give up and flip a coin or make an arbitrary decision. (Why?) Thus the minimum (optimal) error probability over all possible rules is $\min(\epsilon, 1 - \epsilon)$.

The astute reader will notice that having introduced conditional pmf's $p_{Y|X}$, the example considered the alternative pmf $p_{X|Y}$. The two are easily related by Bayes' rule (3.49).

A generalization of the simple binary detection problem provides the typical form of a *statistical classification* system. Suppose that Nature selects a "class" H, a random variable described by a pmf $p_H(h)$, which is no longer assumed to be binary. Once the class is selected, Nature then generates a random "observation" X according to a pmf $p_{X|H}$. For example, the class might be a medical condition and the observations the results of blood pressure, patient's age, medical history, and other information regarding the patient's health. Alternatively, the class might be an "input signal" put into a noisy channel which has the observation X as an "output signal." The question is: given the observation $X = x$, what is the best guess $\hat{H}(x)$ of the unseen class? If by "best" we adopt the criterion that the best guess is the one that minimizes the error probability $P_e = \Pr(\hat{H}(X) \neq H)$, then the optimal classifier is again the MAP rule $\arg\max_u p_{H|X}(u|x)$. More generally we might assign a *cost* $C_{y,h}$ resulting if the true class is h and we guess y. Typically it is assumed that $C_{h,h} = 0$, that is, the cost is zero if our guess is correct. (In fact it can be shown that this assumption involves no real loss of generality.) Given a classifier (classification rule, decision rule) $\hat{h}(x)$, the *Bayes risk* is then defined as

$$B(\hat{h}) = \sum_{x,h} C_{\hat{h}(x),h} p_{H,X}(h,x) , \tag{3.68}$$

which reduces to the probability of error if the cost function is given by

$$C_{y,h} = 1 - \delta_{y,h} . \tag{3.69}$$

The optimal classifier in the sense of minimizing the Bayes risk is then found by observing that the inequality

$$B(\hat{h}) = \sum_x p_X(x) \sum_h C_{\hat{h}(x),h} p_{H|X}(h|x)$$
$$\geq \sum_x p_X(x) \min_y \left(\sum_h C_{y,h} p_{H|X}(h|x) \right),$$

which lower bound is achieved by the classifier

$$\hat{h}(x) = \operatorname*{argmin}_y \left(\sum_h C_{y,h} p_{H|X}(h|x) \right), \tag{3.70}$$

the *minimum average Bayes risk classifier.* This reduces to the MAP detection rule when $C_{y,h} = 1 - \delta_{y,h}$.

3.9 Additive noise

The next examples of the use of conditional distributions treat the distributions arising when one random variable (thought of as a "noise" term) is added to another, independent random variable (thought of as a "signal" term). This is an important example of a derived distribution problem that yields an interesting conditional probability. The problem also suggests a valuable new tool which will provide a simpler way of solving many similar derived distributions – the characteristic function of random variables.

Discrete additive noise

Consider two independent random variables X and W and form a new random variable $Y = X + W$. This could be a description of how errors are actually caused in a noisy communication channel connecting a binary information source to a user. In order to apply the detection and classification signal processing methods, we must first compute the appropriate conditional probabilities of the output Y given the input X. To do this we begin by computing the joint pmf of X and Y using the inverse image formula:

$$p_{X,Y}(x, y) = \Pr(X = x, Y = y) = \Pr(X = x, X + W = y)$$
$$= \sum_{\alpha,\beta:\alpha=x,\alpha+\beta=y} p_{X,W}(\alpha, \beta) = p_{X,W}(x, y - x)$$
$$= p_X(x) p_W(y - x). \tag{3.71}$$

Note that this formula only makes sense if $y - x$ is one of the values in the range space of W. Thus from the definition of conditional pmf's,

$$p_{Y|X}(y|x) = \frac{p_{X,Y}(x,y)}{p_X(x)} = p_W(y-x), \tag{3.72}$$

an answer that should be intuitive: given the input is x, the output will equal a certain value y if and only if the noise exactly makes up the difference, i.e., $W = y - x$. Note that the marginal pmf for the output Y can be found by summing the joint probability:

$$p_Y(y) = \sum_x p_{X,Y}(x,y) = \sum_x p_X(x)p_W(y-x), \tag{3.73}$$

a formula that is known as a *discrete convolution* or *convolution sum.*

Anyone familiar with convolutions knows that they can be unpleasant to evaluate, so we postpone further consideration to the next section and turn to the continuous analog.

The above development assumed ordinary arithmetic, but it is worth pointing out that for discrete random variables sometimes other types of arithmetic are appropriate, e.g., modulo 2 arithmetic for binary random variables. The binary example of Section 3.8 can be considered as an additive noise example if we define a random variable W which is independent of X and has a pmf $p_W(w) = \epsilon^w(1-\epsilon)^{1-w}$; $w = 0, 1$ and where $Y = X + W$ is interpreted as modulo 2 arithmetic, that is, as $Y = X \oplus W$. This additive noise definition is easily seen to yield the conditional pmf of (3.63) and the output pmf via a convolution. To be precise,

$$\begin{aligned} p_{X,Y}(x,y) &= \Pr(X = x, Y = y) = \Pr(X = x, X \oplus W = y) \\ &= \sum_{\alpha,\beta:\alpha=x,\alpha\oplus\beta=y} p_{X,W}(\alpha,\beta) = p_{X,W}(x, y \oplus x) \\ &= p_X(x)p_W(y \oplus x) \end{aligned} \tag{3.74}$$

and hence

$$p_{Y|X}(y|x) = \frac{p_{X,Y}(x,y)}{p_X(x)} = p_W(y \oplus x) \tag{3.75}$$

and

$$p_Y(y) = \sum_x p_{X,Y}(x,y) = \sum_x p_X(x)p_W(y \oplus x), \tag{3.76}$$

a modulo 2 convolution.

Continuous additive noise

An entirely analogous formula arises in the continous case. Again suppose that X is a random variable, a *signal*, with pdf f_X, and that W is a random variable, the *noise*, with pdf f_W. The random variables X and W are assumed to be independent. Form a new random variable Y, an observed signal plus noise. The problem is to find the conditional pdf's $f_{Y|X}(y|x)$ and $f_{X|Y}(x|y)$. The operation of producing an output Y from an input signal X is called an *additive noise channel* in communications systems. The channel is completely described by $f_{Y|X}$. The second pdf, $f_{X|Y}$ will prove useful later when we try to estimate X given an observed value of Y.

Independence of X and W implies that the joint pdf is $f_{X,W}(x,w) = f_X(x)f_W(w)$. To find the joint pdf $f_{X,Y}$, first evaluate the joint cdf and then take the appropriate derivative. The cdf is a straightforward derived distribution calculation:

$$\begin{aligned}
F_{X,Y}(x,y) &= \Pr(X \le x, Y \le y) \\
&= \Pr(X \le x, X + W \le y) \\
&= \int\int_{\alpha,\beta:\alpha\le x,\alpha+\beta\le y} f_{X,W}(\alpha,\beta)\,d\alpha\,d\beta \\
&= \int_{-\infty}^{x} d\alpha \int_{-\infty}^{y-\alpha} d\beta f_X(\alpha) f_W(\beta) \\
&= \int_{-\infty}^{x} d\alpha f_X(\alpha) F_W(y-\alpha).
\end{aligned}$$

Taking the derivatives yields

$$f_{X,Y}(x,y) = f_X(x)f_W(y-x)$$

and hence

$$f_{Y|X}(y|x) = f_W(y-x). \tag{3.77}$$

The marginal pdf for the sum $Y = X + W$ is then found as

$$f_Y(y) = \int f_{X,Y}(x,y)\,dx = \int f_X(x)f_W(y-x)\,dx, \tag{3.78}$$

a convolution integral of the pdf's f_X and f_W, analogous to the convolution sum found when adding independent discrete random variables. Thus the evaluation of the pdf of the sum of two independent continuous random variables is of the same form as the evaluation of the output of a linear system with an input signal f_X and an impulse response f_W. We will later see an easy way to accomplish this using transforms. The pdf $f_{X|Y}$ follows from Bayes' rule:

$$f_{X|Y}(x|y) = \frac{f_X(x) f_W(y-x)}{\int f_X(\alpha) f_W(y-\alpha)\, d\alpha}. \tag{3.79}$$

It is instructive to work through the details of the previous example for the special case of Gaussian random variables. For simplicity the means are assumed to be zero and hence it is assumed that f_X is $\mathcal{N}(0, \sigma_X)$, that f_W is $\mathcal{N}(0, \sigma_Y^2)$, and that as in the example X and W are independent and $Y = X + W$. From (3.77)

$$f_{Y|X}(y|x) = f_W(y-x) = \frac{e^{-(y-x)^2/2\sigma_W^2}}{\sqrt{2\pi\sigma_W^2}} \tag{3.80}$$

from which the conditional pdf can be immediately recognized as being Gaussian with mean x and variance σ_W^2, that is, as $\mathcal{N}(x, \sigma_W^2)$.

To evaluate the pdf $f_{X|Y}$ using Bayes' rule, we begin with the denominator f_Y of (3.54) and write

$$\begin{aligned} f_Y(y) & \\ &= \int_{-\infty}^{\infty} f_{Y|X}(y|\alpha) f_X(\alpha)\, d\alpha \\ &= \int_{-\infty}^{\infty} \frac{\exp\left(-\frac{1}{2\sigma_W^2}(y-\alpha)^2\right)}{\sqrt{2\pi\sigma_W^2}} \frac{\exp\left(-\frac{1}{2\sigma_X^2}\alpha^2\right)}{\sqrt{2\pi\sigma_X^2}}\, d\alpha \\ &= \frac{1}{2\pi\sigma_X\sigma_W} \int_{-\infty}^{\infty} \exp\left(-\frac{1}{2}\left[\frac{y^2 - 2\alpha y + \alpha^2}{\sigma_W^2} + \frac{\alpha^2}{\sigma_X^2}\right]\right) d\alpha \\ &= \frac{\exp\left(-\frac{y^2}{2\sigma_W^2}\right)}{2\pi\sigma_X\sigma_W} \int_{-\infty}^{\infty} \exp\left(-\frac{1}{2}\left[\alpha^2\left(\frac{1}{\sigma_X^2} + \frac{1}{\sigma_W^2}\right) - \frac{2\alpha y}{\sigma_W^2}\right]\right) d\alpha. \end{aligned} \tag{3.81}$$

This convolution of two Gaussian "signals" can be accomplished using an old trick called "complete the square." Call the integral in the square brackets at the end of the above equation I and note that integrand resembles

$$\exp\left(-\frac{1}{2}\left(\frac{\alpha - m}{\sigma}\right)^2\right).$$

We know from (B.13) in Appendix B that this integrates to

$$\int_{-\infty}^{\infty} \exp\left(-\frac{1}{2}\left(\frac{\alpha - m}{\sigma}\right)^2\right) d\alpha = \sqrt{2\pi\sigma^2} \tag{3.82}$$

since a Gaussian pdf integrates to 1. The trick is to modify I to resemble this integral with an additional factor. Compare the two exponents:

$$-\frac{1}{2}\left[\alpha^2\left(\frac{1}{\sigma_X^2}+\frac{1}{\sigma_W^2}\right)-\frac{2\alpha y}{\sigma_W^2}\right]$$

with

$$-\frac{1}{2}\left(\frac{\alpha-m}{\sigma}\right)^2=-\frac{1}{2}\left[\frac{\alpha^2}{\sigma^2}-2\frac{\alpha m}{\sigma^2}+\frac{m^2}{\sigma^2}\right].$$

The exponent from I will equal the left two terms of the expanded exponent in the known integral if we choose

$$\frac{1}{\sigma^2}=\frac{1}{\sigma_W^2}+\frac{1}{\sigma_X^2}$$

or, equivalently,

$$\sigma^2=\frac{\sigma_X^2\sigma_W^2}{\sigma_X^2+\sigma_W^2}, \tag{3.83}$$

and if we choose

$$\frac{y}{\sigma_W^2}=\frac{m}{\sigma^2}$$

or, equivalently,

$$m=\frac{\sigma^2}{\sigma_W^2}y. \tag{3.84}$$

Using (3.83)–(3.84) we have that

$$\alpha^2\left(\frac{1}{\sigma_X^2}+\frac{1}{\sigma_W^2}\right)-\frac{2\alpha y}{\sigma_W^2}=\left(\frac{\alpha-m}{\sigma}\right)^2-\frac{m^2}{\sigma^2},$$

where the addition of the leftmost term is "completing the square." With this identification and again using (3.83)–(3.84) we have that

$$\begin{aligned} I &= \int_{-\infty}^{\infty}\exp\left(-\frac{1}{2}\left[\left(\frac{\alpha-m}{\sigma^2}\right)^2-\frac{m^2}{\sigma^2}\right]\right)d\alpha \\ &= \sqrt{2\pi\sigma^2}\exp\left(\frac{m^2}{2\sigma^2}\right), \end{aligned} \tag{3.85}$$

which implies that

$$f_Y(y) = \frac{\exp\left(-\frac{1}{2}\frac{y^2}{\sigma_W^2}\right)}{2\pi\sigma_X\sigma_W}\sqrt{2\pi\sigma^2}\exp\left(\frac{m^2}{2\sigma^2}\right)$$
$$= \frac{\exp\left(-\frac{1}{2}\frac{y^2}{\sigma_X^2+\sigma_W^2}\right)}{\sqrt{2\pi(\sigma_X^2+\sigma_W^2)}}. \tag{3.86}$$

In other words, f_Y is $\mathcal{N}(0, \sigma_X^2 + \sigma_W^2)$ and we have shown that the sum of two zero-mean independent Gaussian random variables is another zero-mean Gaussian random variable with variance equal to the sum of the variances of the two random variables being added.

Finally we turn to the a posteriori probability $f_{X|Y}$. From Bayes' rule and a lot of algebra

$$f_{X|Y}(x|y) = f_{Y|X}(y|x)f_X(x)/f_Y(y)$$
$$= \frac{\exp\left(-\frac{1}{2\sigma_W^2}(y-x)^2\right)}{\sqrt{2\pi\sigma_W^2}}\frac{\exp\left(-\frac{1}{2\sigma_X^2}x^2\right)}{\sqrt{2\pi\sigma_X^2}}\Bigg/\frac{\exp\left(-\frac{1}{2}\frac{y^2}{\sigma_X^2+\sigma_W^2}\right)}{\sqrt{2\pi(\sigma_X^2+\sigma_W^2)}}$$
$$= \frac{\exp\left(-\frac{1}{2}\left[\frac{y^2-2yx+x^2}{\sigma_W^2}+\frac{x^2}{\sigma_X^2}-\frac{y^2}{\sigma_X^2+\sigma_W^2}\right]\right)}{\sqrt{2\pi\sigma_X^2\sigma_W^2/(\sigma_X^2+\sigma_W^2)}}$$
$$= \frac{\exp\left(-\frac{1}{2\sigma_X^2\sigma_W^2/(\sigma_X^2+\sigma_W^2)}(x-y\sigma_X^2/(\sigma_X^2+\sigma_W^2))^2\right)}{\sqrt{2\pi\sigma_X^2\sigma_W^2/(\sigma_X^2+\sigma_W^2)}}. \tag{3.87}$$

In words: $f_{X|Y}(x|y)$ is a Gaussian pdf

$$\mathcal{N}\left(\frac{\sigma_X^2}{\sigma_X^2+\sigma_W^2}y, \frac{\sigma_X^2\sigma_W^2}{\sigma_X^2+\sigma_W^2}\right).$$

The mean of a conditional distribution is called a *conditional mean* and the variance of a conditional distribution is called a *conditional variance*.

Continuous additive noise with discrete input

Additive noise provides a situation in which mixed distributions having both discrete and continuous parts naturally arise. Suppose that the signal X is binary, say with pmf $p_X(x) = p^x(1-p)^{1-x}$. The noise term W is assumed to be a continuous random variable described by pdf $f_W(w)$, independent of X, with variance σ_W^2. The observation is defined by $Y = X + W$. In this case the joint distribution is not defined by a joint pmf or a joint pdf, but by a combination of the two. Some thought may lead to the reasonable guess that the continuous observation given the discrete signal should be describable by a conditional pdf $f_{Y|X}(y|x) = f_W(y-x)$, where the conditional pdf is of the elementary variety, the given event has nonzero probability. To prove that this is in fact correct, consider the elementary conditional probability $\Pr(Y \le y|X = x)$, for $x = 0, 1$. This is recognizable as the conditional cdf for Y given $X = x$, so that the desired conditional density is given by

$$f_{Y|X}(y|x) = \frac{d}{dy} \Pr(Y \le y|X = x). \tag{3.88}$$

The required probability is evaluated using the independence of X and W as

$$\begin{aligned}\Pr(Y \le y|X = x) &= \Pr(X + W \le y|X = x) = \Pr(x + W \le y|X = x) \\ &= \Pr(W \le y - x) = F_W(y - x).\end{aligned}$$

Differentiating with respect to y gives

$$f_{Y|X}(y|x) = f_W(y - x). \tag{3.89}$$

The joint distribution is described in this case by a combination of a pmf and a pdf. For example, to compute the joint probability that $X \in F$ and $Y \in G$ is accomplished by

$$\begin{aligned}\Pr(X \in F \text{ and } Y \in G) &= \sum_F p_X(x) \int_G f_{Y|X}(y|x)\, dy \\ &= \sum_F p_X(x) \int_G f_W(y - x)\, dy.\end{aligned} \tag{3.90}$$

Choosing $F = \Re$ yields the output distribution

$$\begin{aligned}\Pr(Y \in G) &= \sum p_X(x) \int_G f_{Y|X}(y|x)\, dy \\ &= \sum p_X(x) \int_G f_W(y - x)\, dy.\end{aligned}$$

Choosing $G = (-\infty, y]$ provides a formula for the cdf $F_Y(y)$, which can be differentiated to yield the output pdf

$$f_Y(y) = \sum p_X(x) f_{Y|X}(y|x) = \sum p_X(x) f_W(y - x), \tag{3.91}$$

a mixed discrete convolution involving a pmf and a pdf (and exactly the formula one expects in this mixed situation given the pure discrete and continuous examples).

Continuing the parallel with the pure discrete and continuous cases, one might expect that Bayes' rule could be used to evaluate the conditional distribution in the opposite direction. Since X is discrete this is the conditional pmf:

$$p_{X|Y}(x|y) = \frac{f_{Y|X}(y|x)p_X(x)}{f_Y(y)} = \frac{f_{Y|X}(y|x)p_X(x)}{\sum_\alpha p_X(\alpha)f_{Y|X}(y|\alpha)}. \tag{3.92}$$

Observe that unlike previously treated conditional pmf's, this one is *not* an elementary conditional probability since the conditioning event does not have nonzero probability. Thus it cannot be defined in the original manner, but must be justified in the same way as conditional pdf's, that is, by rewriting the joint distribution (3.90) as

$$\begin{aligned} \Pr(X \in F \text{ and } Y \in G) &= \int_G dy f_Y(y) \Pr(X \in F|Y = y) \\ &= \int_G dy f_Y(y) \sum_F p_{X|Y}(x|y), \end{aligned} \tag{3.93}$$

so that $p_{X|Y}(x|y)$ does indeed play the role of a mass of conditional probability, that is,

$$\Pr(X \in F|Y = y) = \sum_F p_{X|Y}(x|y). \tag{3.94}$$

Applying these results to the specific case of the binary input and Gaussian noise, the conditional pmf of the binary input given the noisy observation is

$$\begin{aligned} p_{X|Y}(x|y) &= \frac{f_W(y-x)p_X(x)}{f_Y(y)} \\ &= \frac{f_W(y-x)p_X(x)}{\sum_\alpha p_X(\alpha)f_W(y-\alpha)}; \; y \in \Re, x \in \{0,1\}. \end{aligned} \tag{3.95}$$

This formula now permits the analysis of a classical problem in communications, the detection of a binary signal in Gaussian noise.

3.10 Binary detection in Gaussian noise

The derivation of the MAP detector or classifier extends immediately to the situation of a binary input random variable and independent Gaussian noise just treated. As in the purely discrete case, the MAP detector $\hat{X}(y)$ of X given $Y = y$ is given by

$$\hat{X}(y) = \operatorname*{argmax}_x p_{X|Y}(x|y) = \operatorname*{argmax}_x \frac{f_W(y-x)p_X(x)}{\sum_\alpha p_X(\alpha)f_W(y-\alpha)}. \tag{3.96}$$

Since the denominator of the conditional pmf does not depend on x (only on y), given y the denominator has no effect on the maximization

$$\hat{X}(y) = \underset{x}{\operatorname{argmax}}\, p_{X|Y}(x|y) = \underset{x}{\operatorname{argmax}}\, f_W(y-x)p_X(x).$$

Assume for simplicity that X is equally likely to be 0 or 1 so that the rule becomes

$$\hat{X}(y) = \underset{x}{\operatorname{argmax}}\, p_{X|Y}(x|y) = \underset{x}{\operatorname{argmax}}\, \frac{1}{\sqrt{2\pi\sigma_W^2}} \exp\left(-\frac{1}{2}\frac{(x-y)^2}{\sigma_W^2}\right).$$

The constant in front of the pdf does not affect the maximization. In addition, the exponential is a mononotically decreasing function of $|x-y|$, so that the exponential is maximized by minimizing this magnitude difference, i.e.,

$$\hat{X}(y) = \underset{x}{\operatorname{argmax}}\, p_{X|Y}(x|y) = \underset{x}{\operatorname{argmin}}\, |x-y|, \tag{3.97}$$

which yields a final simple rule: see if $x = 0$ or 1 is *closer* to y as the best guess of x. This choice yields the MAP detection and hence the minimum probability of error. In our example this yields the rule

$$\hat{X}(y) = \begin{cases} 0 & y < 0.5 \\ 1 & y > 0.5 \end{cases}. \tag{3.98}$$

Because the optimal detector chooses the x that minimizes the Euclidean distance $|x-y|$ to the observation y, it is called a *minimum distance* detector or rule. Because the guess can be computed by comparing the observation to a threshold (the value midway between the two possible values of x), the detector is also called a *threshold detector*.

Assumptions have been made to keep things fairly simple. The reader is invited to work out what happens if the random variable X is biased and if its alphabet is taken to be $\{-1,1\}$ instead of $\{0,1\}$. It is instructive to sketch the conditional pmf's for these cases.

Having derived the optimal detector, it is reasonable to look at the resulting, minimized, probability of error. This can be found using conditional probability:

$$\begin{aligned}
P_e &= \Pr(\hat{X}(Y) \neq X) \\
&= \Pr(\hat{X}(Y) \neq 0|X = 0)p_X(0) + \Pr(\hat{X}(Y) \neq 1|X = 1)p_X(1) \\
&= \Pr(Y > 0.5|X = 0)p_X(0) + \Pr(Y < 0.5|X = 1)p_X(1) \\
&= \Pr(W + X > 0.5|X = 0)p_X(0) + \Pr(W + X < 0.5|X = 1)p_X(1) \\
&= \Pr(W > 0.5|X = 0)p_X(0) + \Pr(W + 1 < 0.5|X = 1)p_X(1) \\
&= \Pr(W > 0.5)p_X(0) + \Pr(W < -0.5)p_X(1)
\end{aligned}$$

where we have used the independence of W and X. These probabilities can be stated in terms of the Φ-function of (2.78) as in (2.82), which combined with the assumption that X is uniform and (2.84) yields

$$P_e = \frac{1}{2}\left(1 - \Phi\left(\frac{0.5}{\sigma_W}\right) + \Phi\left(-\frac{0.5}{\sigma_W}\right)\right) = \Phi\left(-\frac{1}{2\sigma_W}\right). \tag{3.99}$$

3.11 Statistical estimation

Discrete conditional probabilities were seen to provide a method for guessing an unknown class from an observation: if all incorrect choices have equal costs so that the overall optimality criterion is to minimize the probability of error, then the optimal classification rule is to guess that the class $X = k$, where $p_{X|Y}(k|y) = \max_z p_{X|Y}(x|y)$, the maximum a-posteriori or MAP decision rule. There is an analogous problem and solution in the continuous case, but the result does not have as strong an interpretation as in the discrete case. A more complete analogy will be derived in the next chapter.

As in the discrete case, suppose that a random variable Y is observed and the goal is to make a good guess $\hat{X}(Y)$ of another random variable X that is jointly distributed with Y. Unfortunately in the continuous case it does not make sense to measure the quality of such a guess by the probability of its being correct because now that probability is usually zero. For example, if Y is formed by adding a Gaussian signal X to an independent Gaussian noise W to form an observation $Y = X + W$ as in the previous section, then no rule is going to recover X perfectly from Y. Nonetheless, intuitively there should be reasonable ways to make such guesses in continuous situations. Since X is continuous, such guesses are referred to as "estimation" or "prediction" of X rather than as "classification" or "detection" as used in the discrete case. In the statistical literature the general problem is referred to as "regression".

One approach is to mimic the discrete approach on intuitive grounds. If the best guess in the classification problem of a random variable X given an observation Y is the MAP classifier $\hat{X}_{\text{MAP}}(y) = \text{argmax}_x p_{X|Y}(x|y)$, then a natural analog in the continuous case is the so-called MAP estimator defined by

$$\hat{X}_{\text{MAP}}(y) = \text{argmax}_x f_{X|Y}(x|y), \tag{3.100}$$

the value of x maximizing the conditional pdf given y. The advantage of this estimator is that it is easy to describe and provides an immediate application of conditional pdf's paralleling that of classification for discrete conditional probability. The disadvantage is that we cannot argue that this estimate is "optimal" in the sense of optimizing some specified criterion, it is essentially an ad hoc (but reasonable) rule. As an example of its use, consider the Gaussian signal plus noise of the previous section. There it was found that the pdf $f_{X|Y}(x|y)$ is Gaussian with mean $y\sigma_X^2/(\sigma_X^2 + \sigma_W^2)$. Since the Gaussian density has its peak at its mean, in this case the MAP estimate of X given $Y = y$ is given by the conditional mean $y\sigma_X^2/(\sigma_X^2 + \sigma_W^2)$.

Knowledge of the conditional pdf is all that is needed to define another estimator: the *maximum likelihood* or ML estimate of X given $Y = y$ is defined as the value of x that maximizes the conditional pdf $f_{Y|X}(y|x)$, the pdf with the roles of input and output reversed from that of the MAP estimator. Thus

$$\hat{X}_{\text{ML}}(y) = \underset{x}{\text{argmax}}\, f_{Y|X}(y|x). \tag{3.101}$$

Thus in the Gaussian case treated above, $\hat{X}_{\text{ML}}(Y) = y$.

The main interest in the ML estimator in some applications is that it is sometimes simpler. It also does not require any assumption on the input probabilities. The MAP estimator depends strongly on f_X; the ML estimator does not depend on it at all. In the special case where the input pdf f_X is uniform and the conditional pdf $f_{Y|X}(y|x)$ is 0 wherever $f_X(x) = 0$, then maximizing $f_{Y|X}(y|x)$ over x is equivalent to maximizing $f_{X|Y}(x|y) = f_{Y|X}(y|x)f_X(x)/f_Y(y)$ over x so that the MAP estimator and the ML estimator are the same.

3.12 Characteristic functions

We have seen that summing two random variables produces a new random variable whose pmf or pdf is found by convolving the two pmf's or pdf's of the original random variables. Anyone with an engineering background will likely have had experience with convolution and know from experience that convolutions can be somewhat messy to evaluate. To make matters worse, if one wishes to sum additional independent random variables to the existing sum, say form $Y_N = \sum_{k=1}^{N} h_{N-k} X_k$ from an iid collection $\{X_k\}$, then the result will be an N-fold convolution, a potential nightmare in all but the simplest of cases. As in other engineering applications such as circuit design, convolutions can be avoided by Fourier transform methods. In this subsection we describe the method as an alternative approach for the examples to come. We begin with the discrete case.

Historically the transforms used in probability theory have been slightly different from those in traditional Fourier analysis. For a discrete random variable with pmf p_X, define the *characteristic function* M_X of the random variable (or of the pmf) as

$$M_X(ju) = \sum_x p_X(x) e^{jux}, \tag{3.102}$$

where u is usually assumed to be real. Recalling the definition (2.34) of the expectation of a function g defined on a sample space, choosing $g(\omega) = e^{juX(\omega)}$ shows that the characteristic function can be more simply defined as

$$M_X(ju) = E[e^{juX}]. \tag{3.103}$$

Thus characteristic functions, like probabilities, can be viewed as special cases of expectations.

This transform, which is also referred to as an *exponential transform* or *operational transform*, bears a strong resemblance to the discrete-parameter Fourier transform

$$\mathcal{F}_\nu(p_X) = \sum_x p_X(x) e^{-j2\pi\nu x} \tag{3.104}$$

and the z-transform

$$\mathcal{Z}_z(p_X) = \sum_x p_X(x) z^x. \tag{3.105}$$

In particular, $M_X(ju) = \mathcal{F}_{-u/2\pi}(p_X) = \mathcal{Z}_{e^{ju}}(p_X)$. As a result, all of the properties of characteristic functions follow immediately from (are equivalent to) similar properties from Fourier or z-transforms. As with Fourier and z-transforms, the original pmf p_X can be recovered from the transform

M_X by suitable inversion. For example, given a pmf $p_X(k)$; $k \in \mathcal{Z}_N$,

$$\frac{1}{2\pi}\int_{-\pi/2}^{\pi/2} M_X(ju)e^{-iuk}\,du = \frac{1}{2\pi}\int_{-\pi/2}^{\pi/2}\left(\sum_x p_X(x)e^{jux}\right)e^{-iuk}\,du$$
$$= \sum_x p_X(x)\frac{1}{2\pi}\int_{-\pi/2}^{\pi/2} e^{ju(x-k)}\,du$$
$$= \sum_x p_X(x)\delta_{k-x} = p_X(k). \tag{3.106}$$

Consider again the problem of summing two independent random variables X and W with pmf's p_X and p_W and characteristic functions M_X and M_W, respectively. If $Y = X + W$, as before we can evaluate the characteristic function of Y as

$$M_Y(ju) = \sum_y p_Y(y)e^{juy}$$

where from the inverse image formula

$$p_Y(y) = \sum_{x,w:x+w=y} p_{X,W}(x,w)$$

so that

$$M_Y(ju) = \sum_y\left(\sum_{x,w:x+w=y} p_{X,W}(x,w)\right)e^{juy}$$
$$= \sum_y\left(\sum_{x,w:x+w=y} p_{X,W}(x,w)e^{juy}\right)$$
$$= \sum_y\left(\sum_{x,w:x+w=y} p_{X,W}(x,w)e^{ju(x+w)}\right)$$
$$= \sum_{x,w} p_{X,W}(x,w)e^{ju(x+w)}$$

where the last equality follows because the double sum for all y, x and w is the sum for all x and w. This last sum factors, however, as

$$M_Y(ju) = \sum_{x,w} p_X(x)p_W(w)e^{jux}e^{juw}$$
$$= \sum_x p_X(x)e^{jux}\sum_w p_W(w)e^{juw}$$
$$= M_X(ju)M_W(ju), \tag{3.107}$$

which shows that the transform of the pmf of the sum of independent random variables is simply the product of the transforms.

Iterating (3.107) several times gives an extremely useful result that we state formally as a theorem. It can be proved by repeating the above argument, but we shall later see a shorter proof.

Theorem 3.1 *If* $\{X_i;\ i = 1, \ldots, N\}$ *are independent random variables with characteristic functions* M_{X_i}*, then the characteristic function of the random variable* $Y = \sum_{i=1}^{N} X_i$ *is*

$$M_Y(ju) = \prod_{i=1}^{N} M_{X_i}(ju). \tag{3.108}$$

If the X_i *are independent and identically distributed with common characteristic function* M_X*, then*

$$M_Y(ju) = M_X^N(ju). \tag{3.109}$$

As a simple example, the characteristic function of a binary random variable X with parameter $p = p_X(1) = 1 - p_X(0)$ is easily found to be

$$M_X(ju) = \sum_{k=0}^{1} e^{juk} p_X(k) = (1-p) + pe^{ju}. \tag{3.110}$$

If $\{X_i;\ i = 1, \ldots, n\}$ are independent Bernoulli random variables with identical distributions and $Y_n = \sum_{k=1}^{n} X_i$, then $M_{Y_n}(ju) = [(1-p) + pe^{ju}]^n$ and hence

$$\begin{aligned} M_{Y_n}(ju) &= \sum_{k=0}^{n} p_{Y_n}(k) e^{juk} = ((1-p) + pe^{ju})^n \\ &= \sum_{k=0}^{n} \left[\binom{n}{k} (1-p)^{n-k} p^k \right] e^{juk}, \end{aligned}$$

where we have invoked the binomial theorem in the last step. For the equality to hold, however, we have from the uniqueness of transforms that $p_{Y_n}(k)$ must be the square bracketed term, that is, the binomial pmf

$$p_{Y_n}(k) = \binom{n}{k} (1-p)^{n-k} p^k;\ k \in \mathcal{Z}_{n+1}. \tag{3.111}$$

As in the discrete case, convolutions can be avoided by transforming the densities involved. The derivation is exactly analogous to the discrete case, with integrals replacing sums in the usual way.

For a continous random variable X with pmf f_X, define the *characteristic function* M_X of the random variable (or of the pdf) as

$$M_X(ju) = \int f_X(x)e^{jux}\,dx. \tag{3.112}$$

As in the discrete case, this can be considered as a special case of expectation for continuous random variables as defined in (2.34) so that

$$M_X(ju) = E\left[e^{juX}\right]. \tag{3.113}$$

The characteristic function is related to the the continuous-parameter Fourier transform

$$\mathcal{F}_\nu(f_X) = \int f_X(x)e^{-j2\pi\nu x}\,dx \tag{3.114}$$

and the Laplace transform

$$\mathcal{L}_s(f_X) = \int f_X(x)e^{-sx}\,dx \tag{3.115}$$

by $M_X(ju) = \mathcal{F}_{-u/2\pi}(f_X) = \mathcal{L}_{-ju}(f_X)$. As a result, all of the properties of characteristic functions of densities follow immediately from (are equivalent to) similar properties from Fourier or Laplace transforms. For example, given a well-behaved density $f_X(x)$; $x \in \Re$ with characteristic function $M_X(ju)$,

$$f_X(x) = \frac{1}{2\pi}\int_{-\infty}^{\infty} M_X(ju)e^{-jux}\,du. \tag{3.116}$$

Consider again the problem of summing two independent random variables X and Y with pdf's f_X and f_W with characteristic functions M_X and M_W, respectively. As in the discrete case it can be shown that

$$M_Y(ju) = M_X(ju)M_W(ju). \tag{3.117}$$

Rather than mimic the proof of the discrete case, however, we postpone the proof to a more general treatment of characteristic functions in Chapter 4.

As in the discrete case, iterating (3.117) several times yields the following result, which now includes both discrete and continous cases.

Theorem 3.2 *If* $\{X_i;\ i = 1, \ldots, N\}$ *are independent random variables with characteristic functions* M_{X_i}*, then the characteristic function of the random variable* $Y = \sum_{i=1}^{N} X_i$ *is*

$$M_Y(ju) = \prod_{i=1}^{N} M_{X_i}(ju). \tag{3.118}$$

If the X_i are independent and identically distributed with common characteristic function M_X, then

$$M_Y(ju) = M_X^N(ju). \tag{3.119}$$

As an example of characteristic functions and continuous random variables, consider the Gaussian random variable. The evaluation requires a bit of effort, either by using the "complete the square" technique of calculus or by looking up in published tables. Assume that X is a Gaussian random variable with mean m and variance σ^2. Then

$$\begin{aligned} M_X(ju) = E(e^{juX}) &= \int_{-\infty}^{\infty} \frac{1}{(2\pi\sigma^2)^{1/2}} e^{-(x-m)^2/2\sigma^2} e^{jux}\, dx \\ &= \int_{-\infty}^{\infty} \frac{1}{(2\pi\sigma^2)^{1/2}} e^{-(x^2-2mx-2\sigma^2 jux+m^2)/2\sigma^2}\, dx \\ &= \left\{\int_{-\infty}^{\infty} \frac{1}{(2\pi\sigma^2)^{1/2}} e^{-(x-(m+ju\sigma^2))^2/2\sigma^2}\, dx\right\} e^{jum-u^2\sigma^2/2} \\ &= e^{jum-u^2\sigma^2/2}. \end{aligned} \tag{3.120}$$

Thus the characteristic function of a Gaussian random variable with mean m and variance σ_X^2 is

$$M_X(ju) = e^{jum-u^2\sigma^2/2}. \tag{3.121}$$

If $\{X_i;\ i = 1, \ldots, n\}$ are independent Gaussian random variables with identical densities $\mathcal{N}(m, \sigma^2)$ and $Y_n = \sum_{k=1}^n X_i$, then

$$M_{Y_n}(ju) = [e^{jum-u^2\sigma^2/2}]^n = e^{ju(nm)-u^2(n\sigma^2)/2}, \tag{3.122}$$

which is the characteristic function of a Gaussian random variable with mean nm and variance $n\sigma^2$.

The following maxim should be kept in mind whenever faced with sums of independent random variables:

- When given a derived distribution problem involving the sum of independent random variables, first find the characteristic function of the sum by taking the product of the characteristic functions of the individual random variables. Then find the corresponding probability function by inverting the transform. This technique is valid if the random variables are independent – they do not have to be identically distributed.

3.13 Gaussian random vectors

A random vector is said to be *Gaussian* if its density is Gaussian, that is, if its distribution is described by the multidimensional pdf explained in Chapter 2. The component random variables of a Gaussian random vector are said to be *jointly Gaussian* random variables. Note that the symmetric matrix Λ of the k-dimensional vector pdf has $k(k+1)/2$ parameters and that the vector m has k parameters. On the other hand, the k marginal pdf's together have only $2k$ parameters. Again we note the impossibility of constructing joint pdf's without more specification than the marginal pdf's alone. However, the marginals suffice to describe the entire vector if we also know that the vector has independent components, e.g., the vector is iid. In this case the matrix Λ is diagonal.

Although difficult to describe, Gaussian random vectors have several nice properties. One of the most important of these properties is that linear or affine operations on Gaussian random vectors produce Gaussian random vectors. This result can be demonstrated with a modest amount of work using multidimensional characteristic functions, the extension of transforms from scalars to vectors.

The multidimensional characteristic function of a distribution is defined as follows: Given a random vector $\mathbf{X} = (X_0, \ldots, X_{n-1})$ and a vector parameter $\mathbf{u} = (u_0, \ldots, u_{n-1})$, the n-dimensional characteristic function $M_{\mathbf{X}}(j\mathbf{u})$ is defined by

$$\begin{aligned} M_{\mathbf{X}}(j\mathbf{u}) &= M_{X_0,\ldots,X_{n-1}}(ju_0, \ldots, ju_{n-1}) = E\left[e^{j\mathbf{u}^t\mathbf{X}}\right] \\ &= E\left[\exp\left(j\sum_{k=0}^{n-1} u_k X_k\right)\right]. \end{aligned} \tag{3.123}$$

It can be shown using multivariable calculus (problem 3.57) that a Gaussian random vector with mean vector $\mathbf{m}$ and covariance matrix Λ has characteristic function

$$\begin{aligned} M_{\mathbf{X}}(j\mathbf{u}) &= e^{j\mathbf{u}^t\mathbf{m} - \mathbf{u}^t\Lambda\mathbf{u}/2} \\ &= \exp\left(j\sum_{k=0}^{n-1} u_k m_k - 1/2\sum_{k=0}^{n-1}\sum_{m=0}^{n-1} u_k \Lambda(k,m) u_m\right). \end{aligned} \tag{3.124}$$

Observe that the Gaussian characteristic function has the same form as the Gaussian pdf – an exponential function of quadratic and linear terms. However, unlike the pdf, the characteristic function depends on the covariance matrix directly, whereas the pdf contains the inverse of the covariance matrix. Thus the Gaussian characteristic function is in some sense simpler than the Gaussian pdf. As a further consequence of the direct dependence on the covariance matrix, it is interesting to note that, unlike the Gaussian pdf, the characteristic function is well defined even if Λ is only nonnegative definite and not strictly positive definite. Previously we gave a definition of a Gaussian random vector in terms of its pdf. Now we can give an alternate, more general (in the sense that a strictly positive definite covariance matrix is not required) definition of a Gaussian random vector and hence random process):

- A random vector is Gaussian if and only if it has a characteristic function of the form of (3.124).

This may seem strange at first thought – how can we define a vector to be Gaussian by virtue of having a characteristic function of a certain form while allowing that the pdf might not exist. If the pdf does not exist, then of what is the characteristic function the transform? The answer is that in general it is the transform of the distribution, which may exist even if the pdf does not. We are effectively defining a Gaussian distribution here and not a Gaussian pdf. We shall later see that if a distribution is Gaussian but the pdf does not exist, then it is an example of what is called a *singular* distribution and that if the covariance matrix is singular, it is not invertible.

3.14 Simple random processes

In this section several examples of random processes defined on simple probability spaces are given to illustrate the basic definition of an infinite collection of random variables defined on a single space. In the next section more complicated examples are considered by defining random variables on a probability space which is the output space for another random process, a setup that can be viewed as signal processing.

Examples

[3.22] Consider the binary probability space $(\Omega, \mathcal{F}, P)$ with $\Omega = \{0, 1\}$, $\mathcal{F}$ the usual event space, and P induced by the pmf $p(0) = \alpha$ and $p(1) = 1 - \alpha$, where α is some constant, $0 \leq \alpha \leq 1$. Define a random process on this space as follows:

$$X(t, \omega) = \cos(\omega t) = \begin{cases} \cos(t), \ t \in \Re & \text{if } \omega = 1 \\ 1, \ t \in \Re & \text{if } \omega = 0 \end{cases}.$$

Thus if a 1 occurs a cosine is sent forever, and if a 0 occurs a constant 1 is sent forever.

This process clearly has continuous time and at first glance it might appear to have continuous amplitude also, but only two waveforms are possible, a cosine and a constant. Thus the alphabet at each time contains at most two values with nonzero probability and these possible values change with time. Hence this process is in fact a discrete amplitude process and random vectors drawn from this source are described by pmf's. We can consider the alphabet of the process to be either $\Re^{\mathcal{T}}$ or $[-1,1]^{\mathcal{T}}$, among other possibilities. Fix time at $t = \pi/2$. Then $X(\pi/2)$ is a random variable with pmf

$$p_{X(\pi/2)}(x) = \begin{cases} \alpha & \text{if } x = 1 \\ 1-\alpha & \text{if } x = 0 \end{cases}.$$

The reader should try other instances of time. What happens at $t = 0, 2\pi, 4\pi, 2m\pi \ldots$?

[3.23] Consider a probability space $(\Omega, \mathcal{F}, P)$ with $\Omega = \Re$, $\mathcal{F} = \mathcal{B}(\Re)$, the Borel field, and probability measure P induced by the pdf

$$f(r) = \begin{cases} 1 & \text{if } r \in [0,1] \\ 0 & \text{otherwise} \end{cases}.$$

Again define the random process $\{X(t)\}$ by $X(t,\omega) = \cos(\omega t); t \in \Re$.

Again the process is continuous time, but now it has mixed alphabet because an uncountable infinity of waveforms is possible corresponding to all angular frequencies between 0 and 1 so that $X(t,\omega)$ is a continuous random variable except at $t = 0$, where $X(0,\omega) = 1$ is a discrete random variable. If you calculate the pdf of the random variable $X(t)$ you see that it varies as a function of time (Problem 3.33).

[3.24] Consider the probability space of Example [3.23], but cut it down to the unit interval so that $([0,1), \mathcal{B}([0,1)), P)$ is the probability space, where P is the probability measure induced by the pdf $f(r) = 1; r \in [0,1)$. (So far this is just another model for the same thing.) For $n = 1, 2 \ldots, X_n(\omega) = b_n(\omega) =$ the nth digit binary expansion of ω, that is

$$\omega = \sum_{n=1}^{\infty} b_n 2^{-n}$$

or equivalently $\omega = 0.b_1 b_2 b_3 \ldots$ in binary.

The process $\{X_n; n = 1, 2 \ldots\}$ is a one-sided discrete alphabet random process with alphabet $\{0, 1\}$. It is important to understand that Nature has selected ω at the beginning of time, but the observer has no way to determining Ω completely without waiting until the end of time. Nature reveals only one bit of ω per unit time, so the observer can only get an improved *estimate* of ω as time goes on. This is an excellent example of how a random process can be modeled by selecting only a single outcome, yet the observer sees a process that evolves forever.

In this example our change in the sample space to $[0, 1]$ from $\Re$ was done for convenience. By restricting the sample space we did not have to define the random variable outside of the unit interval (as we would have had to do to provide a complete description).

At times it is necessary to extend the definition of a random process to include vector-valued functions of time so that the random process is a function of three arguments instead of two. The most important extension is to complex-valued random processes, i.e., vectors of length 2. We will not make such extensions frequently but we will include an example at this time.

[3.25] Random rotations.
Given the same probability space as in Example [3.24], define a complex-valued random process $\{\mathbf{X}_n\}$ as follows: Let α be a fixed real parameter and define

$$\mathbf{X}_n(\omega) = e^{jn\alpha}e^{j2\pi\omega} = e^{j(n\alpha+2\pi\omega)}; \ n = 1, 2, 3, \ldots$$

This process, called the random rotations process, is a discrete time continuous (complex) alphabet one-sided random process. Note that an alternative description of the same process would be to define Ω as the unit circle in the complex plane together with its Borel field and to define a process $Y_n(\omega) = c^n\omega$ for some fixed $c \in \Omega$; this representation points out that successive values of Y_n are obtained by rotating the previous value through an angle determined by c.

Note that the joint pdf of the complex components of X_n varies with time, n, as does the pdf in Example [3.23] (Problem 3.36).

[3.26] Again consider the probability space of Example [3.24]. We define a random process recursively on this space as follows: define $X_0 = \omega$ and

$$\begin{aligned} X_n(\omega) &= 2X_{n-1}(\omega) \bmod 1 \\ &= \begin{cases} 2X_{n-1}(\omega) & \text{if } 0 \le X_{n-1}(\omega) < 1/2 \\ 2X_{n-1}(\omega) - 1 & \text{if } 1/2 \le X_{n-1}(\omega) < 1 \end{cases}, \end{aligned}$$

where r mod 1 is the fractional portion of r. In other words, if $X_{n-1}(\omega) = x$ is in [0,1/2), then $X_n(\omega) = 2x$. If $X_{n-1}(\omega) = x$ is in [1/2,1), then $X_n(\omega) = 2x - 1$.

[3.27] Given the same probability space as in Example [3.26], define $X(t, \omega) = \cos(t + 2\pi\omega), t \in \Re$. The resulting random process $\{X(t)\}$ is continuous time and continuous amplitude and is called a random phase process since all of the possible waveforms are shifts of one another. Note that the pdf of $X(t, \omega)$ does *not* depend on time (Problem 3.37).

[3.28] Take any one of the foregoing (real) processes and quantize or clip it; that is, define a binary quantizer q by

$$q(r) = \begin{cases} a & \text{if } r \geq 0 \\ b & \text{if } r < 0 \end{cases}$$

and define the process $Y(t, \omega) = q(X(t, \omega))$, all t. (Typically $b = -a$.) This is a common form of signal processing – converting a continuous alphabet random process into a discrete alphabet random process by means of quantization.

This process is discrete alphabet and is either continuous or discrete time, depending on the original X process. In any case $Y(t)$ has a binary pmf that, in general, varies with time.

[3.29] Say we have two random variables U and V defined on a common probability space $(\Omega, \mathcal{F}, P)$. Then

$$X(t) = U \cos(2\pi f_0 t + V)$$

defines a random process on the same probability space for any fixed parameter f_0.

All the foregoing random processes are well defined. The processes inherit probabilistic descriptions from the underlying probability space. The techniques of derived distributions can be used to compute probabilities involving the outputs since, for example, any problem involving a single sample time is simply a derived distribution for a single random variable, and any problem involving a finite collection of sample times is a single derived distribution problem for a random vector. Several examples are explored in the problems at the end of the chapter.

3.15 Directly given random processes

3.15.1 The Kolmogorov extension theorem

Consistency of distributions of random vectors of various dimensions plays a far greater role in the theory and practice of random processes than simply a means of checking the correctness of a computation. We have thus far argued that a *necessary* condition for a set of random vector distributions to describe collections of samples taken from a random process is that the distributions be *consistent*, e.g., given marginals and joints we must be able to compute the marginals from the joints. The Kolmogorov extension theorem states that consistency is also *sufficient* for a family of finite-dimensional vector distributions to describe a random process, that is, for there to exist a well-defined random process that agrees with the given family of finite-dimensional distributions. We state the theorem without proof as the proof is far beyond the assumed mathematical prerequisites for this book. (The interested reader is referred to [56, 7, 32].) Happily, however, it is often straightforward, if somewhat tedious, to demonstrate that the conditions of the theorem hold and hence that a proposed model is well-defined.

Theorem 3.3 *Kolmogorov extension theorem*
Suppose that one is given a consistent family of finite-dimensional distributions $P_{X_{t_0},X_{t_1},\ldots,X_{t_{k-1}}}$ for all positive integers k and all possible sample times $t_i \in \mathcal{T}$; $i = 0, 1, \ldots, k-1$. Then there exists a random process $\{X_t;\ t \in \mathcal{T}\}$ that is consistent with this family. In other words, to describe a random process completely, it is sufficient to describe a consistent family of finite-dimensional distributions of its samples.

3.15.2 Independent identically distributed random processes

The next example extends the idea of an iid vector to provide one of the most important random process models. Although such processes are simple in that they possess no memory among samples, they are a fundamental building block for more complicated processes as well as being an important example in their own right. In a sense these are the most random of all possible random processes because knowledge of the past does not help predict future behavior.

A discrete time random process $\{X_n\}$ is said to be *independent identically distributed* or *iid* if all finite-dimensional random vectors formed by sampling the process are iid; that is, if for any k and any collection of distinct sample times $t_0, t_1, \ldots, t_{k-1}$, the random vector $(X_{t_0}, X_{t_1}, \ldots, X_{t_{k-1}})$ is iid.

This definition is equivalent to the simpler definition of the introduction to this chapter, but the more general form is adopted because it more closely resembles definitions to be introduced later. Independent identically distributed random processes are often called *Bernoulli processes*, especially in the binary case.

It can be shown with cumbersome but straightforward effort that the random process of [3.24] is in fact iid. In fact, for any given marginal distribution there exists an iid process with that marginal distribution. Although eminently believable, this fact requires the Kolmogorov extension theorem, which states that a consistent family of finite-dimensional distributions implies the existence of a random process described or specified by those distributions. The demonstration of consistency for iid processes is straightforward. Readers are encouraged to convince themselves for the case of n-dimensional distributions reducing to $(n-1)$-dimensional distributions and, more specifically, of two-dimensional distributions implying one-dimensional distributions

3.15.3 Gaussian random processes

A random process is Gaussian if the random vectors $(X_{t_0}, X_{t_1}, \ldots, X_{t_{k-1}})$ are Gaussian. for all positive integers k and all possible sample times $t_i \in \mathcal{T}$; $i = 0, 1, \ldots, k-1$.

In order to describe a Gaussian process and verify the consistency conditions of the Kolmogorov extension theorem, one has to provide the covariance matrices Λ and mean vectors m for all random vectors $(X_{t_0}, X_{t_1}, \ldots, X_{t_{k-1}})$ formed from a finite collection of ordered samples from the process. This is accomplished by providing a *mean function* $m(t)$; $t \in \mathcal{T}$ and a *covariance function* $\Lambda(t, s)$; $t, s \in \mathcal{T}$, which then yield all of the required mean vectors and covariance matrices by sampling, that is, the mean vector for $(X_{t_0}, X_{t_1}, \ldots, X_{t_{k-1}})$ is $(m(t_0), m(t_1), \ldots, m(t_{k-1}))$ and the covariance matrix is $\Lambda = \{\Lambda(t_l, t_j);\ l, j \in \mathcal{Z}_k\}$.

That this family of density functions are in fact consistent is much more difficult to verify than was the case for iid processes, but it requires straightforward brute force in calculus rather than any deep mathematical ideas to to do so.

The Gaussian random process in both discrete and continuous time is virtually ubiquitous in the analysis of random systems. This is both because the model is good for a wide variety of physical phenomena and because it is extremely tractable for analysis.

3.16 Discrete time Markov processes

An iid process is often referred to as a *memoryless* process because of the independence among the samples. Such a process is both one of the simplest random processes and one of the most random. It is simple because the joint pmf's are easily found as products of marginals. It is "most random" because knowing the past (or future) outputs does not help improve the probabilities describing the current output. It is natural to seek straightforward means of describing more complicated processes with memory and to analyze the properties of processes resulting from operations on iid processes. A general approach towards modeling processes with memory is to *filter* memoryless processes, i.e., to perform an operation (a form of signal processing) on an input process which produces an output process that is not iid. In this section we explore several examples of such a construction, all of which provide examples of the use of conditional distributions for describing and investigating random processes. All of the processes considered in this section will prove to be examples of *Markov processes*, a class of random processes possessing a specific form of dependence among current and past samples.

3.16.1 A binary Markov process

Suppose that $\{X_n;\ n = 0, 1, \ldots\}$ is a Bernoulli process with

$$p_{X_n}(x) = \begin{cases} p & x = 1 \\ 1-p & x = 0 \end{cases}, \tag{3.125}$$

where $p \in (0, 1)$ is a fixed parameter. Since the pmf does not depend on n, the subscript is dropped and the pmf abbreviated to p_X. The pmf can also be written as

$$p_X(x) = p^x(1-p)^{1-x};\ x = 0, 1. \tag{3.126}$$

Since the process is assumed to be iid,

$$p_{X^n}(x^n) = \prod_{i=0}^{n-1} p_X(x_i) = p^{w(x^n)}(1-p)^{n-w(x^n)}, \tag{3.127}$$

where $w(x^n)$ is the number of nonzero x_i in x^n, called the Hamming weight of the binary vector x^n.

We consider using $\{X_n\}$ as the input to a device which produces an output binary process $\{Y_n\}$. The device can be viewed as a signal processor or as a linear filter. Since the process is binary, the most natural "linear" operations are those in the binary alphabet using modulo 2 arithmetic as defined in

(3.64)–(3.65). Consider the new random process $\{Y_n;\ n = 0, 1, 2, \ldots\}$ defined by

$$Y_n = \begin{cases} Y_0 & n = 0 \\ X_n \oplus Y_{n-1} & n = 1, 2, \ldots \end{cases}, \tag{3.128}$$

where Y_0 is a binary equiprobable random variable ($p_{Y_0}(0) = p_{Y_0}(1) = 0.5$) assumed to be independent of all of the X_n. This is an example of a linear (modulo 2) recursion or difference equation. The process can also be defined for $n = 1, 2, \ldots$ by

$$Y_n = \begin{cases} 1 & \text{if } X_n \neq Y_{n-1} \\ 0 & \text{if } X_n = Y_{n-1} \end{cases}.$$

This process is called a *binary autoregressive* process.

It should be apparent that Y_n has quite different properties from X_n. In particular, it depends strongly on past values. Since $p < 1/2$, Y_n is more likely to equal Y_{n-1} than it is to differ. If p is small, for example, Y_n is likely to have long runs of 0s and 1s. The sequence of random variables $\{Y_n\}$ is defined on a common experiment, the outputs of the $\{X_n\}$ process and an independent selection of Y_0. Thus $\{Y_n\}$ is itself a random process and all of its joint pmf's $p_{Y^n}(y^n) = \Pr(Y^n = y^n)$ should be derivable from the inverse image formula. We proceed to solve this derived distribution and then to interpret the result.

Using the inverse image formula in the general sense, which involves finding a probability of an event involving Y^n in terms of the probability of an event involving X^n (and, in this case, the initial value Y_0), yields the following sequence of steps:

$$\begin{aligned}
p_{Y^n}(y^n) &= \Pr(Y^n = y^n) \\
&= \Pr(Y_0 = y_0, Y_1 = y_1, Y_2 = y_2, \ldots, Y_{n-1} = y_{n-1}) \\
&= \Pr(Y_0 = y_0, X_1 \oplus Y_0 = y_1, X_2 \oplus Y_1 = y_2, \ldots, X_{n-1} \oplus Y_{n-2} = y_{n-1}) \\
&= \Pr(Y_0 = y_0, X_1 \oplus y_0 = y_1, X_2 \oplus y_1 = y_2, \ldots, X_{n-1} \oplus y_{n-2} = y_{n-1}) \\
&= \Pr(Y_0 = y_0, X_1 = y_1 \oplus y_0, X_2 = y_2 \oplus y_1, \ldots, X_{n-1} = y_{n-1} \oplus y_{n-2}) \\
&= p_{Y_0, X_1, X_2, X_3, \ldots, X_{n-1}}(y_0, y_1 \oplus y_0, y_2 \oplus y_1, \ldots, y_{n-1} \oplus y_{n-2}) \\
&= p_{Y_0}(y_0) \prod_{i=1}^{n-1} p_X(y_i \oplus y_{i-1}).
\end{aligned} \tag{3.129}$$

The derivation used the facts that (1) $a \oplus b = c$ if and only if $a = b \oplus c$, (2) the independence of Y_0, X_1, X_2, $\ldots$, X_{n-1}, and (3) the fact that the X_n are iid. This formula completes the first goal, except possibly plugging in the specific forms of p_{Y_0} and p_X to get

$$p_{Y^n}(y^n) = \frac{1}{2}\prod_{i=1}^{n-1} p^{y_i \oplus y_{i-1}}(1-p)^{1-y_i \oplus y_{i-1}}. \tag{3.130}$$

The marginal pmf's for Y_n can be evaluated by summing out the joints, e.g.,

$$\begin{aligned} p_{Y_1}(y_1) &= \sum_{y_0} p_{Y_0,Y_1}(y_0,y_1) = \frac{1}{2}\sum_{y_0} p^{y_1 \oplus y_0}(1-p)^{1-y_1 \oplus y_0} \\ &= \frac{1}{2};\ \ y_1 = 0,1. \end{aligned}$$

In a similar fashion it can be shown that the marginals for Y_n are all the same:

$$p_{Y_n}(y) = \frac{1}{2};\ y = 0,1;\ n = 0,1,2,\ldots, \tag{3.131}$$

and hence as with X_n the pmf can be abbreviated as p_Y, dropping the subscript.

Observe in particular that unlike the iid $\{X_n\}$ process,

$$p_{Y^n}(y^n) \neq \prod_{i=0}^{n-1} p_Y(y_i) \tag{3.132}$$

(provided $p \neq 1/2$) and hence $\{Y_n\}$ is not an iid process and the joint pmf cannot be written as a product of the marginals. Nonetheless, the joint pmf can be written as a product of simple terms, as has been done in (3.130). From the definition of conditional probability and (3.129)

$$p_{Y_l|Y_0,Y_1,\ldots,Y_{l-1}}(y_l|y_0,y_1,\ldots,y_{l-1}) = \frac{p_{Y^{l+1}}(y^{l+1})}{p_{Y^l}(y^l)} = p_X(y_l \oplus y_{l-1}) \tag{3.133}$$

and (3.129) is then recognizable as the chain rule (3.51) for the joint pmf $p_{Y^n}(y^n)$.

Note that the conditional probability of the current output Y_l given the values for the entire past Y_i; $i = 0,1,\ldots,l-1$ depends *only on the most recent past output* Y_{l-1}*!* This property can be summarized nicely by also deriving the conditional pmf

$$p_{Y_l|Y_{l-1}}(y_l|y_{l-1}) = \frac{p_{Y_{l-1},Y_l}(y_l,y_{l-1})}{p_{Y_{l-1}}(y_{l-1})}, \tag{3.134}$$

which with a little effort resembling the previous derivation can be evaluated as $p^{y_i \oplus y_{i-1}}(1-p)^{1-y_i \oplus y_{i-1}}$. Thus the $\{Y_n\}$ process has the property that

$$p_{Y_i|Y_0,Y_1,\ldots,Y_{i-1}}(y_i|y_0,y_1,\ldots,y_{i-1}) = p_{Y_i|Y_{i-1}}(y_i|y_{i-1}). \tag{3.135}$$

A discrete time random process with this property is called a *Markov process* or *Markov chain*. Such processes are among the most studied random processes with memory.

3.16.2 The binomial counting process

We next turn to filtering a Bernoulli process that is linear in the ordinary sense of real numbers. Now the input processes will be binary, but the output process will have the nonnegative integers as an alphabet. Simply speaking, the output process will be formed by counting the number of heads in a sequence of coin flips.

Let $\{X_n\}$ be iid binary random process with marginal pmf $p_X(1) = p = 1 - p_X(0)$. Define a new one-sided random process $\{Y_n; n = 0, 1, \ldots\}$ by

$$Y_n = \begin{cases} Y_0 = 0 & n = 0 \\ \sum_{k=1}^{n} X_k = Y_{n-1} + X_n & n = 1, 2, \ldots \end{cases}. \tag{3.136}$$

For $n \geq 1$ this process can be viewed as the output of a discrete time time-invariant linear filter with Kronecker delta response h_k given by $h_k = 1$ for $k \geq 0$ and $h_k = 0$ otherwise. From (3.136), each random variable Y_n provides a count of the number of 1s appearing in the X_n process through time n. Because of this counting structure we have that either

$$Y_n = Y_{n-1} \text{ or } Y_n = Y_{n-1} + 1; \; n = 2, 3, \ldots \tag{3.137}$$

In general, a discrete time process that satisfies (3.137) is called a *counting process* since it is nondecreasing, and when it jumps, it is always with an increment of 1. (A continuous alphabet counting process is similarly defined as a process with a nondecreasing output which increases in time in steps of 1.)

To completely describe this process it suffices to have a formula for the joint pmf's

$$p_{Y_1,\ldots,Y_n}(y_1,\ldots,y_n) = p_{Y_1}(y_1) \prod_{l=1}^{n} p_{Y_l|Y_1,\ldots,Y_{l-1}}(y_l|y_1,\ldots,y_{l-1}). \tag{3.138}$$

When we have constructed one process $\{Y_n\}$ from an existing process $\{X_n\}$, we need not worry about consistency since we have defined the new process on an underlying probability space (the output space of the original process), and hence the joint distributions must be consistent if they are correctly computed from the underlying probability measure – the process distribution for the iid process.

Since Y_n is formed by summing n Bernoulli random variables, the pmf for Y_n follows immediately from (3.111); it is the binomial pmf and hence the process is referred to as the *binomial counting process*.

The joint probabilities could be computed using the vector inverse image formula as with the binary Markov source, but instead we focus on the conditional distributions and compute them directly. The same approach could have been used for the binary Markov example.

The conditional pmf's are computed by describing probabilistically the next output Y_n of the process given the previous $n-1$ outputs $Y_1, \ldots, Y_{n-1}$. For the binomial counting process, the next output is formed simply by adding a binary random variable to the old sum. Thus all of the conditional probability mass is concentrated on two values – the last value and the last value plus 1. The conditional pmf's can therefore be expressed as

$$\begin{aligned}
&p_{Y_n|Y_{n-1},\ldots,Y_1}(y_n|y_{n-1},\ldots,y_1)\\
&= \Pr(Y_n = y_n|Y_l = y_l; l = 1,\ldots,y_{n-1})\\
&= \Pr(X_n = y_n - y_{n-1}|Y_l = y_l; l = 1,\ldots,n-1) \qquad (3.139)\\
&= \Pr(X_n = y_n - y_{n-1}|X_1 = y_1, X_i = y_i - y_{i-1}; i = 2,3,\ldots,n-1),
\end{aligned}$$

since from the definition of the Y_n process the conditioning event $\{Y_i = y_i; i = 1,2,\ldots,n-1\}$ is identical to the event $\{X_1 = y_1, X_i = y_i - y_{i-1}; i = 2,3,\ldots,n-1\}$ and, given this event, the event $Y_n = y_n$ is identical to the event $X_n = y_n - y_{n-1}$. In words, the Y_n will assume the given values if and only if the X_n assume the corresponding differences since the Y_n are defined as the sum of the X_n. Now, however, the probability is entirely in terms of the given X_i variables, in particular,

$$\begin{aligned}
&p_{Y_n|Y_{n-1},\ldots,Y_1}(y_n|y_{n-1},\ldots,y_1)\\
&\qquad = p_{X_n|X_{n-1},\ldots,X_2,X_1}(y_n - y_{n-1}|y_{n-1} - y_{n-2},\ldots,y_2 - y_1, y_1). \qquad (3.140)
\end{aligned}$$

So far the development is valid for any process and has not used the fact that the $\{X_n\}$ are iid. If the $\{X_n\}$ are iid, then the conditional pmf's are simply the marginal pmf's since each X_n is independent of past X_k; $k < n$! Thus we have that

$$p_{Y_n|Y_{n-1},\ldots,Y_1}(y_n|y_{n-1},\ldots,y_1) = p_X(y_n - y_{n-1}). \qquad (3.141)$$

and hence from the chain rule the vector pmf is (defining $y_0 = 0$)

$$p_{Y_1,\ldots,Y_n}(y_1,\ldots,y_n) = \prod_{i=1}^{n} p_X(y_i - y_{i-1}), \tag{3.142}$$

providing the desired specification.

To apply this formula to the special case of the binomial counting process, we need only plug in the binary pmf for p_X to obtain the desired specification of the binomial counting process:

$$p_{Y_1,\ldots,Y_n}(y_1,\ldots,y_n) = \prod_{i=1}^{n} p^{(y_i - y_{i-1})}(1-p)^{1-(y_i - y_{i-1})},$$

where

$$y_i - y_{i-1} = 0 \text{ or } 1, i = 1, 2, \ldots, n;\ \ y_0 = 0. \tag{3.143}$$

A similar derivation could be used to evaluate the conditional pmf for Y_n given only its immediate predecessor as:

$$\begin{aligned} p_{Y_n|Y_{n-1}}(y_n|y_{n-1}) &= \Pr(Y_n = y_n | Y_{n-1} = y_{n-1}) \\ &= \Pr(X_n = y_n - y_{n-1} | Y_{n-1} = y_{n-1}). \end{aligned}$$

The conditioning event, however, depends only on values of X_k for $k < n$, and X_n is independent of its past; hence

$$p_{Y_n|Y_{n-1}}(y_n|y_{n-1}) = p_X(y_n - y_{n-1}). \tag{3.144}$$

The same conclusion can be reached by the longer route of using the joint pmf for $Y_1, \ldots, Y_n$ previously computed to find the joint pmf for Y_n and Y_{n-1}, which in turn can be used to find the conditional pmf. Comparison with (3.141) reveals that processes formed by summing iid processes (such as the binomial counting process) have the property that

$$p_{Y_n|Y_{n-1},\ldots,Y_1}(y_n|y_{n-1},\ldots,y_1) = p_{Y_n|Y_{n-1}}(y_n|y_{n-1}) \tag{3.145}$$

or, equivalently,

$$\Pr(Y_n = y_n | Y_i = y_i\,;\ i = 1, \ldots, n-1) = \Pr(Y_n = y_n | Y_{n-1} = y_{n-1}), \tag{3.146}$$

that is, they are Markov processes. Roughly speaking, given the most recent past sample (or the current sample), the remainder of the past does not affect the probability of what happens next. Alternatively stated, given the present, the future is independent of the past.

3.16.3 ★Discrete random walk

As a second example of the preceding development, consider the random walk defined as in (3.136), i.e., by

$$Y_n = \begin{cases} 0 & n = 0 \\ \sum_{k=1}^{n} X_k & n = 1, 2, \ldots \end{cases}, \tag{3.147}$$

where the iid process used has alphabet $\{1, -1\}$ and $\Pr(X_n = -1) = p$. This is another example of an autoregressive process since it can be written in the form of a regression

$$Y_n = Y_{n-1} + X_n,\ n = 1, 2, \ldots \tag{3.148}$$

One can think of Y_n as modeling a drunk on a path who flips a coin at each minute to decide whether to take one step forward or one step backward. In this case the transform of the iid random variables is

$$M_X(ju) = (1-p)e^{ju} + pe^{-ju},$$

and hence using the binomial theorem of algebra we have that

$$\begin{aligned} M_{Y_n}(ju) &= ((1-p)e^{ju} + pe^{-ju})^n \\ &= \sum_{k=0}^{n} \left[\binom{n}{k} (1-p)^{n-k} p^k \right] e^{ju(n-2k)} \\ &= \sum_{k=-n,\ -n+2, \ldots, n-2, n} \left[\binom{n}{(n-k)/2} (1-p)^{(n+k)/2} p^{(n-k)/2} \right] e^{juk}. \end{aligned}$$

Comparison of this formula with the definition of the characteristic function reveals that the pmf for Y_n is given by

$$p_{Y_n}(k) = \binom{n}{(n-k)/2} (1-p)^{(n+k)/2} p^{(n-k)/2},$$

$$k = -n, -n+2, \ldots, n-2, n.$$

Note that Y_n must be even or odd depending on whether n is even or odd. This follows from the nature of the increments.

3.16.4 The discrete time Wiener process

Again consider a process formed by summing an iid process as in (3.136). This time, however, let $\{X_n\}$ be an iid process with zero-mean Gaussian marginal pdf's and variance σ^2. Then the process $\{Y_n\}$ defined by (3.136) is called the *discrete time Wiener process.* The discrete time, continuous alphabet case of summing iid random variables is handled in virtually the same manner as the discrete time, discrete alphabet case, with conditional pdfs replacing conditional pmfs.

The marginal pdf for Y_n is given immediately by (3.122) as $N(0, n\sigma_X^2)$.

To find the joint pdf's we evaluate the pdf chain rule of (3.62):

$$f_{Y_1,\ldots,Y_n}(y_1,\ldots,y_n) = \prod_{l=1}^{n} f_{Y_l|Y_1,\ldots,Y_{l-1}}(y_l|y_1,\ldots,y_{l-1}). \tag{3.149}$$

To find the conditional pdf $f_{Y_n|Y_1,\ldots,Y_{n-1}}(y_n|y_1,\ldots,y_{n-1})$ we compute the conditional cdf $P(Y_n \le y_n|Y_{n-i} = y_{n-i}; i = 1, 2, \ldots, n-1)$. Analogous to the discrete case, from the representation of (3.136) and the fact that the X_n are iid we have that

$$\begin{aligned}
&P(Y_n \le y_n|Y_{n-i} = y_{n-i}; i = 1, 2, \ldots, n-1) \\
&= P(X_n \le y_n - y_{n-1}|Y_{n-i} = y_{n-i}; i = 1, 2, \ldots, n-1) \\
&= P(X_n \le y_n - y_{n-1}) = F_X(y_n - y_{n-1}),
\end{aligned} \tag{3.150}$$

and hence differentiating the conditional cdf to obtain the conditional pdf yields

$$f_{Y_n|Y_1,\ldots,Y_{n-1}}(y_n|y_1,\ldots,y_{n-1}) = \frac{\partial}{\partial y_n} F_X(y_n - y_{n-1}) = f_X(y_n - y_{n-1}), \tag{3.151}$$

the continuous analog of (3.141). Application of the pdf chain rule then yields the continuous analog to (3.142):

$$f_{Y_1,\ldots,Y_n}(y_1,\ldots,y_n) = \prod_{i=1}^{n} f_X(y_i - y_{i-1}). \tag{3.152}$$

Finally suppose that f_X is Gaussian with zero mean and variance σ^2. Then this becomes

$$\begin{aligned}
f_{Y^n}(y^n) &= \frac{\exp\left(-\frac{y_1^2}{2\sigma^2}\right)}{\sqrt{2\pi\sigma^2}} \prod_{i=2}^{n} \frac{\exp\left(-\frac{(y_i - y_{i-1})^2}{2\sigma^2}\right)}{\sqrt{2\pi\sigma^2}} \\
&= (2\pi\sigma^2)^{-n/2} \exp\left(-\frac{1}{2\sigma^2}\left(\sum_{i=2}^{n}(y_i - y_{i-1})^2 + y_1^2\right)\right).
\end{aligned} \tag{3.153}$$

This proves to be a Gaussian pdf with mean vector 0 and a covariance matrix with entries $K_X(m,n) = \sigma^2 \min(m,n)$, $m, n = 1, 2, \ldots$ (Readers are invited to test their matrix manipulation skills and verify this claim.)

As in the discrete alphabet case, a similar argument implies that

$$f_{Y_n|Y_{n-1}}(y_n|y_{n-1}) = f_X(y_n - y_{n-1})$$

and hence from (3.151) that

$$f_{Y_n|Y_1,\ldots,Y_{n-1}}(y_n|y_1,\ldots,y_{n-1}) = f_{Y_n|Y_{n-1}}(y_n|y_{n-1}). \tag{3.154}$$

As in the discrete alphabet case, a process with this property is called a *Markov process*. We can combine the discrete alphabet and continuous alphabet definitions into a common definition: a discrete time random process $\{Y_n\}$ is said to be a *Markov process* if the conditional cdf's satisfy the relation

$$\Pr(Y_n \le y_n|Y_{n-i} = y_{n-i};\ i = 1,2,\ldots) = \Pr(Y_n \le y_n|Y_{n-1} = y_{n-1}) \tag{3.155}$$

for all $y_{n-1}, y_{n-2}, \ldots$ More specifically, $\{Y_n\}$ is frequently called a *first-order* Markov process because it depends on only the most recent past value. An extended definition to nth-order Markov processes can be made in the obvious fashion.

3.16.5 Hidden Markov models

A popular random process model that has proved extremely important in the development of modern speech recognition is formed by adding an iid process to a Markov process, so that the underlying Markov process is "hidden." More generally, instead of adding an iid process one can require that the observed process is conditionally independent given the underlying Markov process. Suppose for example that $\{X_n\}$ is a Markov process with either discrete or continuous alphabet and that $\{W_n\}$ is an iid process, for example an iid Gaussian process. Then the resulting process $Y_n = X_n + W_n$ is an example of a hidden Markov model or, in the language of early information theory, a Markov source. A wide literature exists for estimating the parameters of the underlying Markov source when only the sum process Y_n is actually observed. A conditionally Gaussian hidden Markov model can be equivalently considered as viewing a Markov process through a noisy channel with iid Gaussian noise. Perhaps surprisingly, a hidden Markov source is not itself Markov. It is an example of a "conditionally independent" process since if the sequence of underlying states is known, the observed process is independent.

3.17 ⋆Nonelementary conditional probability

Perhaps the most important form for conditional probabilities is the basic form of $\Pr(Y \in F|X = x)$, a probability measure on a random variable Y given the event that another random variable X takes on a specific value x. We consider a general event $Y \in F$ and not simply $Y = y$ since the latter is usually useless in the continuous case. In general, either or both Y or X might be random vectors.

In the elementary discrete case, such conditional probabilities are easily constructed in terms of conditional pmf's using (3.47): conditional probability is found by summing conditional probability mass over the event, just as is done in the unconditional case. We proposed an analogous approach to continuous probability, but this does not lead to a useful general theory. For example, it assumes that the various pdf's all exist and are well behaved. As a first step towards a better general definition (which will reduce in practice to the constructive pdf definition when it makes sense), we derive a variation of (3.47). Multiply both sides of (3.47) by $p_X(x)$ and sum over an X-event G to obtain

$$\begin{aligned}\sum_{x \in G} P(Y \in F|X = x)p_X(x) &= \sum_{x \in G}\sum_{y \in F} p_{Y|X}(y|x)p_X(x) \\ &= \sum_{x \in G}\sum_{y \in F} p_{X,Y}(x,y) \\ &= P(X \in G, Y \in F) \\ &= P_{X,Y}(G \times F); \ \forall G. \qquad (3.156)\end{aligned}$$

This formula describes the essence of the conditional probability by saying what it does: for any X event G, summing the product of the conditional probability that $Y \in F$ and the marginal probability that $X = x$ over all $x \in G$ yields the joint probability that $X \in G$ and $Y \in F$. If our tentative definition of nonelementary conditional probability is to be useful, it must play a similar role in the continuous case, that is, we should be able to average over conditional probabilities to find ordinary joint probabilities, where now averages are integrals instead of sums. This indeed works since

$$\begin{aligned}\int_{x \in G} dx P(Y \in F|X = x)f_X(x) &= \int_{x \in G} dx \int_{y \in F} dy f_{Y|X}(y|x)f_X(x) \\ &= \int_{x \in G} dx \int_{y \in F} dy f_{X,Y}(x,y \\ &= P(X \in G, Y \in F) \\ &= P_{X,Y}(G \times F); \quad \text{all events } G. \quad (3.157)\end{aligned}$$

Thus the tentative definition of nonelementary conditional probability of (3.53) behaves in the manner that one would like. Using the Stieltjes notation we can combine (3.156) and (3.157) into a single requirement:

$$\int_G P(Y \in F|X = x)\, dF_X(x) = P(X \in G, Y \in F) \\ = P_{X,Y}(G \times F); \forall G, \quad (3.158)$$

which is valid in both the discrete case and in the continuous case when one has a conditional pdf. In advanced probability, (3.158) is taken as the *definition* for the general (nonelementary) conditional probability $P(Y \in F|X = x)$; that is, the conditional probability is defined as any function of x that satisfies (3.158). This is a *descriptive* definition which defines an object by its behavior when integrated, much like the rigorous definition of a Dirac delta function is by its behavior inside an integral. This reduces to the given constructive definitions of (3.47) in the discrete case and (3.53) in the continuous case with a well-behaved pdf. It also leads to a useful general theory even when the conditional pdf is not well defined.

Lastly, we observe that elementary and nonelementary conditional probabilities are related in the natural way. Suppose that G is an event with nonzero probability so that the elementary conditional probability $P(Y \in F|X \in G)$ is well defined. Then

$$P(Y \in F|X \in G) = \frac{P_{X,Y}(G \times F)}{P_X(G)} \\ = \frac{1}{P_X(G)} \int P(Y \in F|X = x)\, dF_X(x).$$

3.18 Problems

1. Given the probability space $(\Re, \mathcal{B}(\Re), m)$, where m is the probability measure induced by the uniform pdf f on $[0, 1]$ (that is, $f(r) = 1$ for $r \in [0, 1]$ and is 0 otherwise), find the pdf's for the following random variables defined on this space:

   (a) $X(r) = r^2$,
   (b) $Y(r) = |r|^{1/2}$,
   (c) $Z(r) = \ln |r|$,
   (d) $V(r) = ar + b$, where a and b are fixed constants.
   (e) Find the pmf for the random variable $W(r) = 3$ if $r \geq 2$ and $W(r) = 1$ otherwise.

2. Do Problem 3.1 for an exponential pdf on the original sample space.

3. Do Problem 3.1(a)–(d) for a Gaussian pdf on the original sample space.
4. A random variable X has a uniform pdf on $[0, 1]$. What is the probability density function for the volume of a cube with sides of length X?
5. A random variable X has a cumulative distribution function $F_X(\alpha)$. What is the cdf of the random variable $Y = aX + b$, where a and b are constants?
6. Use the properties of probability measures to prove the following facts about cdf's: if F is the cdf of a random variable, then
   (a) $F(-\infty) = 0$ and $F(\infty) = 1$.
   (b) $F(r)$ is a monotonically nondecreasing function, that is, if $x \geq y$, then $F(x) \geq F(y)$.
   (c) F is continuous from the right, that is, if ϵ_n, $n = 1, 2, ...$ is a sequence of positive numbers decreasing to zero, then

$$\lim_{n - \infty} F(r + \epsilon_n) = F(r).$$

   Note that continuity from the right is a result of the fact that we defined a cdf as the probability of an event of the form $(-\infty, r]$. If instead we had defined it as the probability of an event of the form $(-\infty, r)$ (as is often done in Eastern Europe), then cdf's would be continuous from the left instead of from the right. When is a cdf continuous from the left? When is it discontinuous?
7. Say we are given an arbitrary cdf F for a random variable and we would like to simulate an experiment by generating one of these random variables as input to the experiment. As is typical of computer simulations, all we have available is a uniformly distributed random variable U; that is, U has the pdf of 3.1. This problem explores a means of generating the desired random variable from U (this method is occasionally used in computer simulations). Given the cdf F, define the *inverse* cdf $F^{-1}(r)$ as the smallest value of $x \in \Re$ for which $F(x) \geq r$. We specify "smallest" to ensure a unique definition since F may have the same value for an interval of x. Find the cdf of the random variable Y defined by $Y = F^{-1}(U)$.
   Suppose next that X is a random variable with cdf $F_X(\alpha)$. What is the distribution of the random variable $Y = F_X(X)$? This mapping is used on individual picture elements (pixels) in an image enhancement technique known as "histogram equalization" to enhance contrast.
8. You are given a random variable U described by a pdf that is 1 on $[0, 1]$. Describe and make a labeled sketch of a function g such that the random variable $Y = g(U)$ has a pdf $\lambda e^{-\lambda x}$; $x \geq 0$.
9. A probability space $(\Omega, \mathcal{F}, P)$ models the outcome of rolling two fair four-sided dice on a glass table and reading their down faces. Hence we can take $\Omega = \{1, 2, 3, 4\}^2$, the usual event space (the power set or, equivalently, the Borel field), and a pmf placing equal probability on all 16 points in the space. On this space we define the following random variables: $W(\omega)$ = the down face on die no. 1; that is, if $\omega = (\omega_1, \omega_2)$, where ω_i denotes the down face on die no. i, then $W(\omega) = \omega_1$.

(We could use the sampling function notation here: $W = \prod_1 \cdot$.) Similarly, define $V(\omega) = \omega_2$, the down face on the second die. Define also $X(\omega) = \omega_1 + \omega_2$, the sum of the down faces, and $Y(\omega) = \omega_2\omega_2$, the product of the down faces. Find the pmf and cdf for the random variables X, Y, W, and V. Find the pmf's for the random vectors (X, Y) and (W, V). Write a formula for the distribution of the random vector (W, V) in terms of its pmf.

Suppose that a greedy scientist has rigged the dice using magnets to ensure that the two dice always yield the same value; that is, we now have a new pmf on Ω that assigns equal values to all points where the faces are the same and zero to the remaining points. Repeat the calculations for this case.

10. A random vector (X, Y) has a pmf $p_{X.Y}(k, j) = P(X = k \text{ and } Y = j)$ of $p_{X.Y}(k, j) = 1/8$ if $(k, j) =$ (0,3), (0,1), (2,1), (2,2), (4,0), (4,1), (4,2), or (4,3) and is 0 otherwise.
    (a) Find the pmf $p_Y(y)$ and the conditional pmf $p_{X|Y}(x \mid y)$. Are X and Y independent?
    (b) Find and sketch the pmf for the random variable $R = \min(X, Y)$. (The sketch must be labeled.)
11. Jeff draws cards from a fair deck of 52 cards *without* replacing a card after he draws it. Let the random variable X be the number of cards he draws until (and including) the first draw of a heart.
    Evaluate the following probabilities: $P(X = 1)$. $P(X = 2)$, and $P(X = 52)$. More generally, find the pmf for $p_X(k) = P(X = k)$, $k = 1 \ldots 52$.
12. A biased 4 sided die is rolled and the down face is a random variable N described by the following pmf:

$$p_N(n) = \begin{cases} n/10 & n = 1, 2, 3, 4 \\ 0 & \text{otherwise} \end{cases}.$$

    Given the random variable N a biased coin with bias $(N + 1)/2N$ is flipped and the random variable X is 1 or zero according to whether the coin shows heads or tails, i.e., the conditional pmf is

$$p_{X|N}(x|n) = \left(\frac{n+1}{2n}\right)^x \left(1 - \frac{n+1}{2n}\right)^{1-x}; \; x = 0, 1.$$

    (a) Find the expectation $E(N)$ and variance σ_N^2 of N.
    (b) Find the conditional pmf $p_{N|X}(n|x)$.
    (c) Find the conditional expectation $E(N|X = 1)$, i.e., the expectation with respect to the conditional pmf $p_{N|X}(n|1)$.
    (d) Find the conditional variance of N given $X = 1$.
    (e) Define the event F as the event that the down face of the die is 1 or 4. Are the events F and $\{X = 1\}$ independent?
13. In a certain region of west Texas, there are an average of ten armadillos per square mile. Denote by N the number of armadillos in an area A. Assume that

N is described by a Poisson pmf

$$p_N(n) = \frac{e^{-\lambda A}(\lambda A)^n}{n!}, \quad n = 0, 1, 2, \ldots$$

for some constant λ. How large a circular region should be selected to ensure that the probability of finding at least one armadillo is at least 0.95?

14. Suppose that X is a binary random variable with outputs $\{a, b\}$ with a pmf $p_X(a) = p$ and $p_X(b) = 1 - p$ and Y is a random variable described by the conditional pdf

$$f_{Y|X}(y|x) = \frac{e^{-(y-x)^2/2\sigma_W^2}}{\sqrt{2\pi\sigma_W^2}}.$$

Describe the MAP detector for X given Y and find an expression for the probability of error in terms of the Q function.
Suppose that $p = 0.5$, but you are free to choose a and b subject only to the constraint that $(a^2 + b^2)/2 = E_b$. Which is a better choice, $a = -b$ or a nonzero with $b = 0$? What can you say about the minimum achievable P_e?

15. The famous ubiquitous operating system *defenestration* is run on 2×10^8 computers. For each of the mutually independent computers, the probability mass function for X, the number of operating system crashes in a day, is given by

$$p_X(k) = \frac{4-k}{10}; \; k = 0, 1, 2, 3.$$

On a day when for a given computer the operating system crashes $X = k$ times, the user has a probability of $1 - 2^{-k}$ of reinstalling the operating system.
(a) Find the mean EX and variance σ_X^2 of X.
(b) Find the mean and variance of Y, the total number of operating system crashes among all of the computers on a given day.
(c) Find the probability that a particular computer has its operating system reinstalled on a given day.
(d) Find the conditional probability for the number of crashes for a particular computer given that the operating system was reinstalled on that computer on that day.
(e) In a given group of 10 computers, what is the probability that exactly three of them had their operating systems reinstalled on a particular day?

16. Two random variables X and Y have uniform probability density functions on $(0, 1)$ and they are independent. Find the probability density function $f_W(w)$ for the random variable $W = (X - Y)^2$ and find the mean of W, $E(W)$.

17. John and Mark are going to play a game. John will draw a number X using an exponential distribution with parameter λ; that is

$$f_X(x) = \lambda e^{-\lambda x}, \quad x \geq 0$$

At the same time, Mark will independently draw a number Y using a Poisson distribution with parameter λ, that is

$$p_Y(y) = e^{-\lambda}\frac{\lambda^y}{y!}, \quad y = 0, 1, 2, \ldots$$

If John's number is larger than Mark's, they draw again. Otherwise, the game stops.

(a) Evaluate the probability $P(X > Y)$.

(b) What is the expected number of draws until the game stops?

18. Consider the two-dimensional probability space $(\Re^2, \mathcal{B}(\Re)^2, P)$, where P is the probability measure induced by the pdf g, which is equal to a constant c in the square $\{(x,y) : x \in [-1/2, 1/2],\ y \in [-1/2, 1/2]\}$ and zero elsewhere.
   (a) Find the constant c.
   (b) Find $P(\{x, y : x < y\})$.
   (c) Define the random variable $U : \Re^2 \to \Re$ by $U(x,y) = x + y$. Find an expression for the cdf $F_U(u) = \Pr(U \leq u)$.
   (d) Define the random variable $V : \Re^2 \to \Re$ by $V(x,y) = xy$. Find the cdf $F_V(\nu)$.
   (e) Define the random variable $W : \Re^2 \to \Re$ by $W(x,y) = \max(x,y)$, that is, the larger of the two coordinate values. Thus $\max(x,y) = x$ if $x \geq y$. Find the cdf $F_W(w)$.
19. Suppose that X and Y are two random variables described by a pdf

$$f_{X,Y}(x,y) = Ce^{-x^2 - y^2 + xy}.$$

   (a) Find C.
   (b) Find the marginal pdf's f_X and f_Y. Are X and Y independent? Are they identically distributed?
   (c) Define the random variable $Z = X - 2Y$. Find the joint pdf $f_{X,Z}$.
20. Let (X, Y) be a random vector with distribution $P_{X,Y}$ induced by the pdf $f_{X,Y}(x,y) = f_X(x) f_Y(y)$, where

$$f_X(x) = f_Y(x) = \lambda e^{-\lambda x}; \ x \geq 0,$$

that is, (X, Y) is described by a product pdf with exponential components.
   (a) Find the pdf for the random variable $U = X + Y$.
   (b) Let the "max" function be defined as in Problem 3.18 and define the "min" function as the smaller of two values; that is, $\min(x,y) = x$ if $x \leq y$. Define the random vector (W, V) by $W = \min(X, Y)$ and $V = \max(X, Y)$. Find the pdf for the random vector (W, V).
21. Let (X, Y) be a random vector with distribution $P_{X,Y}$ induced by a product pdf $f_{X,Y}(x,y) = f_X(x) f_Y(y)$ with $f_X(x) = f_Y(y)$ equal to the Gaussian pdf with $m = 0$. Consider the random vector as representing the real and imaginary parts of a complex-valued measurement. It is often useful to consider instead a magnitude–phase representation vector (R, θ), where R is the magnitude $(X^2 + Y^2)^{1/2}$ and $\theta = \tan^{-1}(Y/X)$ (use the principal value of the inverse tangent). Find the joint pdf of the random vector (R, θ). Find the marginal pdf's of

the random variables R and θ. The pdf of R is called the Rayleigh pdf. Are R and θ independent?

22. A probability space $(\Omega, \mathcal{F}, P)$ is defined as follows: Ω consists of all eight-dimensional binary vectors, e.g., every member of Ω has the form $\omega = (\omega_0, \ldots, \omega_{k-1})$, where ω_i is 0 or 1. $\mathcal{F}$ is the power set, P is described by a pmf which assigns a probability of $1/2^8$ to each of the 2^8 elements in Ω (a uniform pmf).
Find the pmf's describing the following random variables:
(a) $g(\omega) = \sum_{i=0}^{k-1} \omega_i$, i.e., the number of 1s in the binary vector.
(b) $X(\omega) = 1$ if there are an even number of 1s in ω and 0 otherwise.
(c) $Y(\omega) = \omega_j$, i.e., the value of the jth coordinate of ω.
(d) $Z(\omega) = \max_i(\omega_i)$.
(e) $V(\omega) = g(\omega)X(\omega)$, where g and X are as above.

23. Suppose that $(X_0, X_1, \ldots, X_N)$ is an iid random vector with marginal pdf's

$$f_{X_n}(\alpha) = \begin{cases} 1 & 0 \le \alpha < 1 \\ 0 & \text{otherwise} \end{cases}.$$

Define the following random variables:
- $U = X_0^2$
- $V = \max(X_1, X_2, X_3, X_4)$
- $W = \begin{cases} 1 & \text{if } X_1 \ge 2X_2 \\ 0 & \text{otherwise} \end{cases}$
- $Y_n = X_n + X_{n-1}$; $n = 1, \ldots, N$.

(a) Find the pdf or pmf as appropriate for U, V, and W.
(b) Find the cumulative distribution function (cdf) for Y_n.

24. Let f be the uniform pdf on $[0, 1]$. Let (X, Y) be a random vector described by a joint pdf

$$f_{X,Y}(x, y) = f(y)f(x - y); \text{ all } x, y.$$

(a) Find the marginal densities f_X and f_Y. Are X and Y independent?
(b) Find $P(X \ge 1/2 | Y \le 1/2)$.

25. In Example [3.24], find the pmf for the random variable X_n for a fixed n. Find the pmf for the random vector (X_n, X_k) for fixed n and k. Consider both the cases where $n = k$ and where $n \ne k$. Find the probability $\Pr(X_5 = X_{12})$.

26. Let X and Y be two random variables with joint pmf

$$p_{XY}(k, j) = C\frac{k}{j+1}; \; j = 1, \ldots, N; \; k = 1, 2, \ldots, j.$$

(a) Find C.
(b) Find $p_Y(j)$.
(c) Find $p_{X|Y}(k|j)$. Are X and Y independent?
(d) Find $E[1/Y]$.

27. In Example [3.27] of the random phase process, find $\Pr(X(t) \geq 1/2)$.
28. Evaluate the pmf $p_{Y(t)}(y)$ for the quantized process of Example [3.28] for each possible case. (Choose $b = 0$ if the process is nonnegative and $b = -a$ otherwise.)
29. Let $([0,1], \mathcal{B}([0,1]), P)$ be a probability space with pdf $f(\omega) = 1$; $\omega \in [0,1]$. Find a random vector $\{X_t;\ t \in \{1, 2, \ldots, n\}\}$ such that $\Pr(X_t = 1) = \Pr(X_t = 0) = 1/2$ and $\Pr(X_t = 1 \text{ and } X_{t-1} = 1) = 1/8$, for relevant t.
30. Give an example of two equivalent random variables (that is, two random variables having the same distribution) that
    (a) are defined on the same space but are not equal for any $\omega \in \Omega$,
    (b) are defined on different spaces and have different functional forms.
31. Let $(\Re, \mathcal{B}(\Re), m)$ be the probability space of Problem 3.1.
    (a) Define the random process $\{X(t);\ t \in [0, \infty)\}$ by

$$X(t, \omega) = \begin{cases} 1 & \text{if } 0 < t \leq \omega \\ 0 & \text{otherwise} \end{cases}.$$

    Find $\Pr(X(t) = 1)$ as a function of t.
    (b) Define the random process $\{X(t);\ t \in [0, \infty)\}$ by

$$X(t, \omega) = \begin{cases} t/\omega & \text{if } 0 < t \leq \omega \\ 0 & \text{otherwise} \end{cases}.$$

    Find $\Pr(X(t) > x)$ as a function of t for $x \in (0, 1)$.
32. Two continuous random variables X and Y are described by the pdf

$$f_{X,Y}(x, y) = \begin{cases} c & \text{if } |x| + |y| \leq r \\ 0 & \text{otherwise} \end{cases}$$

    where r is a fixed real constant and c is a constant. In other words, the pdf is uniform on a square whose side has length $\sqrt{2}\,r$.
    (a) Evaluate c in terms of r.
    (b) Find $f_X(x)$.
    (c) Are X and Y independent random variables? (Prove your answer.)
    (d) Define the random variable $Z = (|X| + |Y|)$. Find the pdf $f_Z(z)$.
33. Find the pdf of $X(t)$ in Example [3.23] as a function of time. Find the joint cdf of the vector $(X(1), X(2))$.
34. Richard III wishes to trade his kingdom for a horse. He knows that the probability that there are k horses within r feet of him is

$$CH^k \frac{r^{2k} e^{-Hr^2}}{k!}; \quad k = 0, 1, 2, \ldots,$$

    where $H > 0$ is a fixed parameter.
    (a) Let R denote a random variable giving the distance from Richard to the nearest horse. What is the probability density function $f_R(\alpha)$ for R? (C should be evaluated as part of this question.)

(b) Rumors of the imminent arrival of Henry Tudor have led Richard to lower his standards and consider alternative means of transportation. Suppose that the probability density function $f_S(\beta)$ for the distance S to the nearest mule is the same as f_R except that the parameter H is replaced by a parameter M. Assume that R and S are independent random variables. Find an expression for the cumulative distribution function (cdf) for W, the distance to the nearest quadruped (i.e., horse or mule).
Hint: if you did not complete or do not trust your answer to part (b), then find the answer in terms of the cdf's for R and S.

35. Suppose that a random vector $\mathbf{X} = (X_0, \ldots, X_{k-1})$ is iid with marginal pmf

$$p_{X_i}(l) = p_X(l) = \begin{cases} p & \text{if } l = 1 \\ 1 - p & \text{if } l = 0 \end{cases}$$

for all i.
(a) Find the pmf of the random variable $Y = \prod_{i=0}^{k-1} X_i$.
(b) Find the pmf of the random variable $W = X_0 + X_{k-1}$.
(c) Find the pmf of the random vector (Y, W).

36. Find the joint cdf of the complex components of $X_n(\omega)$ in Example [3.25] as a function of time.

37. Find the pdf of $X(t)$ in Example [3.27].

38. A certain communication system outputs a discrete time series $\{X_n\}$ where X_n has pmf $p_X(1) = p_X(-1) = 1/2$. Transmission noise in the form of a random process $\{Y_n\}$ is added to X_n to form a random process $\{Z_n = X_n + Y_n\}$. Y_n has a Gaussian distribution with $m = 0$ and $\sigma = 1$.
(a) Find the pdf of Z_n.
(b) A receiver forms a random process $\{R_n = \text{sgn}(Z_n\}$ where sgn is the sign function $\text{sgn}(x) = 1$ if $x \geq 0$, $\text{sgn}(x) = -1$ if $x < 0$. R_n is output from the receiver as the receiver's estimate of what was transmitted. Find the pmf of R_n and the probability of detection (i.e., $\Pr(R_n = X_n)$).
(c) Is this detector optimal?

39. If X is a Gaussian random variable, find the marginal pdf $f_{Y(t)}$ for the random process $Y(t)$ defined by

$$Y(t) = X \cos(2\pi f_0 t); \quad t \in \Re,$$

where f_0 is a known constant frequency.

40. Let X and Z be the random variables of Problems 3.1 through 3.3. For each assumption on the original density find the cdf for the random vector (X, Z), $F_{X,Z}(x, z)$. Does the appropriate derivative exist? Is it a valid pdf?

41. Let N be a random variable giving the number of molecules of hydrogen in a spherical region of radius r and volume $V = 4\pi r^3/3$. Assume that N is described

by a Poisson pmf

$$p_N(n) = \frac{e^{-\rho V}(\rho V)^n}{n!}, \quad n = 0, 1, 2, \ldots$$

where ρ can be viewed as a limiting density of molecules in space. Say we choose an arbitrary point in deep space as the center of our coordinate system. Define a random variable X as the distance from the origin of our coordinate center to the nearest molecule. Find the pdf of the random variable X, $f_X(x)$.

42. Let V be a random variable with a uniform pdf on $[0, a]$. Let W be a random variable, independent of V, with an exponential pdf with parameter λ, that is,

$$f_W(w) = \lambda e^{-\lambda w}; \quad w \in [0, \infty).$$

Let $p(t)$ be a pulse with value 1 when $0 \leq t \leq 1$ and 0 otherwise. Define the random process $\{X(t);\, t \in [0, \infty)\}$ by

$$X(t) = Vp(t - W),$$

(This is a model of a square pulse that occurs randomly in time with a random amplitude.) Find for a fixed time $t > 1$ the cdf $F_{X(t)}(\alpha) = \Pr(X(t) \leq \alpha)$. You must specify the values of the cdf for all possible real values α. Show that there exists a pmf p with a corresponding cdf F_1, a pdf f with a corresponding cdf F_1, a pdf f with a corresponding cdf F_2, and a number $\beta_t \in (0, 1)$ such that

$$F_{X(t)}(\alpha) = \beta_t F_1(\alpha) + (1 - \beta_t) F_2(\alpha).$$

Give expressions for p, f, and β_t.

43. Prove the following facts about characteristic functions:
   (a) $|M_X(ju)| \leq 1$
   (b) $M_X(0) = 1$
   (c) $|M_X(ju)| \leq M_X(0) = 1$
   (d) If a random variable X has a characteristic function $M_X(ju)$, if c is a fixed constant, and if a random variable Y is defined by $Y = X + c$, then

$$M_Y(ju) = e^{juc} M_X(ju).$$

44. Suppose that X is a random variable described by an exponential pdf

$$f_X(\alpha) = \lambda e^{-\lambda \alpha}; \; \alpha \geq 0.$$

($\lambda > 0$.) Define a function q which maps nonnegative real numbers into integers by $q(x) =$ the largest integer less than or equal to x. In other words

$$q(x) = k \text{ if } k \leq x < k + 1, \; k = 0, 1, \ldots$$

(This function is often denoted by $q(x) = \lfloor x \rfloor$.) The function q is a form of quantizer; it rounds its input *downward* to the nearest integer below the input.

Define the following two random variables: the quantizer output

$$Y = q(X)$$

and the quantizer error

$$\epsilon = X - q(X).$$

Note: by construction ϵ can only take on values in $[0, 1)$.

(a) Find the pmf $p_Y(k)$ for Y.

(b) Derive the probability density function for ϵ. (You may find the "divide and conquer" formula useful here, e.g., $P(G) = \sum_i P(G \cap F_i)$, where $\{F_i\}$ is a partition.)

45. Suppose that $(X_1, \ldots, X_N)$ is a random vector described by a product pdf with uniform marginal pdf's

$$f_{X_n}(\alpha) = \begin{cases} 1 & |\alpha| \le \frac{1}{2} \\ 0 & \text{otherwise} \end{cases}.$$

Define the following random variables:

- $U = X_3^2$
- $V = \min(X_1, X_2)$
- $W = n$ if n is the smallest integer for which $X_n \ge 1/4$ and $W = 0$ if there is no such n.

(a) Find pdf's or pmf's for U, V, and W.

(b) What is the joint pdf $f_{X_1, X_3, X_5}(\alpha, \beta, \gamma)$?

46. The joint probability density function of X and Y is

$$f_{X,Y}(\alpha, \beta) = C, \ \ |\alpha| \le 1, \ 0 \le \beta \le 1.$$

Define a new random variable

$$U = \frac{Y}{X^2}.$$

(U is taken to be 0 if $X = 0$.)

(a) Find the constant C and the marginal probability density functions $f_X(\alpha)$ and $f_Y(\beta)$.

(b) Find the probability density function $f_U(\gamma)$ for U.

(c) Suppose that U is quantized into $q(U)$ by defining

$$q(U) = i \text{ for } d_{i-1} \le U < d_i; \ i = 1, 2, 3,$$

where the interval $[d_0, d_3)$ equals the range of possible values of U. Find the quantization levels d_i, $i = 0, 1, 2, 3$ such that $q(U)$ has a uniform probability mass function.

(d) Find the expectations $E(X^3 U)$ and $E(q(U))$.

47. Let (X, Y) be a random vector described by a product pdf $f_{XY}(x, y) = f_X(x) f_Y(y)$. Let F_X and F_Y denote the corresponding marginal cdf's.
    (a) Prove

$$P(X > Y) = \int_{-\infty}^{\infty} F_Y(x) f_X(x)\, dx = 1 - \int_{-\infty}^{\infty} f_Y(x) F_X(x)\, dx.$$

    (b) Assume, in addition, that X and Y are identically distributed, i.e., have the same pdf. Based on the result of (a) calculate the probability $P(X > Y)$. (*Hint:* you should be able to derive or check your answer based on symmetry.)

48. You have two coins and a spinning pointer U. The coins are fair and unbiased, and the pointer U has a uniform distribution over $[0, 1)$. You flip both coins and spin the pointer. A random variable X is defined as follows.
    If the first coin is "heads," then:

$$X = \begin{cases} 1 & \text{if the 2nd coin is "heads"} \\ 0 & \text{otherwise} \end{cases}.$$

    If the first coin is "tails," then $X = U + 2$.
    Define another random variable:

$$Y = \begin{cases} 2U & \text{if the 1st coin is "heads"} \\ 2U + 1 & \text{otherwise} \end{cases}.$$

    (a) Find $F_X(x)$.
    (b) Find $\Pr(\frac{1}{2} \leq X \leq 2)$.
    (c) Sketch the pdf of Y and label important values.
    (d) Design an optimal detection rule to estimate U if you are given only Y. What is the probability of error?
    (e) State how to, or explain why it is not possible to:
        i. Generate a binary random variable Z, $p_Z(1) = p$, given U.
        ii. Generate a continuous, uniformly distributed random variable given Z.

49. The random vector $W = (W_0, W_1, W_2)$ is described by the pdf $f_W(x, y, z) = C|z|$, for $x^2 + y^2 \leq 1, |z| \leq 1$.
    (a) Find C.
    (b) Determine whether the following variables are independent and justify your position:
    - W_0 and W_1
    - W_0 and W_2
    - W_1 and W_2
    - W_0 and W_1 and W_2

    (c) Find $\Pr(W_2 > \frac{1}{3})$.
    (d) Find $F_{W_0,W_2}(0, 0)$.
    (e) Find the cdf of the vector W.
    (f) Let $V = \Pi_{i=0}^{2} W_i$. Find $\Pr(V \geq 0)$.

(g) Find the pdf of M, where $M = \min(W_1^2 + W_2^2, W_3^2)$.

50. Suppose that X and Y are random variables and that the joint pmf is

$$p_{X,Y}(k,j) = c2^{-k}2^{(j-k)};\ k = 0,1,2,\ldots;\ j = k, k+1,\ldots.$$

(a) Find c.
(b) Find the pmf's $p_X(j)$ and $p_Y(j)$.
(c) Find the conditional pmf's $p_{X|Y}(k|j)$ and $p_{Y|X}(j|k)$.
(d) Find the probability that $Y \geq 2X$.

51. Suppose that $\mathbf{X} = (X_0, X_1, \ldots, X_{k-1})$ is a random vector (k is some large number) with joint pdf

$$f_{\mathbf{X}}(\mathbf{x}) = \begin{cases} 1 & \text{if } 0 \leq x_i \leq 1;\ i = 0,\ldots,k-1 \\ 0 & \text{otherwise} \end{cases}.$$

Define the random variables $V = X_0 + X_{10}$ and $W = \max(X_0, X_{10})$.
Define the random vector $\mathbf{Y}$:

$$Y_n = 2^n X_n;\ n = 0,\ldots,k-1.$$

(a) Find the joint pdf $f_{V,W}(v,w)$.
(b) Find the probabilities $\Pr(W \leq 1/2)$, $\Pr(V \leq 1/2)$, and $\Pr(W \leq 1/2 \text{ and } V \leq 1/2)$.
(c) Are W and V independent?
(d) Find the (joint) pdf for $\mathbf{Y}$.

52. The random process described in Example [3.26] is an example of a class of processes that was somewhat of a fad in scientific circles during the 1980s, it is *chaotic*. (See, e.g., *Chaos* by James Gleick [28].) Suppose as in Example [3.26] $X_0(\omega) = \omega$ is chosen at random according to a uniform distribution on $[0,1)$, that is, the pdf is

$$f_{X_0}(\alpha) = \begin{cases} 1 & \text{if } \alpha \in [0,1) \\ 0 & \text{otherwise} \end{cases}.$$

As in the example, the remainder of the process is defined recursively by

$$X_n(\omega) = 2X_{n-1}(\omega) \bmod 1,\ n = 1,2,\ldots$$

Note that if the initial value X_0 is known, the remainder of the process is also known.
Find a nonrecursive expression for $X_n(\omega)$, that is, write $X_n(\omega)$ directly as a function of ω, e.g., $X_n(\omega) = g(\omega) \bmod 1$.
Find the pdf $f_{X_1}(\alpha)$ and $f_{X_n}(\alpha)$.
Hint: after you have found f_{X_1}, try induction.

53. Another random process which resembles that of the previous process but which

is not chaotic is to define X_0 in the same way, but define X_n by

$$X_n(\omega) = (X_{n-1}(\omega) + X_0(\omega)) \bmod 1.$$

Here X_1 is equivalent to that of the previous problem, but the subsequent X_n are different. As in the previous problem, find a direct formula for X_n in terms of ω (e.g., $X_n(\omega) = h(\omega) \bmod 1$) and find the pdf $f_{X_n}(\alpha)$.

54. The Mongol general Subudai is expecting reinforcements from Chenggis Kahn before attacking King Bela of Hungary. The probability mass function describing the number N of *tumens* (units of 10,000 men) that he will receive is

$$p_N(k) = cp^k;\ k = 0, 1, \ldots$$

If he receives $N = k$ tumens, then his probability of losing the battle will be 2^{-k}. This can be described by defining the random variable W which will be 1 if the battle is won, 0 if the battle is lost, and defining the conditional probability mass function

$$p_{W|N}(m|k) = \Pr(W = m|N = k) = \begin{cases} 2^{-k} & m = 0 \\ 1 - 2^{-k} & m = 1 \end{cases}.$$

(a) Find c.
(b) Find the (unconditional) pmf $p_W(m)$, that is, what is the probability that Subudai will win or lose?
(c) Suppose that Subudai is informed that definitely $N < 10$. What is the new (conditional) pmf for N? (That is, find $\Pr(N = k|N < 10)$.)

55. Suppose that $\{X_n;\ n = 0, 1, 2, \ldots\}$ is a binary Bernoulli process, that is, an iid process with marginal pmf's

$$p_{X_n}(k) = \begin{cases} p & \text{if } k = 1 \\ 1 - p & \text{if } k = 0 \end{cases}$$

for all n. Suppose that $\{W_n;\ n = 0, 1, \ldots\}$ is another binary Bernoulli process with parameter ϵ, that is,

$$p_{W_n}(k) = \begin{cases} \epsilon & \text{if } k = 1 \\ 1 - \epsilon & \text{if } k = 0 \end{cases}.$$

We assume that the two random processes are completely independent of each other (that is, any collection of samples of X_n is independent from any collection of W_n). We form a new random process $\{Y_n;\ n = 0, 1, \ldots\}$ by defining

$$Y_n = X_n \oplus W_n,$$

where the $\oplus$ operation denotes mod 2 addition. This setup can be thought of as taking an input digital signal X_n and sending it across a binary channel to a receiver. The binary channel can cause an error between the input X_n and output

Y_n with probability ϵ. Such a communication channel is called an *additive noise* channel because the output is the input plus an independent noise process (where "plus" here means mod 2).

(a) Find the output marginal pmf $p_{Y_n}(k)$.
(b) Is $\{Y_n\}$ Bernoulli? That is, is it an iid process?
(c) Find the conditional pmf $p_{Y_n|X_n}(j|k)$.
(d) Find the conditional pmf $p_{X_n|Y_n}(k|j)$.
(e) Find an expression for the probability of error $\Pr(Y_n \neq X_n)$.
(f) Suppose that the receiver is allowed to think about what the best guess for X_n is, given it receives a value Y_n. In other words, if you are told that $Y_n = j$, you can form an estimate or guess of the input X_n by some function of j, say $\hat{X}(j)$. Given this estimate your new probability of error is given by

$$P_e = \Pr(\hat{X}(Y_n) \neq X_n).$$

What decision rule $\hat{X}(j)$ yields the smallest possible P_e? What is the resulting P_e?

56. Suppose that we have a pair of random variables (X, Y) with a mixed discrete and continuous distribution. Y is a binary $\{0, 1\}$ random variable described by a pmf $p_Y(1) = 0.5$. Conditioned on $Y = y$, X is continuous with a Gaussian distribution with variance σ^2 and mean y, that is,

$$f_{X|Y}(x|y)(x|y) = \frac{\exp\left(-(x-y)^2/2\sigma^2\right)}{\sqrt{2\pi\sigma^2}}; \; x \in \Re; \; y = 0, 1 \, .$$

This can be thought of as the result of communicating a binary symbol (a "bit") over a noisy channel, which adds Gaussian noise with zero mean and variance σ^2 to the bit. In other words, $X = Y + W$, where W is a Gaussian random variable, independent of Y. What is the optimum (minimum error probability) decision for Y given the observation X? Write an expression for the resulting error probability.

57. Find the multidimensional Gaussian characteristic function of Equation (3.124) by completing the square in the exponent of the defining multidimensional integral.

4
Expectation and averages

4.1 Averages

In engineering practice we are often interested in the average behavior of measurements on random processes. The goal of this chapter is to link the two distinct types of averages that are used – long-term time averages taken by calculations on an actual physical realization of a random process and averages calculated theoretically by probabilistic calculus at some given instant of time, averages that are called *expectations*. As we shall see, both computations often (but by no means always) give the same answer. Such results are called *laws of large numbers* or *ergodic theorems*.

At first glance from a conceptual point of view, it seems unlikely that long-term *time* averages and instantaneous probabilistic averages would be the same. If we take a long-term time average of a particular realization of the random process, say $\{X(t, \omega_0);\ t \in \mathcal{T}\}$, we are averaging for a *particular* ω which we cannot know or choose; we do not use probability in any way and we are ignoring what happens with other values of ω. Here the averages are computed by summing the sequence or integrating the waveform over t while ω_0 stays fixed. If, on the other hand, we take an instantaneous probabilistic average, say at the time t_0, we are taking a probabilistic average and summing or integrating over ω for the random variable $X(t_0, \omega)$. Thus we have two averages, one along the time axis with ω fixed, the other along the ω axis with time fixed. It seems that there should be no reason for the answers to agree. Taking a more practical point of view, however, it seems that the time and probabilistic averages must be the same in many situations. For example, suppose that you measure the percentage of time that a particular noise voltage exceeds 10 volts. If you make the measurement over a sufficiently long period of time, the result should be a reasonably good estimate of the probability that the noise voltage exceeds 10 volts at any given instant of time – a probabilistic average value.

To proceed further, for simplicity we concentrate on a random process with discrete alphabet and discrete time. Other cases are considered by converting appropriate sums into integrals. Let $\{X_n\}$ be an arbitrary discrete alphabet, discrete time process. Since the process is random, we cannot predict accurately its instantaneous or short-term behavior – we can only make probabilistic statements. Based on experience with coins, dice, and roulette wheels, however, one expects that the long-term average behavior can be characterized with more accuracy. For example, if one flips a fair coin, short sequences of flips are unpredictable. However, if one flips long enough, one would expect to have an average of about 50% of the flips result in heads. This is a time average of an instantaneous function of a random process – a type of counting function that we will consider extensively. It is obvious that there are many functions that we can average: the average value, the average power, etc. We will proceed by defining one particular average, the value of the sample average of a random process, which is formulated as

$$S_n = n^{-1} \sum_{i=0}^{n-1} X_i; \quad n = 1, 2, 3, \ldots$$

We will investigate the behavior of S_n for large n, i.e., for a long-term time average. Thus, for example, if the random process $\{X_n\}$ is the coin-flipping model, the binary process with alphabet $\{0, 1\}$, then S_n is the number of 1s divided by the total number of flips – the fraction of flips that produced a 1. As noted before, S_n should be close to 50% for large n if the coin is fair.

Note that, as in Example [3.7], for each n, S_n is a random variable that is defined on the same probability space as the random process $\{X_n\}$. This is made explicit by writing the ω dependence:

$$S_n(\omega) = \frac{1}{n} \sum_{k=0}^{n-1} X_k(\omega).$$

In more direct analogy to Example [3.7], we can consider the $\{X_n\}$ as coordinate functions on a sequence space, say $(\Re^{\mathcal{Z}}, \mathcal{B}(\Re^{\mathcal{Z}}), m)$, where m is the distribution of the process, in which case S_n is defined directly on the sequence space. The form of definition is simply a matter of semantics or convenience. Observe, however, that in any case $\{S_n; n = 1, 2, \ldots\}$ is itself a random process since it is an indexed family of random variables defined on a probability space.

For the discrete alphabet random process that we are considering, we can rewrite the sum in another form by grouping together all equal terms:

$$S_n(\omega) = \sum_{a \in A} a r_a^{(n)}(\omega) \tag{4.1}$$

where A is the range space of the discrete alphabet random variable X_n and $r_a^{(n)}(\omega) = n^{-1}$ [number of occurrences of the letter a in $\{X_i(\omega),\ i = 0, 1, 2, \ldots, n-1\}$]. The random variable $r_a^{(n)}$ is called the nth-order *relative frequency* of the symbol a. For the binary coin-flipping example we have considered, $A = \{0, 1\}$, and $S_n(\omega) = r_1^{(n)}(\omega)$, the average number of heads in the first n flips. In other words, for the binary coin-flipping example, the sample average and the relative frequency of heads are the same quantity. More generally, the reader should note that $r_a^{(n)}$ can always be written as the sample average of the indicator function for a, $1_a(x)$:

$$r_a^{(n)} = n^{-1} \sum_{i=0}^{n-1} 1_a(X_i),$$

where

$$1_a(x) = \begin{cases} 1 & \text{if } x = a \\ 0 & \text{otherwise} \end{cases}.$$

Note that $1_{\{a\}}$ is a more precise, but more clumsy, notation for the indicator function of the singleton set $\{a\}$. We shall use the shorter form here.

Let us now assume that all of the marginal pmf's of the given process are the same, say $p_X(x)$, $x \in A$. Based on intuition and gambling experience, one might suspect that as n goes to infinity, the relative frequency of a symbol a should go to its probability of occurrence, $p_X(a)$. To continue the example of binary coin-flipping, the relative frequency of heads in n tosses of a fair coin should tend to $1/2$ as $n \to \infty$. If these statements are true, that is, if in some sense,

$$r_a^{(n)} \underset{n \to \infty}{\to} p_X(a), \tag{4.2}$$

then it follows that in a similar sense

$$S_n \underset{n \to \infty}{\to} \sum_{a \in A} a p_X(a), \tag{4.3}$$

the same expression as (4.1) with the relative frequency replaced by the pmf. The formula on the right is an example of an *expectation* of a random variable, a weighted average with respect to a probability measure. The formula should be recognized as a special case of the definition of expectation of (2.34), where the pmf is p_X and $g(x) = x$, the identity function. The previous plausibility argument motivates the study of such weighted averages because they will characterize the limiting behavior of time averages in the same way that probabilities characterize the limiting behavior of relative frequencies.

Limiting statements of the form of (4.2) and (4.3) are called *laws of large numbers* or *ergodic theorems.* They relate long-run sample averages or time-average behavior to probabilistic calculations made at any given instant of time. It is obvious that such laws or theorems do not always hold. If the coin we are flipping wears in a known fashion with time so that the probability of a head changes, then one could hardly expect that the relative frequency of heads would equal the probability of heads at time zero.

In order to make precise statements and to develop conditions under which the laws of theorems do hold, we first need to develop the properties of the quantity on the right-hand side of (4.2) and (4.3). In particular, we cannot at this point make any sense out of a statement like "$\lim_{n\to\infty} S_n = \sum_{a\in A} a p_X(a)$," since we have no definition for such a limit of random variables or functions of random variables. It is obvious, however, that the usual definition of a limit used in calculus will not do, because S_n is a random variable albeit a random variable whose "randomness" decreases in some sense with increasing n. Thus the limit must be defined in some fashion that involves probability. Such limits are deferred to a later section. We begin by looking at the definitions and calculus of expectations.

4.2 Expectation

Given a discrete alphabet random variable X specified by a pmf p_X, define the *expected value, probabilistic average,* or *mean* of X by

$$E(X) = \sum_{x\in A} x p_X(x). \tag{4.4}$$

The expectation is also denoted by EX or $E[X]$ or by an overbar, as $\overline{X}$. The expectation is also sometimes called an *ensemble average* to denote averaging across the ensemble of sequences that is generated for different values of ω at a given instant of time.

The astute reader might note that we have really provided two definitions of the expectation of X. The definition of (4.4) has already been noted to be a special case of (2.34) with pmf p_X and function $g(x) = x$. Alternatively, we could use (2.34) in a more fundamental form and consider $g(\omega) = X(\omega)$ is a function defined on an underlying probability space described by a pmf

p or a pdf f, in which case (2.34) or (2.57) provide a different formula for finding the expectation in terms of the original probability function:

$$E(X) = \sum_{\omega} X(\omega)p(\omega) \tag{4.5}$$

if the original space is discrete, or

$$E(X) = \int X(r)f(r)\,dr \tag{4.6}$$

if it is continuous. Are these two versions consistent? The answer is yes, as will be proved soon by the fundamental theorem of expectation. The equivalence of these forms is essentially a change of variables formula.

The mean of a random variable is a weighted average of the possible values of the random variable with the pmf used as a weighting. Before continuing, observe that we can define an analogous quantity for a continuous random variable possessing a pdf: if the random variable X is described by a pdf f_X, then we define the *expectation* of X by

$$EX = \int x f_X(x)\,dx, \tag{4.7}$$

where we have replaced the sum by an integral. Analogous to the discrete case, this formula is a special case of (2.57) with pdf $f = f_X$ and g being the identity function. We can also use (2.57) to express the expectation in terms of an underlying pdf, say f, as

$$EX = \int X(r)f(r)\,dr.$$

The equivalence of these two formulas will be considered when the fundamental theorem of expectation is treated.

Although the integral does not have the intuitive motivation involving a relative frequency converging to a pmf that the earlier sum did, we shall see that it plays the analogous role in the laws of large numbers. Roughly speaking, this is because continuous random variables can be approximated by discrete random variables arbitrarily closely by very fine quantization. Through this procedure, the integrals with pdf's are approximated by sums with pmf's and the discrete alphabet results imply the continuous alphabet results by taking appropriate limits. Because of the direct analogy, we shall develop the properties of expectations for continuous random variables along with those for discrete alphabet random variables. Note in passing that, analogous to using the Stieltjes integral as a unified notation for sums and integrals when computing probabilities, the same thing can be done for expectations. If F_X is the cdf of a random variable X, define

$$EX = \int x\,dF_X(x) = \begin{cases} \sum_x x p_X(x) & \text{if } X \text{ discrete} \\ \int x f_X(x)\,dx & \text{if } X \text{ has a pdf} \end{cases}.$$

In a similar manner, we can define the expectation of a mixture random variable having both continuous and discrete parts in a manner analogous to (3.36).

Examples

The following examples provide some typical expectation computations.

[4.1] As a slight generalization of the fair coin flip, consider the more general binary pmf with parameter p; that is, $p_X(1) = p$ and $p_X(0) = 1 - p$. In this case

$$EX = \sum_{i=0}^{1} x p_X(x) = 0(1-p) + 1p = p.$$

It is interesting to note that in this example, as is generally true for discrete random variables, EX is not necessarily in the alphabet of the random variable, i.e., $EX \neq 0$ or 1 unless $p = 0$ or 1.

[4.2] A more complicated discrete example is a geometric random variable. In this case

$$EX = \sum_{k=1}^{\infty} k p_X(k) = \sum_{k=1}^{\infty} k p(1-p)^{k-1},$$

a sum evaluated in (2.48) as $1/p$.

[4.3] As an example of a continuous random variable, assume that X is a uniform random variable on $[0,1]$, that is, its density is one on $[0,1]$. Here

$$EX = \int_0^1 x f_X(x)\,dx = \int_0^1 x\,dx = \frac{1}{2},$$

an integral evaluated in (2.67).

[4.4] If X is an exponential random variable with parameter λ, then from (2.71)

$$\int_0^{\infty} r\lambda e^{-\lambda r}\,dr = \frac{1}{\lambda}. \tag{4.8}$$

In some cases expectations can be found virtually by inspection. For example, if X has an even pdf f_X – that is, if $f_X(-x) = f_X(x)$ for all $x \in \Re$ –

then if the integral exists, $EX = 0$, since $xf_X(x)$ is an odd function and hence has a zero integral. The assumption that the integral exists is necessary because not all even functions are integrable. For example, suppose that we have a pdf $f_X(x) = c/x^2$ for all $|x| \geq 1$, where c is a normalization constant. Then it is not true that EX is zero, even though the pdf is even, because the Riemann integral

$$\int_{x:\,|x|\geq 1} \frac{x}{x^2}\,dx$$

does not exist. (The puzzled reader should review the definition of indefinite integrals. Their existence requires that the limit

$$\lim_{T\to\infty} \lim_{S\to\infty} \int_{-T}^{S} x f_X(x)\,dx$$

exists regardless of how T and S tend to infinity; in particular, the existence of the limit with the constraint $T = S$ is not sufficient for the existence of the integral. These limits do not exist for the given example because $1/x$ is not integrable on $[1, \infty)$.) Nonetheless, it is convenient to set EX to 0 in this example because of the obvious intuitive interpretation.

Sometimes the pdf is an even function about some nonzero value, that is, $f_X(m + x) = f_X(m - x)$, where m is some constant. In this case, it is easily seen that if if the expectation exists, then $EX = m$, as the reader can quickly verify by a change of variable in the integral defining the expectation. The most important example of this is the Gaussian pdf, which is even about the constant m.

The same conclusions also obviously hold for an even pmf.

4.3 Functions of random variables

In addition to the expectation of a given random variable, we will often be interested in the expectations of other random variables formed as functions of the given one. In the beginning of the chapter we introduced the relative frequency function, $r_a^{(n)}$, which counts the relative number of occurrences of the value a in a sequence of n terms. We are interested in its expected value and in the expected value of the indicator function that appears in the expression for $r_a^{(n)}$. More generally, given a random variable X and a function $g : \Re \to \Re$, we might wish to find the expectation of the random variable $Y = g(X)$. If X corresponds to a voltage measurement and g is a simple squaring operation, $g(X) = X^2$, then $g(X)$ provides the instantaneous energy across a unit resistor. Its expected value, then, represents the probabilistic average energy. More generally than the square of a random variable, the *moments* of a random variable X are defined by $E[X^k]$ for $k = 1, 2, \ldots$ The mean is the first moment, the square is the second moment, and so on. If the random variable is complex-valued, then often the *absolute moments* $E[|X|^k]$ are of interest. Moments are often useful as general parameters of a distribution, providing information on its shape without requiring the complete pdf or pmf. Some distributions are completely characterized by a few moments (e.g., the Gaussian). It is often useful to consider moments of a "centralized" random variable formed by removing its mean. The kth centralized moment is defined by $E[(X - E(X))^k]$. The kth centralized absolute moment is defined by $E[|X - E(X)|^k]$.

Of particular interest is the second centralized moment or *variance* $\sigma^2 \triangleq E[(X - E(X))^2]$. Other functions that are of interest are indicator functions of a set, $1_F(x) = 1$ if $x \in F$ and 0 otherwise, so that $1_F(X)$ is a binary random variable indicating whether or not the value of X lies in F, and complex exponentials e^{juX}.

Expectations of functions of random variables were defined in this chapter in terms of the derived distribution for the new random variable. In chapter 3, however, they were defined in terms of the original pmf or pdf in the underlying probability space, a formula not requiring that the new distribution be derived. We next show that the two formulas are consistent. First consider finding the expectation of Y by using derived distribution techniques to find the probability function for Y. Then use the definition of expectation to evaluate EY. Specifically, if X is discrete, the pmf for Y is found as before as

$$p_Y(y) = \sum_{x:\, g(x)=y} p_X(x),\ y \in A_Y.$$

EY is then found as

$$EY = \sum_{A_Y} y p_Y(y).$$

Although it is straightforward to find the probability function for Y, it can be a nuisance if it is being found only as a step in the evaluation of the expectation $EY = Eg(X)$. A second and easier method of finding EY is normally used. Looking at the formula for EX, it seems intuitively obvious that $E(g(X))$ should result if x is replaced by $g(x)$. This can be proved by the following simple procedure. The expectation of Y is found directly from the pmf for X by starting with the pmf for Y, then substituting for its expression in terms of the pmf of X and reordering the summation:

$$\begin{aligned} EY &= \sum_{A_Y} y p_Y(y) = \sum_{A_Y} y \left(\sum_{x:\, g(x)=y} p_X(x) \right) \\ &= \sum_{A_Y} \left(\sum_{x:\, g(x)=y} g(x) p_X(x) \right) = \sum_{A_X} g(x) p_X(x). \end{aligned}$$

This little bit of manipulation is given the fancy name of the *fundamental theorem of expectation*. It is a very useful formula in that it allows one to compute expectations of functions of random variables without needing to perform the (usually more difficult) derived distribution operations.

A similar proof holds for the case of a discrete random variable defined on a continuous probability space described by a pdf. The proof is left as an exercise (Problem 4.4).

A similar change of variables argument with integrals in place of sums yields the analogous pdf result for continuous random variables. As is customary, however, we have only provided the proof for the simple discrete case. For the details of the continuous case, we refer the reader to books on integration or analysis. The reader should be aware that such integral results will have additional technical assumptions (almost always satisfied) required to guarantee the existence of the various integrals. We summarize the results below.

Theorem 4.1 *Fundamental theorem of expectation.*
Let a random variable X be described by a cdf F_X, which is in turn described by either a pmf p_X or a pdf f_X. Given any measurable function $g : \Re \to \Re$, the resulting random variable $Y = g(X)$ has expectation

$$\begin{aligned} EY = E(g(X)) &= \int y \, dF_{g(X)}(y) \\ &= \int g(x) \, dF_X = \begin{cases} \sum_x g(x) p_X(x) & \textit{if discrete with a pmf} \\ \int_x g(x) f_X(x) \, dx & \textit{if continuous with a pdf} \end{cases}. \end{aligned}$$

The qualification "measurable" is needed in the theorem to guarantee the existence of the expectation. Measurability is satisfied by almost any function that you can think of and, for all practical purposes, the requirement can be neglected.

Examples

As a simple example of the use of this formula, consider a random variable X with a uniform pdf on $[-1/2, 1/2]$. Define the random variable $Y = X^2$, that is $g(r) = r^2$. We can use the derived distribution formula (3.40) to write

$$f_Y(y) = y^{-(1/2)} f_X\left(y^{(1/2)}\right); \quad y \geq 0,$$

and hence

$$f_Y(y) = y^{-(1/2)}; \quad y \in (0, 1/4],$$

where we have used the fact that $f_X(y^{(1/2)})$ is 1 only if the nonnegative argument is less than $(1/2)$ or $y \leq \frac{1}{4}$. We can then find EY as

$$EY = \int y f_Y(y)\, dy = \int_0^{1/4} y^{1/2}\, dy = \frac{(1/4)^{3/2}}{3/2} = \frac{1}{12}.$$

Alternatively, we can use the theorem to write

$$EY = E(X^2) = \int_{-1/2}^{1/2} x^2\, dx = 2\,\frac{(1/2)^3}{3} = \frac{1}{12}.$$

Note that the result is the same for each method. However, the second calculation is much simpler, especially if one considers the work that was required in Chapter 3 to derive the density formula for the square of a random variable.

[4.5] A second example generalizes an observation of Chapter 2 and shows that expectations can be used to express probabilities (and hence that probabilities can be considered as special cases of expectation). Recall that the indicator function of an event F is defined by

$$1_F(x) = \begin{cases} 1 & \text{if } x \in F \\ 0 & \text{otherwise} \end{cases}.$$

The probability of the event F can be written in the following form which is convenient in certain computations:

$$E1_F(X) = \int 1_F(x)\, dF_X(x) = \int_F dF_X(x) = P_X(F), \tag{4.9}$$

where we have used the universal Stieltjes integral representation of (3.32) to save writing out both sums of pmf's and integrals of pdf's (the reader who is unconvinced by (4.9) should write out the specific pmf and pdf forms). Observe also that finding probability by taking expectations of indicator functions is like finding a relative frequency by taking a sample average of an indicator function.

It is obvious from the fundamental theorem of expectation that the expected value of any function of a random value can be calculated from its probability distribution. The preceding example demonstrates that the converse is also true: the probability distribution can be calculated from a knowledge of the expectation of a large enough set of functions of the random variable. The example provides the result for the set of all indicator functions. The choice is not unique, as shown by the following example.

[4.6] Let $g(x)$ be the complex function e^{jux} where u is an arbitrary constant. For a cdf F_X, define

$$E(g(X)) = E(e^{juX}) = \int e^{jux}\, dF_X(x).$$

This expectation is immediately recognizable as the characteristic function of the random variable (or its distribution), providing a shorthand definition

$$M_X(ju) = E[e^{juX}].$$

In addition to its use in deriving distributions for sums of independent random variables, the characteristic function can be used to compute moments of a random variable (as the Fourier transform can be used to find moments of a signal). For example, consider the discrete case and take a derivative of the characteristic function $M_X(ju)$ with respect to u:

$$\frac{d}{du}M_X(ju) = \frac{d}{du}\sum_x p_X(x)e^{jux} = \sum_x p_X(x)(jx)e^{jux}$$

and evaluate the derivative at $u = 0$ to find that

$$M_X{}'(0) = \frac{d}{du}M_X(ju)|_{u=0} = jEX.$$

Thus the mean of a random variable can be found by differentiating the characteristic function and setting the argument to 0 as

$$EX = \frac{M_X{}'(0)}{j}. \tag{4.10}$$

Repeated differentiation can be used to show more generally that the kth moment can be found as

$$E[X^k] = j^{-k}M^{(k)}(0) = j^{-k}\frac{d^k}{du^k}M_X(ju)|_{u=0}. \tag{4.11}$$

If one needs several moments of a given random variable, it is usually easier to do one integration to find the characteristic function and then several differentiations than it is to do the several integrations necessary to find the moments directly. Note that if we make the substitution $w = ju$ and differentiate with respect to w, instead of u,

$$\frac{d^k}{dw^k} M_X(w)|_{w=0} = E(X^k).$$

Because of this property, characteristics function with $ju = w$ are called *moment-generating functions.* From the defining sum or integral for characteristic functions in Example [4.6], the moment-generating function may not exist for all $w = v + ju$, even when it exists for all $w = ju$ with u real. This is a variation on the idea that a Laplace transform might not exist for all complex frequencies $s = \sigma + j\omega$ even when it exists for all $s = j\omega$ with ω real, that is, when the Fourier transform exists.

Example [4.6] illustrates an obvious extension of the fundamental theorem of expectation. In [4.6] the complex function is actually a vector function of length 2. Thus it is seen that the theorem is valid for vector functions, $g(x)$, as well as for scalar functions, $g(x)$. The expectation of a vector is simply the vector of expected values of the components.

As a simple example, recall from (3.110) that the characteristic function of a binary random variable X with parameter $p = p_X(1) = 1 - p_X(0)$ is

$$M_X(ju) = (1-p) + pe^{ju}. \tag{4.12}$$

It is easily seen that

$$\frac{M_X{}'(0)}{j} = p = E[X], \quad -M_X{}''(0) = -M_X^{(2)}(0)p = E[X^2].$$

As another example, consider $\mathcal{N}(m, \sigma^2)$, the Gaussian pdf with mean m and variance σ^2. Differentiating easily yields

$$\frac{M_X{}'(0)}{j} = m = E[X], \quad -M_X{}''(0) = \sigma_X^2 + m^2 = E[X^2].$$

The relationship between the characteristic function of a distribution and the moments of a distribution becomes particularly striking when the characteristic function is sufficiently nice near the origin to possess a Taylor series expansion. The Taylor series of a function $f(u)$ about the point $u = 0$

has the form

$$f(u) = \sum_{k=0}^{\infty} u^k \frac{f^{(k)}(0)}{k!}$$

$$= f(0) + uf^{(1)}(0) + u^2 \frac{f^{(2)}(0)}{2} + \text{terms in } u^k \ ; k \geq 3, \quad (4.13)$$

where the derivatives

$$f^{(k)}(0) = \frac{d^k}{du^k} f(u)|_{u=0} \ ;$$

are assumed to exist, that is, the function is assumed to be analytic at the origin. Combining the Taylor series expansion with the moment-generating property (4.11) yields

$$M_X(ju) = \sum_{k=0}^{\infty} u^k \frac{M_X^{(k)}(0)}{k!} = \sum_{k=0}^{\infty} (ju)^k \frac{E(X^k)}{k!}$$

$$= 1 + juE(X) - u^2 \frac{E(X^2)}{2} + o(u^2)/2, \quad (4.14)$$

where $o(u^2)$ contains higher-order terms that go to zero as $u \to 0$ faster than u^2.

This result has an interesting implication: knowing all of the moments of the random variable is equivalent to knowing the behavior of the characteristic function near the origin. If the characteristic function is sufficiently well behaved for the Taylor series to be valid over the entire range of u rather than just in the area around 0, then knowing all of the moments of a random variable is sufficient to know the transform. Since the transform in turn implies the distribution, this guarantees that knowing *all* of the moments of a random variable completely describes the distribution. This is true, however, only when the distribution is sufficiently "nice," that is, when the technical conditions are met that ensure the existence of all of the required derivatives and of the convergence of the Taylor series.

The approximation of the first three terms of (4.14) plays an important role in the central limit theorem, so it is worth pointing out that it holds under even more general conditions than having an analytic function. In particular, if X has a second moment so that $E[X^2] < \infty$, then

$$M_X(ju) = 1 + juE(X) - \frac{u^2 E(X^2)}{2} + o(u^2). \quad (4.15)$$

See, for example, Breiman's treatment of characteristic functions [7].

The most important application of the characteristic function is its use in deriving properties of sums of independent random variables, as was seen in (3.109).

4.4 Functions of several random variables

Thus far expectations have been considered for functions of a single random variable, but it will often be necessary to treat functions of multiple random variables such as sums, products, maxima, and minima. For example, given random variables U and V defined on a common probability space we might wish to find the expectation of $Y = g(U, V)$. The fundamental theorem of expectation has a natural extension (which is proved in the same way).

Theorem 4.2 *Fundamental theorem of expectation for functions of several random variables*
Given random variables $X_0, X_1, \ldots, X_{k-1}$ described by a cdf $F_{X_0,X_1,\ldots,X_{k-1}}$ and given a measurable function $g : \Re^k \to \Re$,

$$E[g(X_0, \ldots, X_{k-1})]$$
$$= \int g(x_0, \ldots, x_{k-1})\, dF_{X_0,\ldots,X_{k-1}}(x_0, \ldots, x_{k-1})$$
$$= \begin{cases} \displaystyle\sum_{x_0,\ldots,x_{k-1}} g(x_0, \ldots, x_{k-1}) p_{X_0,\ldots,X_{k-1}}(x_0, \ldots, x_{k-1}) \\ or \\ \displaystyle\int g(x_0, \ldots, x_{k-1}) f_{X_0,\ldots,X_{k-1}}(x_0, \ldots, x_{k-1})\, dx_0 \ldots dx_{k-1} \end{cases}.$$

We will consider correlation, covariance, multidimensional characteristic functions, and differential entropy as examples of the expectation of several random variables. First, however, we develop some simple and important properties of expectation that will be needed.

4.5 Properties of expectation

Expectation possesses several basic properties that will prove useful. We now present these properties and prove them for the discrete case. The continuous results follow by using integrals in place of sums.

Property 1. If X is a random variable such that $\Pr(X \geq 0) = 1$, then $EX \geq 0$.

Proof $\Pr(X \geq 0) = 1$ implies that the pmf $p_X(x) = 0$ for $x < 0$. If $p_X(x)$ is nonzero only for nonnegative x, then the sum defining the expectation contains only terms $xp_X(x) \geq 0$, and hence them sum which equals EX is nonnegative. Note that Property 1 parallels Axiom 2.1 of probability. In other words, the nonnegativity of probability measures implies Property 1.□

Property 2. If X is a random variable such that for some fixed number r, $\Pr(X = r) = 1$, then $EX = r$. Thus the expectation of a constant equals the constant.

Proof $\Pr(X = r) = 1$ implies that $p_X(r) = 1$. Thus the result follows from the definition of expectation. Observe that Property 2 parallels Axiom 2.2 of probability. That is, the normalization of the total probability to 1 leaves the constant unscaled in the result. If total probability were different from 1, the expectation of a constant as defined would be a different, scaled value of the constant. □

Property 3. Expectation is linear; that is, given two random variables X and Y and two real constants a and b,

$$E(aX + bY) = aEX + bEY.$$

Proof Let $g(x, y) = ax + by$, where a and b are constants. In this case the fundamental theorem of expectation for functions of several (here two) random variables implies that

$$\begin{aligned} E[aX + bY] &= \sum_{x,y} (ax + by) p_{X,Y}(x, y) \\ &= a \sum_x x \sum_y p_{X,Y}(x, y) + b \sum_y y \sum_x p_{X,Y}(x, y). \end{aligned}$$ □

Using the consistency of marginal and joint pmf's of (3.13)–(3.14) this becomes

$$\begin{aligned} E[aX + bY] &= a \sum_x x p_X(x) + b \sum_y y p_Y(y) \\ &= aE(X) + bE(Y). \end{aligned} \tag{4.16}$$

Keep in mind that this result has nothing to do with whether or not the random variables are independent.

The linearity of expectation follows from the additivity of probability. In particular, the summing out of joint pmf's to get marginal pmf's in the proof was a direct consequence of Axiom 2.4. The alert reader will probably have noticed the method behind the presentation of the properties of expectation – each follows directly from the corresponding axiom of probability. Furthermore, using (4.9), the converse is true: that is, instead of starting with the axioms of probability, suppose we start by using the properties of expectation as the axioms of expectation. Then the axioms of probability become the derived properties of probability. Thus the first three axioms of probability and the first three properties of expectation are dual; one can start with either and get the other. One might suspect that to get a useful theory based on expectation, one would require a property analogous to Axiom 2.4 of probability, that is, a limiting form of expectation Property 3. This is, in fact, the case, and the fourth basic property of expectation is the countably infinite version of Property 3. When dealing with expectations, however, the fourth property is more often stated as a continuity property. That is, it is stated in a form analogous to Axiom 2.4 of probability given in (2.28). For reference we state the property below without proof.

Property 4. Given an increasing sequence of nonnegative random variables X_n; $n = 0, 1, 2, \ldots$, that is, $X_n \geq X_{n-1}$ for all n (i.e., $X_n(\omega) \geq X_{n-1}(\omega)$ for all $\omega \in \Omega$), which converge to a limiting random variable $X = \lim_{n\to\infty} X_n$, then

$$E\left(\lim_{n\to\infty} X_n\right) = \lim_{n\to\infty} EX_n.$$

Thus as with probabilities, one can in certain cases exchange the order of limits and expectation. The cases include but are not limited to those of Property 4. Property 4 is called the *monotone convergence theorem* and is one of the basic properties of integration as well as expectation. This theorem is discussed in Appendix B along with another important limiting result, the dominated convergence theorem.

In fact, the four properties of expectation can be taken as a definition of an integral (the Stieltjes integral) and used to develop the general Lebesgue theory of integration. In other words, the theory of expectation is really just a specialization of the theory of integration. The duality between probability and expectation is just a special case of the duality between measure theory and the theory of integration.

4.6 Examples

4.6.1 Correlation

We next introduce the idea of correlation or expectation of products of random variables which will lead to the development of a property of expectation that is special to independent random variables. A weak form of this property will be seen to provide a weak form of independence that will later be useful in characterizing certain random processes. Correlations will later be seen to play a fundamental role in many signal processing applications. Suppose we have two independent random variables X and Y and we have two functions or measurements on these random variables, say $g(X)$ and $h(Y)$, where $g : \Re \to \Re$, $h : \Re \to \Re$, and $E[g(X)]$ and $E[h(Y)]$ exist and are finite. Consider the expected value of the product of these two functions, called the *correlation* between $g(X)$ and $h(Y)$. As we shall consider in more detail later, if we are considering complex-valued random variables or functions, the correlation of $g(X)$ and $h(Y)$ is defined by $E[g(X)h(Y)^*]$, where the asterix denotes the complex conjugate. For the time being, however, we focus on the simpler case of real-valued random variables and functions.

Applying the two-dimensional vector case of the fundamental theorem of expectation to discrete random variables results in

$$\begin{aligned} E[g(X)h(Y)] &= \sum_{x,y} g(x)h(y)p_{X,Y}(x,y) \\ &= \sum_x \sum_y g(x)h(y)p_X(x)p_Y(y) \\ &= \left(\sum_x g(x)p_X(x)\right)\left(\sum_y h(y)p_Y(y)\right) \\ &= E[g(X)]E[h(Y)]. \end{aligned}$$

A similar manipulation with integrals shows the same to be true for random variables possessing pdf's. Thus we have proved the following result, which we state formally as a lemma.

Lemma 4.1 *For any two independent random variables X and Y,*

$$E[g(X)h(Y)] = E[g(X)]E[h(Y)] \tag{4.17}$$

for all functions g and h with finite expectation.

By stating that the functions have finite expectation we implicitly assume them to be measurable, i.e., to have a distribution with respect to which we can evaluate an expectation. Measurability is satisfied by almost all functions so that the qualification can be ignored for all practical purposes.

To cite the most important example, if g and h are identity functions ($h(r) = g(r) = r$), then the independence of X and Y implies that

$$E[XY] = (EX)(EY), \tag{4.18}$$

that is, the correlation of X and Y is the product of the means, in which case the two random variables are said to be *uncorrelated*. (The term *linear independence* is sometimes used as a synonym for uncorrelated.)

We have shown that if two discrete random variables are independent, then they are also uncorrelated. Note that independence implies not only that two random variables are uncorrelated but also that all *functions* of the random variables are uncorrelated – a much stronger property. In particular, two uncorrelated random variables need not be independent. For example, consider two random variables X and Y with the joint pmf

$$p_{X,Y}(x,y) = \begin{cases} 1/4 & \text{if } (x,y) = (1,1) \text{ or } (-1,1) \\ 1/2 & \text{if } (x,y) = (0,0) \end{cases}.$$

A simple calculation shows that $E(XY) = (1/4)(1-1) + (1/2)(0) = 0$ and $(EX)(EY) = (0)((1/2)) = 0$, and hence the random variables are uncorrelated. They are not, however, independent. For example, $\Pr(X = 0|Y = 0) = 1$ whereas $\Pr(X = 0) = 1/2$. As another example, consider the case where $p_X(x) = 1/3$ for $x = -1, 0, 1$ and $Y = X^2$. X and Y are uncorrelated but not independent.

As another example, suppose that U has a uniform pdf on $[0,1]$ define the random variables $X = \cos U$ and $Y = \sin U$. Then X and Y are not independent, but they are uncorrelated. The proof is left as an exercise, but note that $X^2 + Y^2 = 1$ so knowledge of X strongly affects the distribution for Y.

Thus uncorrelation does not imply independence. If, however, all possible functions of the two random variables are uncorrelated – that is, if (4.17) holds – then they must be independent. To see this in the discrete case, just consider all possible indicator functions of of points, $1_a(x)$. In particular, the function $1_a(x)$ is 1 if $x = a$ and zero otherwise. Let $g = 1_a$ and $h = 1_b$ for a in the range space of X and b in the range space of Y. It follows from (4.17) and (4.9) that $p_{X,Y}(a,b) = p_X(a)p_Y(b)$. A similar argument extends the result to continuous random variables and pdf's. Obviously the result holds for all a and b. Thus the two random variables are independent. It can now be seen that (4.17) provides a necessary and sufficient condition for independence, a fact we formally state as a theorem.

Theorem 4.3 *Two random variables X and Y are independent if and only if $g(X)$ and $h(Y)$ are uncorrelated for all functions g and h with finite expectations, that is, if (4.17) holds. More generally, random variables X_i; $i = 1, \ldots, n$, are mutually independent if and only if for all functions g_i; $i = 1, \ldots, n$ the random variables $g_i(X_i)$ are uncorrelated.*

This theorem is useful as a means of showing that two random variables are *not* independent. If we can find *any* functions g and h such that $E(g(X)h(Y)) \neq (Eg(X))(Eh(Y))$, then the random variables are not independent. The theorem also provides a simple and general proof of the fact that the characteristic function of the sum of independent random variables is the product of the characteristic functions of the random variables being summed.

Corollary 4.1 *Given a sequence of mutually independent random variables* $X_1, X_2, \ldots, X_n$, *define*

$$Y_n = \sum_{i=1}^{n} X_i.$$

Then

$$M_{Y_n}(ju) = \prod_{i=1}^{n} M_{X_i}(ju).$$

Proof Successive application of Theorem 4.3, which states that functions of independent random variables are uncorrelated, yields

$$E\left(e^{juY_n}\right) = E\left(\exp\left(ju\sum_{i=1}^{n} X_i\right)\right) = E\left(\prod_{i=1}^{n} e^{juX_i}\right)$$
$$= \prod_{i=1}^{n} E\left(e^{juX_i}\right) = \prod_{i=1}^{n} M_{X_i}(ju). \qquad \square$$

4.6.2 Covariance

The idea of uncorrelation can be stated conveniently in terms of another quantity, which we now define. Given two random variables X and Y, define their *covariance*, $\mathrm{COV}(X, Y)$ by

$$\mathrm{COV}(X, Y) \triangleq E[(X - EX)(Y - EY)].$$

In the more general case of complex-valued random variables, the covariance is defined by

$$\mathrm{COV}(X, Y) \triangleq E[(X - EX)(Y - EY)^*].$$

As you can see, the *covariance* of two random variables equals the *correlation* of the two "centralized" random variables, $X - EX$ and $Y - EY$, that are formed by subtracting the means from the respective random variables. Keeping in mind that EX and EY are constants, it is seen that centralized random variables are zero-mean random variables: $E(X - EX) = E(X) - E(EX) = EX - EX = 0$. Expanding the product in the definition, the covariance can be written in terms of the correlation and means of the random variables. Again remembering that EX and EY are constants, we get

$$\begin{aligned} \mathrm{COV}(X,Y) &= E[XY - YEX - XEY + (EX)(EY)] \\ &= E(XY) - (EY)(EX) - (EX)(EY) + (EX)(EY) \\ &= E(XY) - (EX)(EY). \end{aligned} \tag{4.19}$$

Thus the covariance is the correlation minus the product of the means. Using this fact and the definition of uncorrelated, we have the following statement:

Corollary 4.2 *Two random variables X and Y are uncorrelated if and only if their covariance is zero; that is, if* $\mathrm{COV}(X,Y) = 0$.

If we set $Y = X$, the result is the correlation of X with itself, $E(X^2)$; this is called the *second moment* of the random variable X. The covariance $\mathrm{COV}(X,X)$ is called the *variance* of the random variable and is given the special notation σ_X^2. The square root of the variance, $\sigma_X = \sqrt{\sigma_X^2}$, is called the *standard deviation* of X. From the definition of covariance and (4.18),

$$\sigma_X^2 = E[(X - EX)^2] = E(X^2) - (EX)^2.$$

By the first property of expectation, the variance is nonnegative, yielding the simple but powerful inequality

$$|EX| \leq [E(X^2)]^{\frac{1}{2}}, \tag{4.20}$$

a special case of the *Cauchy–Schwarz inequality* (see Problem 4.20 with the random variable Y set equal to the constant 1).

4.6.3 Covariance matrices

The fundamental theorem of expectation of functions of several random variables can also be extended to vector or even matrix functions g of random vectors. There are two primary examples, the covariance matrix treated here and the multivariable characteristic functions treated next.

Suppose that $X = (X_0, X_1, \ldots, X_{n-1})$ is a given n-dimensional random vector. The *mean vector* $m = (m_0, m_1, \ldots, m_{n-1})^t$ is defined as the vector of the means, i.e., $m_k = E(X_k)$ for all $k = 0, 1, \ldots, n-1$. This can be written more conveniently as a single vector expectation

$$m = E(X) = \int_{\Re^n} f_X(x) x \, dx \tag{4.21}$$

where the random vector X and the dummy integration vector x are both

n-dimensional and the integral of a vector is simply the vector of the integrals of the individual components. Similarly we could define for each $k, l = 0, 1, \ldots, n-1$ the covariance $K_X(k, l) = E[(X_k - m_k)(X_l - m_l)]$ and then collect these together to form the *covariance matrix*

$$K = \{K_X(k, l);\ k = 0, 1, \ldots, n-1; l = 0, 1, \ldots, n-1\}.$$

Alternatively, we can use the outer product notation of linear algebra and the fundamental theorem of expectation to write

$$K = E[(X - m)(X - m)^t] = \int_{\Re^n} (x - m)(x - m)^t \, dx, \tag{4.22}$$

where the outer product of a vector a with a vector b, ab^t, has (k, j) entry equal to $a_k b_j$.

In particular, by straightforward but tedious multiple integration, it can be shown that the mean vector and the covariance matrix of a Gaussian random vector are indeed the mean and covariance, i.e., using the fundamental theorem of expectation

$$m = E(X) = \int_{\Re^n} x \frac{\exp\left(-\frac{1}{2}(x - m)^t K^{-1}(x - m)\right)}{(2\pi)^{n/2}\sqrt{\det K}} \, dx \tag{4.23}$$

$$\begin{aligned} K &= E[(X - m)(X - m)^t] \\ &= \int_{\Re^n} (x - m)(x - m)^t \frac{\exp\left(-\frac{1}{2}(x - m)^t K^{-1}(x - m)\right)}{(2\pi)^{n/2}\sqrt{\det K}} \, dx. \end{aligned} \tag{4.24}$$

When considering more general complex-valued random vectors, the transpose becomes a conjugate transpose so that the covariance is defined by

$$K = E[(X - m)(X - m)^*] = \int_{\Re^n} (x - m)(x - m)^* \, dx, \tag{4.25}$$

4.6.4 Multivariable characteristic functions

Multivariable characteristic functions provide another example of the extension of the theorem to vector functions g of random vectors. In fact we implicitly assumed this extension in the evaluation of the characteristic function of a Gaussian random variable (since e^{juX} is a complex function of X and hence a vector function) and of the multidimensional characteristic function of a Gaussian random vector in (3.124): if a Gaussian random vector X has a mean vector m and covariance matrix Λ, then

$$M_X(ju) = \exp\left(ju^t m - \frac{1}{2}u^t \Lambda u\right)$$
$$= \exp\left[j\sum_{k=0}^{n-1} u_k m_k - \frac{1}{2}\sum_{k=0}^{n-1}\sum_{m=0}^{n-1} u_k \Lambda(k,m) u_m\right].$$

This representation for the characteristic function yields the proof of the following important result.

Theorem 4.4 *Let X be a k-dimensional Gaussian random vector with mean m_X and covariance matrix Λ_X. Let Y be the new random vector formed by an affine (linear plus a constant) operation of X:*

$$Y = HX + b, \tag{4.26}$$

where H is an $n \times k$ matrix and b is an n-dimensional vector. Then Y is a Gaussian random vector of dimension n with mean

$$m_Y = Hm_X + b \tag{4.27}$$

and covariance matrix

$$\Lambda_Y = H\Lambda_X H^t. \tag{4.28}$$

Proof The characteristic function of Y is found by direct substitution of the expression for Y in terms of X into the definition of a characteristic function. A little matrix algebra, and (3.124) yield

$$M_Y(ju) = E\left[e^{ju^t Y}\right] = E\left[e^{ju^t(HX+b)}\right]$$
$$= e^{ju^t b} E\left[e^{j(H^t u)^t X}\right] = e^{ju^t b} M_X(jH^t u)$$
$$= e^{ju^t b} e^{ju^t Hm - (1/2)(H^t u)^t \Lambda_X (H^t u)}$$
$$= e^{ju^t(Hm+b)} e^{-(1/2)u^t H\Lambda_X H^t u}.$$

It can be seen by reference to (3.124) that the resulting characteristic function is the transform of a Gaussian random vector pdf with mean vector $Hm + b$ and covariance matrix $H\Lambda H^t$. □

The following observation is trivial, but it emphasizes a useful fact. Suppose that X is a Gaussian random vector of dimension, say, k, and we form a new vector Y by subsampling X, that is, by selecting a subset of $(X_0, X_1, \ldots, X_{k-1})$, such as $Y_i = X_{l(i)}$, $i = 0, 1, \ldots, m < k$. Then we can write $Y = AX$, where A is a matrix that has $A_{l(i),l(i)} = 1$ for $i = 0, 1, \ldots, m < k$ and 0s everywhere else. The preceding result implies immediately that Y is Gaussian and shows how to compute the mean and covariance. Thus any subvector of a Gaussian vector is also a Gaussian vector. This could also be proved by a derived distribution and messy multidimensional integrals, but the previous result provides a nice shortcut.

The next observation is that given a Gaussian random vector X we could take H to be an all zero matrix and b to be a constant vector, which means that constant vectors can be considered as special cases of Gaussian vectors. Clearly they are trivial in the sense that the covariance is zero. They provide an example of *singular* Gaussian distributions. A more interesting example is considered next.

Suppose that X is a scalar Gaussian random variable with 0 mean and variance $\sigma_X^2 > 0$. We form a two-dimensional random vector $Y = HX$ where $H = (a, b)^t$ for two constants a and b. Then from the previous discussion Y will be Gaussian with covariance

$$K_Y = \sigma_X^2 \begin{bmatrix} a^2 & ba \\ ab & b^2 \end{bmatrix}.$$

The determinant of K_Y is zero, which means the matrix is not invertible, Y does not have a pdf, and the distribution for Y is singular. More generally, suppose that X is a k-dimensional nonsingular Gaussian random vector, i.e., K_X is an invertible matrix and X has a pdf. Let H be an $n \times k$ matrix where now $n > k$. Then $Y = HX + b$ will be Gaussian, but as with the previous simple example, the covariance K_Y will be singular and Y will have a singular Gaussian distribution. It can be shown that all singular Gaussian distributions have one of the preceding forms; either they are a constant or they are formed by taking a linear transformation of a nonsingular Gaussian random vector of lower dimension.

Singular distributions do occasionally arise in practice, although they are little treated in the literature. One can think of them (at least in the Gaussian case) as having a nonsingular distribution in a lower-dimensional subspace.

4.6.5 Differential entropy of a Gaussian vector

Suppose that $X = (X_0, X_1, \ldots, X_{n-1})$ is a Gaussian random vector described by a pdf f_X specified by a mean vector m and a covariance matrix

K_X. The *differential entropy* of a continuous vector X is defined by

$$h(X) = -\int f_X(x) \log f_X(x)\, dx$$
$$= -\int f_{X_0,X_1,\ldots,X_{n-1}}(x_0, x_1, \ldots, x_{n-1}) \times \tag{4.29}$$
$$\log f_{X_0,X_1,\ldots,X_{n-1}}(x_0, x_1, \ldots, x_{n-1})\, dx_0\, dx_1 \ldots\, dx_{n-1}$$

where the units are called "bits" if the logarithm is base 2 and "nats" if the logorithm is base e.[1] The differential entropy plays a fundamental role in Shannon information theory for continuous alphabet random processes. See, for example, Cover and Thomas [12]. It will also prove a very useful aspect of a random vector when considering linear prediction or estimation. Here we use the fundamental theorem of expectation for functions of several variables to evaluate the differential entropy $h(X)$ of a Gaussian random vector.

Plugging in the density for the Gaussian pdf and using natural logarithms results in

$$h(X) = -\int f_X(x) \ln f_X(x)\, dx$$
$$= \frac{1}{2} \ln\left((2\pi)^n \det K\right) +$$
$$\frac{1}{2}\int (x-m)^t K^{-1}(x-m) \frac{\exp\left(-(1/2)(x-m)^t K^{-1}(x-m)\right)}{(2\pi)^{n/2}\sqrt{\det K}}\, dx.$$

The final term can be easily evaluated by a trick. From linear algebra we can write for any n-dimensional vector a and $n \times n$ matrix K

$$a^t A a = \mathrm{Tr}(A a a^t), \tag{4.30}$$

where Tr is the trace or sum of diagonals of the matrix. Thus using the

[1] Generally the natural log is used in mathematical derivations, whereas the base 2 log is used for intuitive interpretations and explanations.

linearity of expectation we can rewrite the previous equation as

$$\begin{aligned} h(X) &= \frac{1}{2}\ln((2\pi)^n \det K) + \frac{1}{2}E\left((X-m)^t K^{-1}(X-m)\right) \\ &= \frac{1}{2}\ln((2\pi)^n \det K) + \frac{1}{2}E\left(\mathrm{Tr}[K^{-1}(X-m)(X-m)^t)\right] \\ &= \frac{1}{2}\ln((2\pi)^n \det K) + \frac{1}{2}\mathrm{Tr}[K^{-1}E\left((X-m)(X-m)^t\right)] \\ &= \frac{1}{2}\ln((2\pi)^n \det K) + \frac{1}{2}\mathrm{Tr}[K^{-1}K] \\ &= \frac{1}{2}\ln((2\pi)^n \det K) + \frac{1}{2}\mathrm{Tr}[I] \\ &= \frac{1}{2}\ln((2\pi)^n \det K) + \frac{n}{2} = \frac{1}{2}\ln((2\pi e)^n \det K). \end{aligned} \tag{4.31}$$

4.7 Conditional expectation

Expectation is essentially a weighted integral or summation with respect to a probability distribution. If one uses a conditional distribution, then the expectation is also conditional. For example, suppose that (X,Y) is a random vector described by a joint pmf $p_{X,Y}$. The ordinary expectation of Y is defined as usual by $EY = \sum y p_Y(y)$. Suppose, however, that one is told that $X = x$ and hence one has the conditional (a-posteriori) pmf $p_{Y|X}$. Then one can define the *conditional expectation of* Y *given* $X = x$ by

$$E(Y|x) = \sum_{y \in A_Y} y p_{Y|X}(y|x) \tag{4.32}$$

that is, the usual expectation, but with respect to the pmf $p_{Y|X}(\cdot|x)$. So far, this is an almost trivial generalization. Perhaps unfortunately, however, (4.32) is not in fact what is usually defined as conditional expectation. The actual definition might appear to be only slightly different, but there is a fundamental difference and a potential for confusion because of the notation. As we have defined it so far, the conditional expectation of Y given $X = x$ is a function of the independent variable x, say $g(x)$. In other words,

$$g(x) = E(Y|x).$$

If we take any function $g(x)$ of x and replace the independent variable x by a random variable X, we get a new random variable $g(X)$. If we simply replace the independent variable x in $E(Y|x)$ by the random variable X, the resulting quantity is a random variable and is denoted by $E(Y|X)$. It is this random variable that is defined as the *conditional expectation* of Y given X. The previous definition $E(Y|x)$ can be considered as a sample value of the random variable $E(Y|X)$. Note that we can write the definition as

$$E(Y|X) = \sum_{y \in A_Y} y p_{Y|X}(y|X), \tag{4.33}$$

but the reader must beware the dual use of X: in the subscript it denotes as usual the *name* of the random variable, in the argument it denotes the random variable itself, i.e., $E(Y|X)$ is a function of the random variable X and hence is itself a random variable.

Since $E(Y|X)$ is a random variable, we can evaluate its expectation using the fundamental theorem of expectation. The resulting formula has wide application in probability theory. Taking this expectation we have that

$$\begin{aligned} E[E(Y|X)] &= \sum_{x \in A_X} p_X(x) E(Y|x) = \sum_{x \in A_X} p_X(x) \sum_{y \in A_Y} y p_{Y|X}(y|x) \\ &= \sum_{y \in A_Y} y \sum_{x \in A_X} p_{X,Y}(x,y) = \sum_{y \in A_Y} y p_Y(y) = EY, \end{aligned}$$

a result known as *iterated expectation* or *nested expectation*. Roughly speaking it states that if we wish to find the expectation of a random variable Y, then we can first find its conditional expectation with respect to another random variable, $E(Y|X)$, and then take the expectation of the resulting random variable to obtain

$$EY = E[E(Y|X)]. \tag{4.34}$$

In the next section we shall see an interpretation of conditional expectation as an estimator of one random variable given another. A simple example now, however, helps point out how this result can be useful.

Suppose that one has a random process $\{X_k;\ k = 0, 1, \ldots\}$, with identically distributed random variables X_n, and a random variable N that takes on positive integer values. Suppose also that the random variables X_k are all independent of N. Suppose that one defines a new random variable

$$Y = \sum_{k=0}^{N-1} X_k,$$

that is, the sum of a random number of random variables. How does one evaluate the expectation EY? Finding the derived distribution is daunting, but iterated expectation comes to the rescue. Iterated expectation states that $EY = E[E(Y|N)]$, where $E(Y|N)$ is found by evaluating $E(Y|n)$ and replacing n by N. But given $N = n$, the random variable Y is simply $Y = \sum_{k=0}^{n-1} X_k$. The distribution of the X_k is not affected by the fact that $N = k$

since the X_k are independent of N. Hence by the linearity of expectation,

$$E(Y|n) = \sum_{k=0}^{n-1} EX_k,$$

where the identically distributed assumption implies that the EX_k are all equal, say EX. Thus $E(Y|n) = nEX$ and hence $E(Y|N) = NEX$. Then iterated expectation implies that

$$EY = E(NEX) = (EN)(EX), \tag{4.35}$$

the product of the two means. Try finding this result without using iterated expectation! As a particular example, if the random variables are Bernoulli random variables with parameter p and N has a Poisson distribution with parameter λ, then $\Pr(X_i = 1) = p$ for all i and $EN = \lambda$ and hence $EY = p\lambda$.

Iterated expectation has a more general form. Just as constants can be pulled out of ordinary expectations, quantities depending on the variable on which the expectation is conditioned can be pulled out of conditional expectations. We state and prove this formally.

Lemma 4.2 *General iterated expectation*
Suppose the X, Y are discrete random variables and that $g(X)$ and $h(X, Y)$ are functions of these random variables. Then

$$E[g(X)h(X,Y)] = E\left(g(X)E[h(X,Y)|X]\right). \tag{4.36}$$

Proof

$$\begin{aligned} E[g(X)h(X,Y)] &= \sum_{x,y} g(x)h(x,y)p_{X,Y}(x,y) \\ &= \sum_{x} p_X(x)g(x)\sum_{y} h(x,y)p_{Y|X}(y|x) \\ &= \sum_{x} p_X(x)g(x)E[h(X,Y)|x] \\ &= E\left(g(X)E[h(X,Y)|X]\right). \end{aligned}$$ □

As with ordinary iterated expectation, this is primarily an interpretation of an algebraic rewriting of the definition of expectation. Note that if we take $g(x) = 1$ and $h(x, y) = y$, this general form reduces to the previous form.

In a similar vein, one can extend the idea of conditional expectation to continuous random variables by using pdf's instead of pmf's. For example,

$$E(Y|x) = \int y f_{Y|X}(y|x)\, dy,$$

and $E(Y|X)$ is defined by replacing x by X in the above formula. Both iterated expectation and its general form extend to this case by replacing sums by integrals.

4.8 ⋆Jointly Gaussian vectors

Gaussian vectors provide an interesting example of a situation where conditional expectations can be explicitly computed. This in turn provides additional fundamental, if unsurprising, properties of Gaussian vectors. Instead of considering a Gaussian random vector $X = (X_0, X_1, \ldots, X_{N-1})^t$, say, consider instead a random vector

$$U = \begin{pmatrix} X \\ Y \end{pmatrix}$$

formed by concatening two vectors X and Y of dimensions, say, k and m, respectively. For this section we will drop the boldface notation for vectors. If U is Gaussian, then we say that X and Y are *jointly Gaussian*. From Theorem 4.4 it follows that if X and Y are jointly Gaussian, then they are individually Gaussian with, say, means m_X and m_Y, respectively, and covariance matrices K_X and K_Y, respectively. The goal of this section is to develop the conditional second-order moments for Y given X and to show in the process that given X, Y has a Gaussian density. Thus not only is any subcollection of a Gaussian random vector Gaussian, it is also true that the *conditional* density of any subvector of a Gaussian vector given a disjoint subvector (another subvector with no components in common) is Gaussian. This generalizes (3.60) from two jointly Gaussian scalar random variables to two jointly Gaussian random vectors. The idea behind the proof is the same, but the algebra is messier in higher dimensions.

Begin by writing

$$\begin{aligned} K_U &= E[(U - m_U)(U - m_U)^t] \\ &= E\left[\begin{pmatrix} X - m_X \\ Y - m_Y \end{pmatrix} \left((X - m_X)^t \; (Y - m_Y)^t\right)\right] \\ &= \begin{bmatrix} E[(X - m_X)(X - m_X)^t] & E[(X - m_X)(Y - m_Y)^t] \\ E[(Y - m_Y)(X - m_X)^t] & E[(Y - m_Y)(Y - m_Y)^t] \end{bmatrix} \\ &= \begin{bmatrix} K_X & K_{XY} \\ K_{YX} & K_Y \end{bmatrix}, \end{aligned} \tag{4.37}$$

where K_X and K_Y are ordinary covariance matrices and the matrices $K_{XY} = K_{YX}^t$ are called *cross-covariance* matrices. We shall also denote K_U

by $K_{(X,Y)}$, where the subscript is meant to emphasize that it is the covariance of the cascade vector of both X and Y in distinction to K_{XY}, the cross-covariance of X and Y.

The key to recognizing the conditional moments and densities is the following admittedly unpleasant matrix equation, which can be proved with a fair amount of brute force linear algebra:

$$\begin{bmatrix} K_X & K_{XY} \\ K_{YX} & K_Y \end{bmatrix}^{-1} = \begin{bmatrix} K_X^{-1} + K_X^{-1}K_{XY}K_{Y|X}^{-1}K_{YX}K_X^{-1} & -K_X^{-1}K_{XY}K_{Y|X}^{-1} \\ -K_{Y|X}^{-1}K_{YX}K_X^{-1} & K_{Y|X}^{-1} \end{bmatrix}, \tag{4.38}$$

where $K_{Y|X} \triangleq K_Y - K_{YX}K_X^{-1}K_{XY}$. The determined reader who wishes to verify the above should do the block matrix multiplication

$$\begin{bmatrix} a & b \\ c & d \end{bmatrix} \triangleq \begin{bmatrix} K_X^{-1} + K_X^{-1}K_{XY}K_{Y|X}^{-1}K_{YX}K_X^{-1} & -K_X^{-1}K_{XY}K_{Y|X}^{-1} \\ -K_{Y|X}^{-1}K_{YX}K_X^{-1} & K_{Y|X}^{-1} \end{bmatrix} \times \begin{bmatrix} K_X & K_{XY} \\ K_{YX} & K_Y \end{bmatrix}$$

and show that a is a $k \times k$ identity matrix, d is an $m \times m$ identity matrix, and that c and d contain all zeros so that the right-hand matrix is indeed an identity matrix.

The conditional pdf for Y given X follows directly from the definitions as

$$\begin{aligned} f_{Y|X}(y|x) &= \frac{f_{XY}(x,y)}{f_X(x)} \\ &= \frac{\exp\left(-(1/2)((x-m_X)^t \; (y-m_Y)^t)K_U^{-1}\begin{pmatrix} x-m_X \\ y-m_Y \end{pmatrix}\right)}{\sqrt{(2\pi)^{(k+m)}\det K_U}} \\ &\quad \times \frac{\sqrt{(2\pi)^k \det K_X}}{\exp\left(-(x-m_X)^t K_X^{-1}(x-m_X)/2\right)} \\ &= \frac{1}{\sqrt{(2\pi)^m \det K_U / \det K_X}} \\ &\quad \times \exp\left(-(1/2)((x-m_X)^t \; (y-m_Y)^t)K_U^{-1}\begin{pmatrix} x-m_X \\ y-m_Y \end{pmatrix}\right. \\ &\quad\quad \left. + (1/2)(x-m_X)^t K_X^{-1}(x-m_X)\right) \end{aligned}$$

Again using some brute force linear algebra, it can be shown that the quadratic terms in the exponential can be expressed in the form

$$((x - m_X)^t \ (y - m_Y)^t) K_U^{-1} \begin{pmatrix} x - m_X \\ y - m_Y \end{pmatrix} + (x - m_X)^t K_X^{-1} (x - m_X)$$
$$= (y - m_Y - K_{YX} K_X^{-1}(x - m_X))^t K_{Y|X}^{-1} (y - m_Y - K_{YX} K_X^{-1}(x - m_X)).$$

Defining

$$m_{Y|x} = m_Y + K_{YX} K_X^{-1}(x - m_X) \tag{4.39}$$

the conditional density simplifies to

$$f_{Y|X}(y|x) =$$
$$(2\pi)^{-m/2} \left(\frac{\det K_U}{\det K_X} \right)^{-\frac{1}{2}} \times \exp\left(-(y - m_{Y|x})^t K_{Y|X}^{-1} (y - m_{Y|x})/2 \right), \tag{4.40}$$

which shows that conditioned on $X = x$, Y has a Gaussian density. This means that we can immediately recognize the conditional expectation of Y given X as

$$E(Y|X = x) = m_{Y|x} = m_Y + K_{YX} K_X^{-1}(x - m_X), \tag{4.41}$$

so that the conditional expectation is an affine function of the vector x. We can also infer from the form that $K_{Y|X}$ is the (conditional) covariance

$$K_{Y|X} = E[(Y - E(Y|X = x))(Y - E(Y|X = x))^t|x], \tag{4.42}$$

which unlike the conditional mean does *not* depend on the vector x! Furthermore, since we know how the normalization must relate to the covariance matrix, we have that

$$\det(K_{Y|X}) = \frac{\det(K_U)}{\det(K_X)}. \tag{4.43}$$

These relations completely describe the conditional densities of one subvector of a Gaussian vector given another subvector. We shall see, however, that the importance of these results goes beyond the above evaluation and provides some fundamental results regarding optimal nonlinear estimation for Gaussian vectors and optimal linear estimation in general.

4.9 Expectation as estimation

Suppose that you are asked to guess the value that a random variable Y will take on, knowing the distribution of the random variable. What is the *best* guess or estimate, say $\hat{Y}$? Obviously there are many ways to define a best estimate, but one of the most popular ways to define a cost or distortion resulting from estimating the "true" value of Y by $\hat{Y}$ is to look at the expected value of the square of the error $Y - \hat{Y}$, $E[(Y - \hat{Y})^2]$, the so-called *mean squared error* or *mean square error* or simply *MSE*. Many arguments have been advanced in support of this approach, perhaps the simplest being that if one views the error as a voltage, then the average squared error is the average energy in the error. The smaller the energy, the weaker the signal in some sense. Perhaps a more honest reason for the popularity of the measure is its tractability in a wide variety of problems; it often leads to nice solutions that indeed work well in practice. As an example, we show that the optimal estimate of the value of an unknown random variable in the minimum mean squared error sense is in fact the mean of the random variable, a result that is highly intuitive. Rather than use calculus to prove this result – a tedious approach requiring setting derivatives to zero and then looking at second derivatives to verify that the stationary point is a minimum – we directly prove the global optimality of the result. Suppose that that our estimate is $\hat{Y}$ is some constant a. We will show that this estimate can never have mean squared error smaller than that resulting from using the expected value of Y as an estimate. This is accomplished by a simple sequence of equalities and inequalities. Begin by adding and subtracting the mean, expanding the square, and using the second and third properties of expectation as

$$\begin{aligned} E[(Y-a)^2] &= E[(Y - EY + EY - a)^2] \\ &= E[(Y - EY)^2] + 2E[(Y - EY)(EY - a)] + (EY - a)^2. \end{aligned}$$

The cross-product is evaluated using the linearity of expectation and the fact that EY is a constant as

$$E[(Y - EY)(EY - a)] = (EY)^2 - aEY - (EY)^2 + aEY = 0$$

and hence from Property 1 of expectation,

$$E[(Y-a)^2] = E[(Y - EY)^2] + (EY - a)^2 \geq E[(Y - EY)^2], \tag{4.44}$$

which is the mean squared error resulting from using the mean of Y as an estimate. Thus *the mean of a random variable is the minimum mean squared error (MMSE) estimate of the value of a random variable in the absence of any a-priori information.*

What if one is given a-priori information? For example, suppose that now you are told that $X = x$. What then is the best estimate of Y, say $\hat{Y}(X)$? This problem is easily solved by modifying the previous derivation to use conditional expectation, that is, by using the conditional distribution for Y given X instead of the a-priori distribution for Y. Once again we try to minimize the mean squared error:

$$\begin{aligned}E[(Y-\hat{Y}(X))^2] &= E\left(E[(Y-\hat{Y}(X))^2|X]\right)\\ &= \sum_x p_X(x)E[(Y-\hat{Y}(X))^2|x].\end{aligned}$$

Each of the terms in the sum, however, is just a mean squared error between a random variable and an estimate of that variable with respect to a distribution, here the conditional distribution $p_{Y|X}(\cdot|x)$. By the same argument as was used in the unconditional case, the best estimate is the mean, but now the mean with respect to the conditional distribution, i.e., $E(Y|x)$. In other words, for each x the best $\hat{Y}(x)$ in the sense of minimizing the mean squared error is $E(Y|x)$. Plugging in the random variable X in place of the dummy variable x we have the following interpretation:

- The conditional expectation $E(Y|X)$ of a random variable Y given a random variable X is the minimum mean squared estimate of Y given X.

A direct proof of this result without invoking the conditional version of the result for unconditional expectation follows from general iterated expectation. Suppose that $g(X)$ is an estimate of Y given X. Then the resulting mean squared error is

$$\begin{aligned}E[(Y-g(X))^2] &= E[(Y-E(Y|X)+E(Y|X)-g(X))^2]\\ &= E[(Y-E(Y|X))^2]\\ &\quad +2E[(Y-E(Y|X))(E(Y|X)-g(X))]\\ &\quad +E[(E(Y|X)-g(X))^2].\end{aligned}$$

Expanding the cross term yields

$$\begin{aligned}&E[(Y-E(Y|X))(E(Y|X)-g(X))] =\\ &\qquad E[YE(Y|X)]-E[Yg(X)]-E[E(Y|X)^2]+E[E(Y|X)g(X)].\end{aligned}$$

From the general iterated expectation (4.36), $E[YE(Y|X)] = E[E(Y|X)^2]$ (setting $g(X)$ of the lemma to $E(Y|X)$ and $h(X,Y)=Y$); and $E[Yg(X)] = E[E(Y|X)g(X)]$ (setting $g(X)$ of the lemma to the $g(X)$ used here and $h(X,Y)=Y$).

As with ordinary expectation, the ideas of conditional expectation can be extended to continuous random variables by substituting conditional pdf's for the unconditional pdf's. As is the case with conditional probability, however, this constructive definition has its limitations and only makes sense when the pdf's are well defined. The rigorous development of conditional expectation is, like conditional probability, analogous to the rigorous treatment of the Dirac delta; it is defined by its behavior underneath the integral sign rather than by a construction. When the constructive definition makes sense, the two approaches agree.

The conditional expectation cannot be explicitly evaluated in most cases. However, for the very important case of jointly Gaussian random variables we can immediately identify from (3.60) that

$$E[Y|X] = m_Y + \rho(\sigma_Y/\sigma_X)(X - m_X). \tag{4.45}$$

It is important that this is in fact an affine function of X.

The same ideas extend from scalars to vectors. Suppose we observe a real-valued column vector $X = (X_0, \ldots, X_{k-1})^t$ and we wish to predict or estimate a second random vector $Y = (Y_0, \ldots, Y_{m-1})^t$. Note that the dimensions of the two vectors need not be the same.

The prediction $\hat{Y} = \hat{Y}(X)$ is to be chosen as a function of X which yields the smallest possible mean squared error, as in the scalar case. The mean squared error is defined as

$$\begin{aligned} \epsilon^2(\hat{Y}) &= E(\|Y - \hat{Y}\|^2) \triangleq E[(Y - \hat{Y})^t(Y - \hat{Y})] \\ &= \sum_{i=0}^{m-1} E[(Y_i - \hat{Y}_i)^2]. \end{aligned} \tag{4.46}$$

An estimator or predictor is said to be *optimal* within some class of predictors if it minimizes the mean squared error over all predictors in the given class.

Two specific examples of vector estimation are of particular interest. In the first case, the vector X consists of k consecutive samples from a stationary random process, say $X = (X_{n-1}, X_{n-2}, \ldots, X_{n-k})$ and Y is the one step ahead, next, or "future", sample $Y = X_n$. In this case the goal is to find the best one-step predictor given the finite past. In the second example, Y is a rectangular subblock of pixels in a sampled image intensity raster and X consists of similar subgroups above and to the left of Y. Here the goal is to use portions of an image already coded or processed to predict a new portion of the same image. This vector prediction problem is depicted in Figure 4.1 where subblocks A, B, and C would be used to predict subblock D.

The following theorem shows that the best nonlinear predictor of Y given X is simply the conditional expectation of Y given X. Intuitively, our best guess of an unknown vector is its expectation or mean *given* whatever ob-

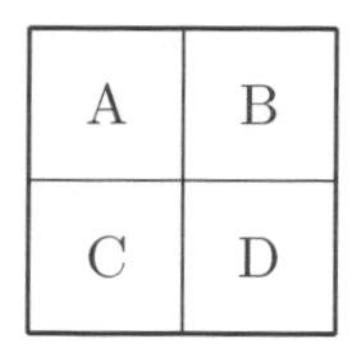

Figure 4.1 Vector Prediction of image subblocks

servations that we have. This extends the interpretation of a conditional expectation as an optimal estimator to the vector case.

Theorem 4.5 *Given two random vectors Y and X, the minimum mean squared error (MMSE)estimate of Y given X is*

$$\hat{Y}(X) = E(Y|X). \tag{4.47}$$

Proof As in the scalar case, the proof does not require calculus or Lagrange minimizations. Suppose that $\hat{Y}$ is the claimed optimal estimate and that $\tilde{Y}$ is some other estimate. We will show that $\tilde{Y}$ must yield a mean squared error no smaller than does $\hat{Y}$. To see this consider

$$\begin{aligned}\epsilon^2(\tilde{Y}) &= E(\|Y - \tilde{Y}\|^2) = E(\|Y - \hat{Y} + \hat{Y} - \tilde{Y}\|^2) \\ &= E(\|Y - \hat{Y}\|^2) + E(\|\hat{Y} - \tilde{Y}\|^2) + 2E[(Y - \hat{Y})^t(\hat{Y} - \tilde{Y})] \\ &\geq \epsilon^2(\hat{Y}) + 2E[(Y - \hat{Y})^t(\hat{Y} - \tilde{Y})].\end{aligned}$$

We will prove that the rightmost term is zero and hence that $\epsilon^2(\tilde{Y}) \geq \epsilon^2(\hat{Y})$, which will prove the theorem. Recall that $\hat{Y} = E(Y|X)$ and hence

$$E[(Y - \hat{Y})|X] = 0.$$

Since $\hat{Y} - \tilde{Y}$ is a deterministic function of X,

$$E[(Y - \hat{Y})^t(\hat{Y} - \tilde{Y})|X] = 0.$$

Then, by iterated expectation applied to vectors, we have

$$E[(Y - \hat{Y})^t(\hat{Y} - \tilde{Y})] = E(E[(Y - \hat{Y})^t(\hat{Y} - \tilde{Y})|X]) = 0$$

as claimed, which proves the theorem. □

As in the scalar case, the conditional expectation is in general a difficult function to evaluate with the notable exception of jointly Gaussian vectors. From (4.40)–(4.43) , recall that for jointly Gaussian vectors Y and X with $K_{(X,Y)} = E[((X^t, Y^t) - (m_X^t, m_Y^t))^t((X^t, Y^t) - (m_X^t, m_Y^t))]$, $K_Y = E[(Y - m_Y)(Y - m_Y)^t]$, $K_X = E[(X - m_X)(X - m_X)^t]$, $K_{XY} = E[(X -$

$m_X)(Y - m_Y)^t]$, $K_{YX} = E[(Y - m_Y)(X - m_X)^t]$ the conditional pdf is is

$$f_{Y|X}(y|x) = (2\pi)^{-m/2}(\det(K_{Y|X}))^{-\frac{1}{2}} \times \exp\left(-\frac{(y - m_{Y|x})^t K_{Y|X}^{-1}(y - m_{Y|x})}{2}\right), \tag{4.48}$$

where

$$\begin{aligned} K_{Y|X} &\triangleq K_Y - K_{YX}K_X^{-1}K_{XY} \\ &= E[(Y - E(Y|X))(Y - E(Y|X))^t|X], \end{aligned} \tag{4.49}$$

$$\det(K_{Y|X}) = \frac{\det(K_{(Y,X)})}{\det(K_X)}, \tag{4.50}$$

and

$$E(Y|X = x) = m_{Y|x} = m_Y + K_{YX}K_X^{-1}(x - m_X), \tag{4.51}$$

and hence the minimum mean squared error estimate of Y given X is

$$E(Y|X) = m_Y + K_{YX}K_X^{-1}(X - m_X), \tag{4.52}$$

which is an affine function of X! The resulting minimum mean squared error is (using iterated expectation)

$$\begin{aligned} &E[(Y - E(Y|X))^t(Y - E(Y|X))] \\ &= E\left(E[(Y - E(Y|X))^t(Y - E(Y|X))|X]\right) \\ &= E\left(E[\mathrm{Tr}[(Y - E(Y|X))(Y - E(Y|X))^t]|X]\right) \\ &= \mathrm{Tr}(K_{Y|X}). \end{aligned} \tag{4.53}$$

In the special case where $X = X^n = (X_0, X_1, \ldots, X_{n-1})$ and $Y = X_n$, the so-called one-step linear prediction problem, the solution takes an interesting form. For this case define the nth-order covariance matrix as the $n \times n$ matrix

$$K_X^{(n)} = E[(X^n - E(X^n))(X^n - E(X^n))^t], \tag{4.54}$$

i.e., the (k, j) entry of $K_X^{(n)}$ is $E[(X_k - E(X_k))(X_j - E(X_j))]$, $k, j = 0, 1, \ldots, n-1$. Then if X^{n+1} is Gaussian, the optimal one-step predictor for X_n given X^n is

$$\begin{aligned} &\hat{X}_n(X^n) = \\ &E(X_n) + E[(X_n - E(X_n))(X^n - E(X^n))^t](K_X^{(n)})^{-1}(X^n - E(X^n)) \end{aligned} \tag{4.55}$$

which has an affine form

$$\hat{X}_n(X^n) = AX^n + b \tag{4.56}$$

where

$$A = r^t(K_X^{(n)})^{-1}, \tag{4.57}$$

$$r = \begin{pmatrix} K_X(n,0) \\ K_X(n,1) \\ \vdots \\ K_X(n,n-1) \end{pmatrix}, \tag{4.58}$$

and

$$b = E(X_n) - AE(X^n). \tag{4.59}$$

The resulting mean squared error is

$$\begin{aligned} \text{MMSE} &= E[(X_n - \hat{X}_n(X^n))^2] \\ &= \text{Tr}(K_Y - K_{YX}K_X^{-1}K_{XY}) \\ &= \sigma_{X_n}^2 - r^t(K_X^{(n)})^{-1}r \end{aligned}$$

or

$$\text{MMSE} = E[(X_n - \hat{X}_n(X^n))^2] = \sigma^2_{X_n|X^n}, \tag{4.60}$$

which from (4.50) can be expressed as

$$\text{MMSE} = \frac{\det(K_X^{(n+1)})}{\det(K_X^{(n)})}, \tag{4.61}$$

a classical result from minimum mean squared error estimation theory.

A special case of interest occurs when the mean of X_n is constant for all n and the covariance function has the property that

$$K_X(k,l) = K_X(k+n, l+n), \text{ all } k, l, n,$$

that is, the covariance function evaluated at two sample times is unchanged if both sample times are shifted by an equal amount. In this case the covariance of two samples depends only on how far apart in time they are and not on the actual absolute time of each. As we shall detail in Section 4.18, a process with this property is said to be *weakly stationary*. In this case with the added assumption (for simplicity) that the constant mean is zero, the estimator simplifies to

$$\hat{X}_n(X^n) = r^t(K_X^{(n)})^{-1}X^n, \tag{4.62}$$

where r is the n-dimensional vector

$$r = \begin{pmatrix} K_X(n) \\ K_X(n-1) \\ \vdots \\ K_X(1) \end{pmatrix}. \tag{4.63}$$

4.10 ⋆Implications for linear estimation

The development of minimum mean squared error estimation for the Gaussian case provides a preview and an approach to the problem of minimum mean squared estimation for the situation of completely general random vectors (not necessarily Gaussian) where only linear or affine estimators are allowed (to avoid the problem of possibly intractable conditional expectations in the non-Gaussian case). This topic will be developed in some detail in a later section, but the key results will be shown to follow directly from the Gaussian case by reinterpreting the results.

The key fact is that the optimal estimator for a vector Y given a vector X when the two are jointly Gaussian was found to be an affine estimator, that is, to have the form

$$\hat{Y}(X) = AX + b.$$

The MMSE was found over all possible estimators as an estimator of this form with $A = K_{YX}K_X^{-1}$ and $b = E(Y) - AE(X)$ with a resulting MMSE $= \mathrm{Tr}(K_Y - K_{YX}K_X^{-1}K_{XY})$. Therefore it is obviously true that this MMSE must be the minimum achievable MSE over all affine estimators, i.e., that for all $k \times m$ matrices A and m-dimensional vectors b it is true that

$$\begin{aligned} \mathrm{MSE}(A,b) &= E\left[\mathrm{Tr}\left((Y - AX - b)(Y - AX - b)^t\right)\right] \\ &\geq \mathrm{Tr}(K_Y - K_{YX}K_X^{-1}K_{XY}) \end{aligned} \tag{4.64}$$

and that equality holds if and only if $A = K_{YX}K_X^{-1}$ and $b = E(Y) - AE(X)$. We shall now see that this version of the result has nothing to do with Gaussianity and that the inequality and solution are true for any distribution (providing of course that K_X is invertible).

Expanding the MSE and using some linear algebra results in

$$\begin{aligned}\text{MMSE}(A,b) &= E\left[\text{Tr}\left((Y-AX-b)(Y-AX-b)^t\right)\right] \\ &= E\left[\text{Tr}\left((Y-m_Y-A(X-m_X)-b+m_Y-Am_X)\right.\right. \\ &\qquad \left.\left.\times\ (Y-m_Y-A(X-m_X)-b+m_Y-Am_X)^t\right)\right] \\ &= \text{Tr}\left(K_Y - AK_{XY} - K_{YX}A^t + AK_XA^t\right) \\ &\qquad + (b-m_Y+Am_X)^t(b-m_Y+Am_X)\end{aligned}$$

where all the remaining cross terms are zero. Regardless of A the final term is nonnegative and hence is bounded below by 0. Therefore the minimum is achieved by the choice

$$b = m_Y - Am_X. \tag{4.65}$$

Thus the inequality we wish to prove becomes

$$\text{Tr}\left(K_Y - AK_{XY} - K_{YX}A^t + AK_XA^t\right) \geq \text{Tr}(K_Y - K_{YX}K_X^{-1}K_{XY}) \tag{4.66}$$

or

$$\text{Tr}\left(K_{YX}K_X^{-1}K_{XY} + AK_XA^t - AK_{XY} - K_{YX}A^t\right) \geq 0. \tag{4.67}$$

Since K_X is a covariance matrix it is Hermitian and since it has an inverse it must be positive definite. It therefore has a well-defined square root $K_X^{(1/2)}$ (see Section A.4) and hence the left-hand side of the previous equation can be expressed as

$$\text{Tr}\left((AK_X^{\frac{1}{2}} - K_{YX}K_X^{-\frac{1}{2}})(AK_X^{\frac{1}{2}} - K_{YX}K_X^{-\frac{1}{2}})^t\right) \tag{4.68}$$

(just expand this expression to verify it is the same as the previous expression). But this has the form $\text{Tr}(BB^t)$ which is just $\sum_i b_{i,i}^2$, which is nonnegative, proving the inequality. Plugging in $A = K_{YX}K_X^{-1}$ achieves the lower bound with equality.

We summarize the result in the following theorem.

Theorem 4.6 *Given random vectors X and Y with $K_{(X,Y)} = E[((X^t,Y^t)-(m_X^t,m_Y^t))^t((X^t,Y^t)-(m_X^t,m_Y^t))]$, $K_Y = E[(Y-m_Y)(Y-m_Y)^t]$, $K_X = E[(X-m_X)(X-m_X)^t]$, $K_{XY} = E[(X-m_X)(Y-m_Y)^t]$, $K_{YX} = E[(Y-m_Y)(X-m_X)^t]$, assume that K_X is invertible (or that it is positive definite). Then*

$$\begin{aligned}\min_{A,b} MMSE(A,b) &= \min_{A,b} \text{Tr}\left((Y-AX-b)(Y-AX-b)^t\right) \\ &= \text{Tr}(K_Y - K_{YX}K_X^{-1}K_{XY})\end{aligned} \tag{4.69}$$

and the minimum is achieved by $A = K_{YX}K_X^{-1}$ *and* $b = E(Y) - AE(X)$.

In particular, this result does not require that the vectors be jointly Gaussian.

As in the Gaussian case, the results can be specialized to the situation where $Y = X_n$, $X = X^n$ and $\{X_n\}$ is a weakly stationary process to obtain that the optimal linear estimator of X_n given $(X_0, \ldots, X_{n-1})$ in the sense of minimizing the mean squared error is

$$\hat{X}_n(X^n) = r^t(K_X^{(n)})^{-1}X^n, \tag{4.70}$$

where r is the n-dimensional vector

$$r = \begin{pmatrix} K_X(n) \\ K_X(n-1) \\ \vdots \\ K_X(1) \end{pmatrix}. \tag{4.71}$$

The resulting minimum mean squared error (called the "linear least squares error") is

$$\text{LLSE} = \sigma_X^2 - r^t(K_X^{(n)})^{-1}r = \frac{\det(K_X^{(n)})}{\det(K_X^{(n-1)})}. \tag{4.72}$$

a classical result of linear estimation theory. Note that the equation with the determinant form does not require a Gaussian density, although a Gaussian density was used to identify the first form with the determinant form (both being $\sigma^2_{X_n|X^n}$ in the Gaussian case).

There are fast algorithms for computing inverses of covariance matrices and hence for computing linear least squares estimators. This is of particular interest in applications such as speech processing which depend on fast computation of linear estimators. For example, the classical Gram–Schmidt orthogonalization procedure of matrix theory leads to the Cholesky decomposition of positive definite matrices. If the process is stationary (and hence the covariance matrix is Toeplitz), then the Levinson-Durbin algorithm can be used. See, e.g., [50] for a discussion and and reference list for these methods along with a general survey of applications of linear prediction methods to speech processing, an area pioneered by Fumitada Itakura and Bishnu S. Atal.

4.11 Correlation and linear estimation

As an example of the application of correlations, we consider a constrained form of the minimum mean squared error estimation problem that provided an application and interpretation for conditional expectation. A problem with the earlier result is that in some applications the conditional expectation will be complicated or unknown, but the simpler correlation might be known or at least possible to approximate from observed data. While the conditional expectation provides the optimal estimator over all possible estimators, the correlation turns out to provide an optimal estimator over a restricted class of estimators.

Suppose again that the value of X is observed and that a good estimate of Y, say $\hat{Y}(X)$, is desired. Once again the quality of an estimator will be measured by the resulting mean squared error, but this time we do not allow the estimator to be an arbitrary function of the observation. It is required to be a *linear* function of the form

$$\hat{Y}(x) = ax + b, \tag{4.73}$$

where a and b are fixed constants which are chosen to minimize the mean squared error. Strictly speaking, this is an *affine* function rather than a linear function; it is linear if $b = 0$. The terminology is common, however, and we will use it.

The goal now is to find a and b which minimizes

$$E[(Y - \hat{Y}(X))^2] = E[(Y - aX - b)^2]. \tag{4.74}$$

Rewriting the formula for the error in terms of the mean-removed random variables yields for any a, b:

$$\begin{aligned} &E\left([Y - (aX + b)]^2\right) \\ &= E\left([(Y - EY) - a(X - EX) - (b - EY + aEX)]^2\right) \\ &= \sigma_Y^2 + a^2\sigma_X^2 + (b - EY + aEX)^2 - 2a\,\mathrm{COV}(X, Y) \end{aligned}$$

since the remaining cross-products are all zero (why?). (Recall that the covariance $\mathrm{COV}(X, Y)$ was defined in Subsection 4.6.2.) Since the first term does not depend on a or b, minimizing the mean squared error is equivalent to minimizing

$$a^2\sigma_X^2 + (b - EY + aEX)^2 - 2a\,\mathrm{COV}(X, Y).$$

First note that the middle term is nonnegative. Once a is chosen, this term will be minimized by choosing $b = EY - aEX$, which makes this term 0. Thus the best a must minimize $a^2\sigma_X^2 - 2a\,\mathrm{COV}(X, Y)$. A little calculus shows that the minimizing a is

$$a = \frac{\text{COV}(X,Y)}{\sigma_X^2} \tag{4.75}$$

and hence the best b is

$$b = EY - \frac{\text{COV}(X,Y)}{\sigma_X^2} EX. \tag{4.76}$$

The connection of second-order moments and linear estimation also plays a fundamental role in the vector analog to the problem of the previous section, that is, in the estimation of a vector Y given an observed vector X. The details are more complicated, but the basic ideas are essentially the same.

Unfortunately the conditional expectation is mathematically tractable only in a few very special cases, e.g., the case of jointly Gaussian vectors. In the Gaussian case the conditional expectation given X is formed by a simple matrix multiplication on X with possibly a constant vector being added; that is, the optimal estimate has a linear form. (As in the scalar case, technically this is an *affine* form and not a linear form if a constant vector is added.) Even when the random vectors are not Gaussian, linear predictors or estimates are important because of their simplicity. Although they are not in general optimal, they play an important role in signal processing. Hence we next turn to the problem of finding the optimal *linear* estimate of one vector given another.

Suppose as before that we are given a k-dimensional vector X and wish to predict an m-dimensional vector Y. We now restrict ourselves to estimates of the form

$$\hat{Y} = AX,$$

where the $m \times k$-dimensional matrix A can be considered as a matrix of k-dimensional row vectors a_k^t; $k = 0, \ldots, m-1$:

$$A = [a_0, a_2, \ldots, a_{m-1}]^t$$

so that if $\hat{Y} = (\hat{Y}_0, \ldots, \hat{Y}_{m-1})^t$, then

$$\hat{Y}_i = a_i^t X$$

and hence

$$\epsilon^2(\hat{Y}) = \sum_{i=1}^{k} E[(Y_i - a_i^t X)^2]. \tag{4.77}$$

The goal is to find the matrix A that minimizes ϵ^2. This can be considered as a function of the estimate $\hat{Y}$ or of the matrix A defining the estimate. We shall provide two separate solutions which are almost, but not quite, equivalent. The first is constructive in nature: a specific A will be given and shown to be optimal. The second development is descriptive: without actually giving the matrix A, we will show that a certain property is necessary and sufficient for the matrix to be optimal. That property is called the *orthogonality principle.* It states that the optimal matrix is the one that causes the error vector $Y - \hat{Y}$ to be orthogonal to (have zero correlation with) the observed vector X. The first development is easier to use because it provides a formula for A that can be immediately computed in many cases. The second development is less direct and less immediately applicable, but it turns out to be more general: the descriptive property can be used to derive A even when the first development is not applicable. The orthogonality principle plays a fundamental role in all of linear estimation theory.

The error $\epsilon^2(A)$ is minimized if each term $E[(Y_i - a_i^t X)^2]$ is minimized over a_i since there is no interaction among the terms in the sum. We can do no better when minimizing a sum of such positive terms than to minimize each term separately. Thus the fundamental problem is the following simpler one: given a random vector X and a random variable (one-dimensional or scalar vector) Y, we seek a vector a that minimizes

$$\epsilon^2(a) = E[(Y - a^t X)^2]. \tag{4.78}$$

One way to find the optimal a is to use calculus, setting derivatives of $\epsilon^2(a)$ to zero and verifying that the stationary point so obtained is a global minimum. As previously discussed, variational techniques can be avoided via elementary inequalities if the answer is known. We shall show that the optimal a is a solution of

$$a^t R_X = E(YX^t), \tag{4.79}$$

so that if the *autocorrelation matrix* defined by

$$R_X = E[XX^t] = \{R_X(k,i) = E(X_k X_i);\ k,i = 0,\ldots,k-1\}$$

is invertible, then the optimal a is given by

$$a^t = E(YX^t)R_X^{-1}. \tag{4.80}$$

To prove this we assume that a satisfies (4.80) and show that for any other vector b

$$\epsilon^2(b) \geq \epsilon^2(a)\,. \tag{4.81}$$

To do this we write

$$\begin{aligned}\epsilon^2(b) &= E[(Y - b^t X)^2] = E[(Y - a^t X + a^t X - b^t X)^2] \\ &= E[(Y - a^t X)^2] + 2E[(Y - a^t X)(a^t X - b^t X)] + E[(a^t X - b^t X)^2].\end{aligned}$$

Of the final terms, the first term is just $\epsilon^2(a)$ and the rightmost term is obviously nonnegative. Thus we have the bound

$$\epsilon^2(b) \geq \epsilon^2(a) + 2E[(Y - a^t X)(a^t - b^t)X]. \tag{4.82}$$

The cross-product term can be written as

$$\begin{aligned}2E[(Y - a^t X)(a^t - b^t)X] &= 2E[(Y - a^t X)X^t(a - b)] \\ &= 2E[(Y - a^t X)X^t](a - b) \\ &= 2\left(E[YX^t] - a^t E[XX^t]\right)(a - b) \\ &= 2\left(E[YX^t] - a^t R_X\right)(a - b) \\ &= 0\end{aligned} \tag{4.83}$$

invoking (4.79). Combining this with (4.82) proves (4.81) and hence optimality. Note that because of the symmetry of autocorrelation matrices and their inverses, we can rewrite (4.80) as

$$a = (E(YX^t)R_X^{-1})^t = R_X^{-1}E[YX]. \tag{4.84}$$

Using the above result to perform a termwise minimization of (4.77) now yields the following theorem describing the optimal linear vector predictor.

Theorem 4.7 *The minimum mean squared error linear predictor of the form* $\hat{Y} = AX$ *is given by any solution* A *of the equation*

$$AR_X = E(YX^t).$$

If the matrix R_X *is invertible, then* A *is uniquely given by*

$$A^t = R_X^{-1}E[XY^t],$$

that is, the matrix A *has rows* a_i^t; $i = 0, 1, \ldots, m$, *with*

$$a_i = R_X^{-1}E[Y_i X].$$

Alternatively,

$$A = E[YX^t]R_X^{-1}. \tag{4.85}$$

Having found the best linear estimate, it is easy to modify the development to find the best estimate of the form

$$\hat{Y}(X) = AX + b, \tag{4.86}$$

where now we allow an additional constant term. This is also often called a linear estimate, although as previously noted it is more correctly called an affine estimate because of the extra constant vector term. As the end result and proof strongly resemble the linear estimate result, we proceed directly to the theorem.

Theorem 4.8 *The minimum mean squared estimate of the form $\hat{Y} = AX + b$ is given by any solution A of the equation*

$$AK_X = E[(Y - E(Y))(X - E(X))^t], \tag{4.87}$$

where the covariance matrix K_X is defined by

$$K_X = E[(X - E(X))(X - E(X))^t] = R_{X-E(X)},$$

and

$$b = E(Y) - AE(X).$$

If K_X is invertible, then

$$A = E[(Y - E(Y))(X - E(X))^t]K_X^{-1}. \tag{4.88}$$

Note that if X and Y have zero-means, then the result reduces to the previous result; that is, affine predictors offer no advantage over linear predictors for zero-mean random vectors.

Proof Let C be any matrix and d any vector (both of suitable dimensions) and note that

$$\begin{aligned}
&E(\|Y - (CX + d)\|^2)\\
&\quad = E(\|(Y - E(Y)) - C(X - E(X)) + E(Y) - CE(X) - d\|^2)\\
&\quad = E(\|(Y - E(Y)) - C(X - E(X))\|^2) + E(\|E(Y) - CE(X) - d\|^2)\\
&\qquad\quad + 2E[Y - E(Y) - C(X - E(X))]^t[E(Y) - CE(X) - d].
\end{aligned}$$

From Theorem 4.7, the first term is minimized by choosing $C = A$, where A is a solution of (4.87); also, the second term is the expectation of the squared norm of a vector that is identically zero if $C = A$ and $d = b$, and similarly for this choice of C and d the third term is zero. Thus

$$E(\|Y - (CX + d)\|^2) \geq E(\|Y - (AX + b)\|^2).$$

We often restrict interest to linear estimates by assuming that the various vectors have zero-mean. This is not always possible, however. For example, groups of pixels in a sampled image intensity raster can be used to predict other pixel groups, but pixel values are always nonnegative and hence have nonzero-means. Hence in some problems affine predictors may be preferable. Nonetheless, we will often follow the common practice of focusing on the linear case and extending when necessary. In most studies of linear prediction it is assumed that the mean is zero, i.e., that any dc value of the process has been removed. If this assumption is not made, linear estimation theory is still applicable but will generally give inferior performance to the use of affine prediction.

4.11.1 The orthogonality principle

Although we have proved the form of the optimal linear predictor of one vector given another, there is another way to describe the result that is often useful for deriving optimal linear predictors in somewhat different situations. To develop this alternative viewpoint we focus on the error vector

$$e = Y - \hat{Y}. \tag{4.89}$$

Rewriting (4.89) as $Y = \hat{Y} + e$ points out that the vector Y can be considered as its estimate plus an error or "noise" term. The goal of an optimal predictor is then to minimize the error energy $e^t e = \sum_{n=0}^{k-1} e_n^2$. If the estimate is linear, then

$$e = Y - AX.$$

As with the basic development for the linear predictor, we simplify things for the moment. We look at the scalar prediction problem of predicting a random variable Y by $\hat{Y} = a^t X$ yielding a scalar error of $e = Y - \hat{Y} = Y - a^t X$. Since we have seen that the overall mean squared error $E[e^t e]$ in the vector case is minimized by separately minimizing each component $E[e_k^2]$, we can later easily extend our results for the scalar case to the vector case.

Suppose that a is chosen optimally and consider the crosscorrelation between an arbitrary error term and the observable vector:

$$\begin{aligned} E[(Y - \hat{Y})X] &= E[(Y - a^t X)X] \\ &= E[YX] - E[X(X^t a)] \\ &= E[YX] - R_X a = 0 \end{aligned}$$

using (4.79).

Thus for the optimal predictor, the error satisfies

$$E[eX] = 0,$$

or, equivalently,

$$E[eX_n] = 0;\ n = 0, \ldots, k. \tag{4.90}$$

When two random variables e and X are such that their expected product $E(eX)$ is 0, they are said to be *orthogonal* and we write

$$e \perp X.$$

We have therefore shown that the optimal linear estimate of a scalar random variable given a vector of observations causes the error to be orthogonal to all of the observables and hence orthogonality of error and observations is a necessary condition for optimality of a linear estimate.

Conversely, suppose that we know a linear estimate a is such that it renders the prediction error orthogonal to all of the observations. Arguing as we have before, suppose that b is any other linear predictor vector and observe that

$$\begin{aligned}\epsilon^2(b) &= E[(Y - b^tX)^2] \\ &= E[(Y - a^tX + a^tX - b^tX)^2] \\ &\geq \epsilon^2(a) + 2E[(Y - a^tX)(a^tX - b^tX)],\end{aligned}$$

where the equality holds if $b = a$. Letting $e = Y - a^tX$ denote the error resulting from an a that makes the error orthogonal with the observations, the rightmost term can be rewritten as

$$2E[e(a^tX - b^tX)] = 2(a^t - b^t)E[eX] = 0.$$

Thus we have shown that $\epsilon^2(b) \geq \epsilon^2(a)$ and hence no linear estimate can outperform one yielding an error orthogonal to the observations and hence such orthogonality is sufficient as well as necessary for optimality.

Since the optimal estimate of a vector Y given X is given by the componentwise optimal predictions given X, we have thus proved the following alternative to Theorem 4.7.

Theorem 4.9 *The orthogonality principle*
A linear estimate $\hat{Y} = AX$ is optimal (in the mean squared error sense) if and only if the resulting errors are orthogonal to the observations, that is, if $e = Y - AX$, then

$$E[e_kX_n] = 0;\ k = 1, \ldots, K;\ n = 1, \ldots, N.$$

4.12 Correlation and covariance functions

We turn now to correlation in the framework of random processes. The notion of an iid random process can be generalized by specifying the component random variables to be merely uncorrelated rather than independent. Although requiring the random process to be uncorrelated is a much weaker requirement, the specification is sufficient for many applications, as will be seen in several ways. In particular, in this chapter, the basic laws of large numbers require only the weaker assumption and hence are more general than they would be if independence were required. To define the class of uncorrelated processes, it is convenient to introduce the notions of autocorrelation functions and covariance functions of random processes.

Given a random process $\{X_t;\, t \in \mathcal{T}\}$, the *autocorrelation function* $R_X(t,s);\, t,s \in \mathcal{T}$ is defined by

$$R_X(t,s) = E(X_t X_s); \;\text{ all } t,s \in \mathcal{T}.$$

If the process is complex-valued instead of real-valued, the definition becomes

$$R_X(t,s) = E(X_t X_s^*); \;\text{ all } t,s \in \mathcal{T}.$$

The *autocovariance function* or simply the *covariance function* $K_X(t,s)$; $t,s,\in \mathcal{T}$ is defined by

$$K_X(t,s) = \mathrm{COV}(X_t, X_s).$$

Observe that (4.18) relates the two functions by

$$K_X(t,s) = R_X(t,s) - (EX_t)(EX_s^*). \tag{4.91}$$

Thus the autocorrelation and covariance functions are equal if the process has zero-mean, that is, if $EX_t = 0$ for all t. The covariance function of a process $\{X_t\}$ can be viewed as the autocorrelation function of the process $\{X_t - EX_t\}$ formed by removing the mean from the given process to form a new process having zero-mean.

The autocorrelation function of a random process is given by the correlation of all possible pairs of samples; the covariance function is the covariance of all possible pairs of samples. Both functions provide a measure of how dependent the samples are and will be seen to play a crucial role in laws of large numbers. Note that both definitions are valid for random processes in either discrete time or continuous time and having either a discrete alphabet or a continuous alphabet.

In terms of the correlation function, a random process $\{X_t;\, t \in \mathcal{T}\}$ is said to be *uncorrelated* if

$$R_X(t,s) = \begin{cases} E(X_t^2) & \text{if } t = s \\ EX_tEX_s & \text{if } t \neq s \end{cases}$$

or, equivalently, if

$$K_X(t,s) = \begin{cases} \sigma_{X_t}^2 & \text{if } t = s \\ 0 & \text{if } t \neq s \end{cases}.$$

The reader should not overlook the obvious fact that if a process is iid or uncorrelated, the random variables are independent or uncorrelated *only if taken at different times.* That is, X_t and X_s will not be independent or uncorrelated when $t = s$, only when $t \neq s$ (except, of course, in such trivial cases as that where $\{X_t\} = \{a_t\}$, a sequence of constants where $E(X_tX_t) = a_ta_t = EX_tEX_t$ and hence X_t is uncorrelated with itself).

Gaussian Processes Revisited

Recall from Chapter 3 that a Gaussian random process $\{X_t;\ t \in \mathcal{T}\}$ is completely described by a mean function $\{m_t;\ t \in \mathcal{T}\}$ and a covariance function $\{\Lambda(t,s);\ t,s \in \mathcal{T}\}$. As one might suspect, the names of these functions come from the fact that they are indeed the mean and covariance functions as defined in terms of expectations, i.e.,

$$m_t = EX_t \tag{4.92}$$

$$\Lambda(t,s) = K_X(t,s). \tag{4.93}$$

The result for the mean follows immediately from our computation of the mean of a Gaussian $N(m,\sigma^2)$ random variable. The result for the covariance can be derived by brute force integration (not too bad if the integrator is well versed in matrix transformations of multidimensional integrals) or looked up in tables somewhere. The computation is tedious and we will simply state the result without proof. The multidimensional characteristic functions to be introduced later can be used to get a relatively simple proof, but again it is not worth the effort to fill in the details.

A more important issue is the properties that were required for a covariance function when the Gaussian process was defined. Recall that it was required that the covariance function of the process be symmetric, i.e., $K_X(t,s) = K_X(s,t)$, and positive definite, i.e., given any positive integer k, any collection of sample times $\{t_0,\ldots,t_{k-1}\}$, and any k real numbers

a_i; $i = 0, \ldots, k-1$ (not all 0), then

$$\sum_{i=0}^{n-1}\sum_{l=0}^{n-1} a_i a_l K_X(t_i, t_l) \geq 0. \tag{4.94}$$

If the process is complex-valued, then the covariance function is Hermitian, i.e., $K_X(t,s) = K_X^*(s,t)$ We now return to these conditions to see if they are indeed necessary conditions for all covariance functions, Gaussian or not.

Symmetry is easy. It immediately follows from the definitions that

$$\begin{aligned} K_X(t,s) &= E[(X_t - EX_t)(X_s - EX_s)] \\ &= E[(X_s - EX_s)(X_t - EX_t)] \\ &= K_X(s,t) \end{aligned} \tag{4.95}$$

and hence clearly all covariance functions of real-valued random processes are symmetric, and so are covariance matrices formed by sampling covariance functions. The Hermitian property for complex-valued random processes follows similarly. To see that positive definiteness is indeed almost a requirement, consider the fact that

$$\begin{aligned} \sum_{i=0}^{n-1}\sum_{l=0}^{n-1} a_i a_l K_X(t_i, t_l) &= \sum_{i=0}^{n-1}\sum_{l=0}^{n-1} a_i a_l E[(X_{t_i} - EX_{t_i})(X_{t_l} - EX_{t_l})] \\ &= E\left(\sum_{i=0}^{n-1}\sum_{l=0}^{n-1} a_i a_l (X_{t_i} - EX_{t_i})(X_{t_l} - EX_{t_l})\right) \\ &= E\left(\left|\sum_{i=0}^{n-1} a_i (X_{t_i} - EX_{t_i})\right|^2\right) \geq 0. \end{aligned} \tag{4.96}$$

Thus any covariance function K_X must at least be nonnegative definite, which implies that any covariance matrix formed by sampling the covariance function must also be nonnegative definite. Thus nonnegative definiteness is necessary for a covariance function and our requirement for a Gaussian process was only slightly stronger than was needed. We will later see how to define a Gaussian process when the covariance function is only nonnegative definite and not necessarily positive definite.

A slight variation on the above argument shows that for any random vector $X = (X_0, \ldots, X_{k-1})^t$, then the covariance matrix $\Lambda = \{\lambda_{i,l};\ i, l \in \mathcal{Z}_k\}$ defined by $\lambda_{i,l} = E[(X_i - EX_i)(X_l - EX_l)]$ must also be symmetric and nonnegative definite. This was the reason for assuming that the covariance matrix for a Gaussian random vector had at least these properties.

We make two important observations before proceeding. First, remember that the four basic properties of expectation have nothing to do with independence. In particular, whether or not the random variables involved are independent or uncorrelated, one can always interchange the expectation operation and the summation operation (Property 3), because expectation is linear. On the other hand, one cannot interchange the expectation operation with the product operation (this is not a property of expectation) unless the random variables involved are uncorrelated, e.g., when they are independent. Second, an iid process is also a discrete time uncorrelated random process with identical marginal distributions. The converse statement is not true in general; that is, the notion of an uncorrelated process is more general than that of an iid process. Correlation measures only a weak pairwise degree of independence. A random process can even be pairwise independent (and hence uncorrelated) but still not be iid (Problem 4.31).

4.13 ⋆The central limit theorem

The characteristic function of a sum of iid Gaussian random variables has been shown to be Gaussian, and linear combinations of jointly Gaussian variables have also been shown to be Gaussian. Far more surprising is that the characteristic function of the sum of many non-Gaussian random variables turns out to be approximately Gaussian if the variables are suitably scaled and shifted. This result is called the *central limit theorem* and is one of the primary reasons for the importance of Gaussian distributions. When a large number of effects are added up with suitable scaling and shifting, the resulting random variable looks Gaussian even if the underlying individual effects are not at all Gaussian. This result is developed in this subsection.

Just as with laws of large numbers, there is no single central limit theorem – there are many versions. The various central limit theorems differ in the conditions of applicability. However, they have a common conclusion: the distribution or characteristic function of the sum of a collection of random variables converges to that of a Gaussian random variable. We will present only the simplest form of central limit theorem, that for iid random variables.

Suppose that $\{X_n\}$ is an iid random process with a common distribution F_X described by a pmf or pdf except that it has a finite mean $EX_n = m$ and finite variance $\sigma^2_{X_n} = \sigma^2$. It will also be assumed that the characteristic function $M_X(ju)$ is well behaved for small u in a manner to be made precise. Consider the "standardized" or "normalized" sum

$$R_n = \frac{1}{n^{1/2}} \sum_{i=0}^{n-1} \frac{X_i - m}{\sigma}. \tag{4.97}$$

By subtracting the means and dividing by the square root of the variance (the standard deviation), the resulting random variable is easily seen to have zero-mean and unit variance; that is,

$$ER_n = 0, \quad \sigma^2_{R_n} = 1,$$

hence the description "standardized," or "normalized." Note that unlike the sample average that appears in the law of large numbers, the sum here is normalized by $n^{-1/2}$ and not n^{-1}.

Using characteristic functions, we have from the independence of the $\{X_i\}$ and Lemma 4.1 that

$$M_{R_n}(ju) = M_{(X-m)/\sigma}\left(\frac{ju}{n^{1/2}}\right)^n. \tag{4.98}$$

We wish to investigate the asymptotic behavior of the characteristic function of (4.98) as $n \to \infty$. This is accomplished by assuming that σ^2 is finite and applying the approximation of (4.15) to $M_{(X-m)/\sigma}\left(ju/n^{-1/2}\right)^n$ and then finding the limiting behavior of the expression. Let $Y = (X - m)/\sigma$. Y has zero-mean and a second moment of 1, and hence from (4.15)

$$M_{(X-m)/\sigma}(jun^{1/2}) = 1 - \frac{u^2}{2n} + o(\frac{u^2}{n}), \tag{4.99}$$

where the rightmost term goes to zero faster than u^2/n. Combining this result with (4.98) produces

$$\lim_{n\to\infty} M_{R_n}(ju) = \lim_{n\to\infty}\left[1 - \frac{u^2}{2n} + o\left(\frac{u^2}{n}\right)\right]^n.$$

From elementary real analysis, however, this limit is

$$\lim_{n\to\infty} M_{R_n}(ju) = e^{-u^2/2},$$

the characteristic function of a Gaussian random variable with zero mean and unit variance! Thus, provided that (4.99) holds, a standardized sum of a family of iid random variables has a transform that converges to the transform of a Gaussian random variable regardless of the actual marginal distribution of the iid sequence.

By taking inverse transforms, the convergence of transforms implies that the cdf's will also converge to a Gaussian cdf (provided some technical conditions are satisfied to ensure that the operations of limits and integration can be exchanged). This does *not* imply convergence to a Gaussian *pdf*, however, because, for example, a finite sum of discrete random variables will not have a pdf (unless one resorts to Dirac delta functions). Given a sequence of random variables R_n with cdf F_n and a random variable R with distribution F, then if $\lim_{n\to\infty} F_n(r) = F(r)$ for all points r at which $F(r)$ is continuous, we say that R_n converges to R *in distribution.* Thus the central limit theorem states that under certain conditions, sums of iid random variables adjusted to have zero-mean and unit variance converge in distribution to a Gaussian random variable with the same mean and variance.

A slight modification of the above development shows that if $\{X_n\}$ is an iid sequence with mean m and variance σ^2, then

$$n^{-1/2}\sum_{k=0}^{n-1}(X_i - m)$$

will have a transform and a cdf converging to those of a Gaussian random variable with mean 0 and variance σ^2. We summarize the central limit theorem that we have established as follows.

Theorem 4.10 *A central limit theorem*
Let $\{X_n\}$ be an iid random process with a finite mean m and variance σ^2. Then

$$n^{-1/2}\sum_{k=0}^{n-1}(X_i - m)$$

converges in distribution to a Gaussian random variable with mean m and variance σ^2.

Intuitively the theorem states that if we sum up a large number of independent random variables and normalize by $n^{-1/2}$ so that the variance of the normalized sum stays constant, then the resulting sum will be approximately Gaussian. For example, a current meter across a resistor will measure the effects of the sum of millions of electrons randomly moving and colliding with each other. Regardless of the probabilistic description of these micro-events, the global current will appear to be Gaussian. Making this precise yields a model of thermal noise in resistors. Similarly, if dust particles are suspended on a dish of water and subjected to the random collisions of millions of molecules, then the motion of any individual particle in two dimensions will appear to be Gaussian. Making this rigorous yields the classic model for what is called "Brownian motion." A similar development in one dimension yields the Wiener process.

Note that in (4.98), if the Gaussian characteristic function is substituted on the right-hand side, a Gaussian characteristic function appears on the left. Thus the central limit theorem says that if you sum up random variables, you approach a Gaussian distribution. Once you have a Gaussian distribution, you "get stuck" there – adding more random variables of the same type (or Gaussian random variables) to the sum does not change the Gaussian characteristic. The Gaussian distribution is an example of an *infinitely divisible* distribution – the nth root of its characteristic function is a distribution of the same type as seen in (4.98). Equivalently stated, the distribution class is invariant under summations.

4.14 Sample averages

In many applications, engineers analyze the average behavior of systems based on observed data. It is important to estimate the accuracy of such estimates as a function of the amount of data available. This and the next sections are a prelude to such analyses. They also provide some very good practice in manipulating expectations and a few results of interest in their own right.

In this section we study the behavior of the arithmetic average of the first n values of a discrete time random process with either a discrete or a continuous alphabet. Specifically, the variance of the average is considered as a function of n.

Suppose we are given a process $\{X_n\}$. The sample average of the first n values of $\{X_n\}$ is $S_n = n^{-1}\sum_{i=0}^{n-1} X_i$. The mean of S_n is found easily using the linearity of expectation (expectation Property 3) as

$$ES_n = E\left[n^{-1}\sum_{i=0}^{n-1} X_i\right] = n^{-1}\sum_{i=0}^{n-1} EX_i. \tag{4.100}$$

Hence the mean of the sample average is the same as the average of the mean of the random variables produced by the process. Suppose now that we assume that the mean of the random variables is a constant, $EX_i = \overline{X}$ independent of i. Then $ES_n = \overline{X}$. In terms of estimation theory, if one estimates an unknown random process mean, $\overline{X}$, by S_n, then the estimate is said to be *unbiased* because the expected value of the estimate is equal to the value being estimated. Obviously an unbiased estimate is not unique, so being unbiased is only one desirable characteristic of an estimate (Problem 4.28).

Next consider the variance of the sample average:

$$\begin{aligned}
\sigma_{S_n}^2 &\triangleq E[(S_n - E(S_n))^2] \\
&= E\left[\left(n^{-1}\sum_{i=0}^{n-1} X_i - n^{-1}\sum_{i=0}^{n-1} EX_i\right)^2\right] \\
&= E\left[\left(n^{-1}\sum_{i=0}^{n-1} (X_i - EX_i)\right)^2\right] \\
&= n^{-2}\sum_{i=0}^{n-1}\sum_{j=0}^{n-1} E[(X_i - EX_i)(X_j - EX_j)].
\end{aligned}$$

The reader should be certain that the preceding operations are well understood, as they are frequently encountered in analyses. Note that expanding the square requires the use of separate dummy indices in order to get all of the cross-products. Once expanded, linearity of expectation permits the interchange of expectation and summation.

Recognizing the expectation in the sum as the covariance function, the variance of the sample average becomes

$$\sigma_{S_n}^2 = n^{-2}\sum_{i=0}^{n-1}\sum_{j=0}^{n-1} K_X(i,j). \tag{4.101}$$

Note that so far we have used none of the specific knowledge of the process, i.e., the above formula holds for general discrete time processes and does not require such assumptions as time-constant mean, time-constant variance, identical marginal distributions, independence, uncorrelated processes, etc. If we now use the assumption that the process is uncorrelated, the covariance becomes zero except when $i = j$, and expression (4.101) becomes

$$\sigma_{S_n}^2 = n^{-2}\sum_{i=0}^{n-1} \sigma_{X_i}^2. \tag{4.102}$$

If we now also assume that the variances $\sigma_{X_i}^2$ are equal to some constant value σ_X^2 for all times i, e.g., the process has identical marginal distributions as for an iid process, then the equations become

$$\sigma_{S_n}^2 = n^{-1}\sigma_X^2. \tag{4.103}$$

Thus, for uncorrelated discrete time random processes with mean and variance not depending on time, the sample average has expectation equal to the (time-constant) mean of the process, and the variance of the sample average tends to zero as $n \to \infty$. Of course we have only specified sufficient conditions. Expression (4.101) goes to zero with n under more general circumstances, as we shall see in detail later.

For now, however, we stick with uncorrelated process with mean and variance independent of time and require only a definition to obtain our first law of large numbers, a result implicit in equation (4.103).

4.15 Convergence of random variables

The preceding section demonstrated a form of convergence for the sequence of random variables, $\{S_n\}$, the sequence of sample averages, that is different from convergence as it is seen for a nonrandom sequence. To review, a nonrandom sequence $\{x_n\}$ is said to converge to the limit x if for every $\epsilon > 0$ there exists an N such that $|x_n - x| < \epsilon$ for every $n > N$. The preceding section did not see S_n converge in this sense. Nothing was said about the *individual* realizations $S_n(\omega)$ as a function of ω. Only the variance of the sequence $\sigma^2_{S_n}$ was shown to converge to zero in the usual ϵ, N sense. The variance calculation probabilistically averages across ω. For any particular ω, the realization S_n may, in fact, *not* converge to zero. It is important that the reader understand this distinction clearly.

Thus, in order to make precise the notion of convergence of sample averages to a limit, we need to make precise the notion of convergence of a sequence of random variables. We have already seen one example of convergence of random variables, the convergence in distribution used in the central limit theorem. In this section we will describe four additional notions of convergence of random variables. These are perhaps the most commonly encountered, but they are by no means an exhaustive list. The common goal is to quantify a useful definition for saying that a sequence of random variables, say Y_n, $n = 1, 2, \ldots$, converges to a random variable Y, which will be considered the *limit* of the sequence. Our main application will be the case where $Y_n = S_n$, a sample average of n samples of a random process, and Y is the expectation of the samples, that is, the limit is a trivial random variable, a constant.

One might wonder why five different notions of convergence of random variables are considered. Unfortunately no single definition is useful in all applications and the definitions that are arguably most natural are the most difficult to work with. Each has its uses, as will be seen.

The most straightforward generalization of the usual idea of a limit to random variables is easy to define, but virtually useless. If for *every* sample point ω we had $\lim_{n\to\infty} Y_n(\omega) = Y(\omega)$ in the usual sense of convergence of numbers, then we could say that Y_n converges *pointwise* to Y, that is, for every sample point in the sample space. Unfortunately it is rarely possible to prove so strong a result, nor is it necessary, nor is it usually true.

A slight variation of this yields a far more important important notion of convergence. A sequence of random variables Y_n, $n = 1, 2, \ldots$, is said to converge to a random variable Y *with probability one* or converge *w.p. 1* if the set of sample points ω such that $\lim_{n\to\infty} Y_n(\omega) = Y(\omega)$ is an event with probability one. Thus a sequence converges with probability one if it converges pointwise on a set of probability one; it can do anything outside that set, e.g., converge to something else or not converge at all. Since the total probability of all such bad sequences is 0, this has no practical significance. Although convergence w.p. 1 is the easiest useful concept of convergence to define, it is the most difficult to use. Most proofs involving convergence with probability one are beyond the mathematical prerequisites and capabilities of this book. Hence we will focus on two other notions of convergence that are perhaps less intuitive to understand, but are far easier to use when proving results. First note, however, that there are many equivalent names for convergence with probability one. It is often called convergence *almost surely* and abbreviated *a.s.* or convergence *almost everywhere* and abbreviated *a.e.* Convergence with probability one will not be considered in any depth here, but some toy examples will be considered in the problems to help get the concept across and a special case of the strong law of large numbers will be proved in an optional section.

Henceforth two definitions of convergence for random variables will be emphasized, both well suited to the type of results developed here (and one that was used in historically the first such results, Bernoulli's weak law of large numbers for iid random processes). The first is *convergence in mean square,* convergence of the type seen in the last section, which leads to a result called a *mean ergodic theorem.* The second is called *convergence in probability,* which is implied by the first and leads to a result called the weak law of large numbers. The second result will follow from the first via a simple but powerful inequality relating probabilities and expectations.

A sequence of random variables Y_n; $n = 1, 2, \ldots$ is said to *converge in mean square* or *converge in quadratic mean* to a random variable Y if $E[|Y_n|^2] < \infty$ all n and

$$\lim_{n\to\infty} E[|Y_n - Y|^2] = 0.$$

This is also written $Y_n \to Y$ in mean square or $Y_n \to Y$ in quadratic mean or by

$$\underset{n \to \infty}{\text{l.i.m.}}\, Y_n = Y,$$

where l.i.m. is an acronym for "limit in the mean." Although it is probably not obvious to the novice, it is important to understand that convergence in mean square *does not* imply convergence with probability one. Examples converging in one sense and not the other may be found in Problem 4.36.

Thus a sequence of random variables converges in mean square to another random variable if the second absolute moment of the difference converges to zero in the ordinary sense of convergence of a sequence of real numbers. This can be given the physical interpretation that the energy in the error goes to zero in the limit. Although the definition encompasses convergence to a random variable with any degree of "randomness," in most applications that we shall encounter the limiting random variable is a degenerate random variable, i.e., a constant. In particular, the sequence of sample averages, $\{S_n\}$, of the preceding section is next seen to converge in this sense.

The final notion of convergence bears a strong resemblance to the notion of convergence with probability one, but the resemblance is a *faux ami*, the two notions are fundamentally different. A sequence of random variables $Y_n;\ n = 1, 2, \ldots$ is said to *converge in probability* to a random variable Y if for every $\epsilon > 0$,

$$\lim_{n \to \infty} \Pr(|Y_n - Y| > \epsilon) = 0.$$

Thus a sequence of random variables converges in probability if the probability that the nth member of the sequence differs from the limit by more than an arbitrarily small ϵ goes to zero as $n \to \infty$. Note that just as with convergence in mean square, convergence in probability is silent on the question of convergence of individual realizations $Y_n(\omega)$. You could, in fact, have no realizations converge individually and yet have convergence in probability. All that convergence in probability tells you is that $\Pr(\omega : |Y_n(\omega) - Y(\omega)| > \epsilon)$ tends to zero with n. Suppose at time n a given subset of Ω satisfies the inequality, at time $n + 2$ still a different subset satisfies the inequality, etc. As long as the subsets have diminishing probability, convergence in probability can occur without convergence of the individual sequences.

Also, as in convergence in the mean square sense, convergence in probability is to a random variable in general. This includes the most interesting case of a degenerate random variable – a constant.

The two notions of convergence – convergence in mean square and convergence in probability – can be related to each other via simple, but important, inequalities. It will be seen that convergence in mean square is the stronger of the two notions; that is, if a random sequence converges in mean square, then it converges in probability, but not necessarily vice versa. The two inequalities are slight variations on each other, but they are stated separately for clarity. Both an elementary and a more elegant proof are presented.

The Tchebychev inequality

Suppose that X is a random variable with mean m_X, and variance σ_X^2. Then

$$\Pr(|X - m_X| \geq \epsilon) \leq \frac{\sigma_X^2}{\epsilon^2}. \tag{4.104}$$

We prove the result here for the discrete case. The continuous case is similar (and can be inferred from the more general proof of the Markov inequality to follow).

Proof The result follows from a sequence of inequalities.

$$\begin{aligned}
\sigma_X^2 &= E[(X - m_X)^2] = \sum_x (x - m_X)^2 p_X(x) \\
&= \sum_{x:|x-m_X|<\epsilon} (x - m_X)^2 p_X(x) + \sum_{x:|x-m_X|\geq\epsilon} (x - m_X)^2 p_X(x) \\
&\geq \sum_{x:|x-m_X|\geq\epsilon} (x - m_X)^2 p_X(x) \\
&\geq \epsilon^2 \sum_{x:|x-mX|\geq\epsilon} p_X(x) = \epsilon^2 \Pr(|X - m_X| \geq \epsilon).
\end{aligned}$$

□

Note that the Tchebychev inequality implies that

$$\Pr(|V - \overline{V}| \geq \gamma\sigma_V) \leq \frac{1}{\gamma^2},$$

that is, the probability that V is farther from its mean by more than γ times its standard deviation (the square root of its variance) is no greater than γ^{-2}.

The Markov Inequality

Given a *nonnegative* random variable U with finite expectation $E[U]$, for any $a > 0$ we have

$$\Pr(U \geq a) = P_U([a, \infty)) \leq \frac{E[U]}{a}.$$

Proof The result can be proved in the same manner as the Tchebychev inequality by separate consideration of the discrete and continuous cases.

Here we give a more general proof. Fix $a > 0$ and set $F = \{u : u \geq a\}$. Let $1_F(u)$ be the indicator of the function F, $1_F(u) = 1$ if $u \geq a$ and 0 otherwise. Then since $F \cap F^c = \emptyset$ and $F \cup F^c = \Omega$, using the linearity of expectation and the fact that $U \geq 0$ with probability one:

$$\begin{aligned} E[U] &= E[U(1_F(U) + 1_{F^c}(U))] \\ &= E[U(1_F(U))] + E[U(1_{F^c}(U))] \\ &\geq E[U(1_F(U))] \geq aE[1_F(U)] \\ &= aP(F). \end{aligned}$$

completing the proof. □

Observe that if a random variable U is nonnegative and has small expectation, say $E[U] \leq \epsilon$, then the Markov inequality with $a = \sqrt{\epsilon}$ implies that

$$\Pr(U \geq \sqrt{\epsilon}) \leq \sqrt{\epsilon}.$$

This can be interpreted as saying that the random variable can take on values greater that $\sqrt{\epsilon}$ no more than $\sqrt{\epsilon}$ of the time.

Before applying this result, we pause to present a second proof of the Markov inequality that has a side result of some interest in its own right. As before assume that $U \geq 0$. Assume for the moment that U is continuous so that

$$E[U] = \int_0^\infty u f_U(u)\, du.$$

Consider the admittedly strange-looking equality

$$u = \int_0^\infty 1_{[\alpha,\infty)}(u)\, d\alpha,$$

which follows since the integrand is 1 if and only if $\alpha \leq u$ and hence integrating 1 as α ranges from 0 to u yields u. Plugging this equality into the previous integral expression for expectation and changing the order of integration yields

$$\begin{aligned} E[U] &= \int_0^\infty \left(\int_0^\infty 1_{[\alpha,\infty)}(u)\, d\alpha \right) f_U(u)\, du \\ &= \int_0^\infty \left(\int_0^\infty 1_{[\alpha,\infty)}(u) f_U(u)\, du \right) d\alpha, \end{aligned}$$

which can be expressed as

$$E[U] = \int_0^\infty \Pr(U \geq \alpha)\, d\alpha. \tag{4.105}$$

This result immediately gives the Markov inequality since for any fixed $a > 0$,

$$E[U] = \int_0^\infty \Pr(U \geq \alpha)\, d\alpha \geq a \Pr(U \geq a).$$

To see this, $\Pr(U \geq \alpha)$ is monotonically nonincreasing with α, so for all $\alpha \leq a$ we must have $\Pr(U \geq \alpha) \geq \Pr(U \geq a)$ (and for other α $\Pr(U \geq \alpha) \geq 0$). Plugging the bound into the integral yields the claimed inequality.

The second proof provides a geometric interpretation of the Markov inequality. The expectation $E[U]$ is given by the integral $\int_0^\infty \Pr(U \geq \alpha)\, d\alpha$, which is the area under the curve $\Pr(U \geq \alpha)$ which is bound below by $a \Pr(U \geq a)$, the area under a rectangle of width a and height $\Pr(U \geq a)$.

Lemma 4.3 *If Y_n converges to Y in mean square, then it also converges in probability.*

Proof From the Markov inequality applied to $|Y_n - Y|^2$, we have for any $\epsilon > 0$

$$\Pr(|Y_n - Y| \geq \epsilon) = \Pr(|Y_n - Y|^2 \geq \epsilon^2) \leq \frac{E(|Y_n - Y|^2)}{\epsilon^2}.$$

The right-hand term goes to zero as $n \to \infty$ by definition of convergence in mean square. □

Although convergence in mean square implies convergence in probability, the reverse statement cannot be made; they are not equivalent. This is shown by a simple counterexample. Let Y_n be a discrete random variable with pmf

$$p_{Y_n}(y) = \begin{cases} 1 - 1/n & \text{if } y = 0 \\ 1/n & \text{if } y = n \end{cases}.$$

Convergence in probability to zero without convergence in mean square is easily verified. In particular, the sequence converges in probability since $\Pr[|Y_n - 0| > \epsilon] = \Pr[Y_n > 0] = 1/n$, which goes to 0 as $n \to \infty$. On the other hand, $E[|Y_n - 0|^2]$ would have to go to 0 for Y_n to converge to 0 in mean square, but it is $E[Y_n^2] = 0(1 - 1/n) + n^2/n = n$, which does not converge to 0 as $n \to \infty$.

It merits noting that a sequence of random variables can converge to more than one mean square limit and to more than one limit in probability, but the limits must be equal with probability one. Suppose for example Y_n converges in mean square to two random variables Y and W. Then using the triangle inequality and the Cauchy–Schwarz inequality (see Problem 4.20) we have that for any n

$$\begin{aligned}
&E(|Y - W|^2) \\
&\qquad \leq E(|Y - Y_n + Y_n - W|^2) \\
&\qquad \leq E\left((|Y - Y_n| + |Y_n - W|)^2\right) \\
&\qquad = E(|Y - Y_n|^2) + 2E(|Y_n - Y||Y_n - W|) + E(|W - Y_n|^2) \\
&\qquad \leq E(|Y - Y_n|^2) + 2(E(|Y - Y_n|^2)E(|W - Y_n|^2))^{1/2} + E(|W - Y_n|^2)
\end{aligned}$$

which goes to zero as $n \to \infty$, so that

$$E(|Y - W|^2) = 0.$$

The Markov inequality then implies that $\Pr(|Y - W| > \epsilon) = 0$ for any $\epsilon > 0$, which in turn implies that $\Pr(|Y - W| = 0) = 1$ or $\Pr(Y = W) = 1$ from the continuity of probability. The similar demonstration for convergence in probability is left as an exercise (see Problem 4.34).

Before leaving the topic of convergence of random variables for the time being, we close this section by showing that both convergence in probability and convergence in mean square imply convergence in distribution, the convergence used in the central limit theorem. Later, in Section 5.8, we shall dig more deeply into what is arguably the most important form of convergence in engineering: mean square convergence.

Lemma 4.4 *If Y_n converges to Y in probability, then it also converges in distribution.*

Proof Consider the cdfs for Y_n and Y:

$$\begin{aligned}
F_{Y_n}(y) &= \Pr(Y_n \leq y) \\
&= \Pr(Y_n \leq y \text{ and } Y \leq y + \epsilon) + \Pr(Y_n \leq y \text{ and } Y > y + \epsilon) \\
F_Y(y + \epsilon) &= \Pr(Y \leq y + \epsilon) \\
&= \Pr(Y \leq y + \epsilon \text{ and } Y_n \leq y) + \Pr(Y \leq y + \epsilon \text{ and } Y_n > y)
\end{aligned}$$

so that

$$\begin{aligned}
&F_{Y_n}(y) - F_Y(y + \epsilon) \\
&\qquad = \Pr(Y_n \leq y \text{ and } Y > y + \epsilon) - \Pr(Y \leq y + \epsilon \text{ and } Y_n > y) \\
&\qquad \leq \Pr(Y_n \leq y \text{ and } Y > y + \epsilon) \\
&\qquad \leq \Pr(|Y - Y_n| \geq \epsilon).
\end{aligned}$$

The final term converges to 0 from the assumption of convergence in probability. Since the leftmost term is nonnegative, this implies that

$$\lim_{n\to\infty} F_{Y_n}(y) \leq F_Y(y+\epsilon).$$

As similar argument shows that also

$$F_Y(y-\epsilon) \leq \lim_{n\to\infty} F_{Y_n}(y).$$

The lemma then follows if y is a continuity point of the cdf F_Y. □

The lemma has the following obvious corollary:

Corollary 4.3 *If a sequence of random variables Y_n converges in mean square to a random variable Y, then it also converges in distribution.*

4.16 Weak law of large numbers

We now have the definitions and preliminaries to prove laws of large numbers showing that sample averages converge to the expectation of the individual samples. The basic (and classical) results hold for uncorrelated random processes with constant variance.

A mean ergodic theorem

Theorem 4.11 *Let $\{X_n\}$ be a discrete time uncorrelated random process such that $EX_n = \overline{X}$ is finite and $\sigma^2_{X_n} = \sigma^2_X < \infty$ for all n; that is, the mean and variance are the same for all sample times. Then*

$$\underset{n\to\infty}{\text{l.i.m.}} \frac{1}{n}\sum_{i=0}^{n-1} X_i = \overline{X},$$

that is, $\frac{1}{n}\sum_{i=0}^{n-1} X_i \to \overline{X}$ *in mean square.*

Proof The proof follows directly from the last section with $S_n = \frac{1}{n}\sum_{i=0}^{n-1} X_i$, $ES_n = EX_i = \overline{X}$. To summarize from (4.103),

$$\begin{aligned}
\lim_{n\to\infty} E[(S_n - \overline{X})^2] &= \lim_{n\to\infty} E[(S_n - ES_n)^2] \\
&= \lim_{n\to\infty} \sigma^2_{S_n} \\
&= \lim_{n\to\infty} \frac{\sigma^2_X}{n} = 0.
\end{aligned}$$

□

This theorem is called *a* mean ergodic theorem because it is a special case of the more general mean ergodic theorem – it holds only for uncorrelated random processes. We shall later consider more general results along this line, but this simple result and the one to follow provide the basic ideas.

Combining Lemma 4.3 with Theorem 4.11 yields the following famous result, one of the original limit theorems of probability theory:

Theorem 4.12 *The weak law of large numbers*
Let $\{X_n\}$ be a discrete time process with finite mean $EX_n = \overline{X}$ and variance $\sigma^2_{X_n} = \sigma^2_X < \infty$ for all n. If the process is uncorrelated, then the sample average $n^{-1}\sum_{i=0}^{n-1} X_i$ converges to $\overline{X}$ in probability.

An alternative means of describing a law of large numbers is to define the limiting time average or sample average of a sequence of random variables $\{X_n\}$ by

$$< X_n >= \lim_{n\to\infty} n^{-1}\sum_{i=0}^{n-1} X_i, \tag{4.106}$$

if the limit exists in any of the manners considered, e.g., in mean square, in probability, or with probability one. Note that ordinarily the limiting time average must be considered as a random variable since it is a function of random variables. Laws of large numbers then provide conditions under which

$$< X_n >= E(X_k), \tag{4.107}$$

which requires that $< X_n >$ not be a random variable, i.e., that it be a constant and not vary with the underlying sample point ω, and that $E(X_k)$ not depend on time, i.e., that it be a constant and not vary with time k.

The best-known (and earliest) application of the weak law of large numbers is to iid processes such as the Bernoulli process. Note that the iid specification is not needed, however. All that is used for the weak law of large numbers is constant means, constant variances, and uncorrelation. The actual distributions could be time-varying and dependent within these constraints. The *weak* law is called weak because convergence in probability is one of the weaker forms of convergence. Convergence of individual realizations of the random process is not assured. This could be very annoying because in many practical engineering applications, we have only one realization to work with (i.e., only one ω), and we need to calculate averages that converge as determined by actual calculations, e.g., with a computer.

The *strong* law of large numbers considers convergence with probability one. Such strong theorems are much harder to prove, but fortunately are satisfied in most engineering situations.

The astute reader may have noticed the remarkable difference in behavior caused by the apparently slight change of division by $\sqrt{n}$ instead of n when normalizing sums of iid random variables. In particular, if $\{X_n\}$ is a zero-mean process with unit variance, then the weighted sum $n^{-(1/2)} \sum_{k=0}^{n-1} X_k$ converges to a Gaussian random variable in some sense because of the central limit theorem, whereas the weighted sum $n^{-1} \sum_{k=0}^{n-1} X_k$ converges to a constant, the mean of the individual random variables!

4.17 ⋆Strong law of large numbers

The strong law of large numbers replaces the convergence in probability of the weak law with convergence with probability one. It will shortly be shown that convergence with probability one implies convergence in probability, so the "strong" law is indeed stronger than the "weak" law. Although the two terms sound the same, they are really quite different. Convergence with probability one applies to individual realizations of the random process, whereas convergence in probability does not. Convergence with probability one is closer to the usual definition of convergence of a sequence of numbers since it says that for each sample point ω, the limiting sample average $\lim_{n\to\infty} n^{-1} \sum_{n=1}^{\infty} X_n$ exists in the usual sense for all ω in a set of probability one. Although convergence with probability one is a more satisfying notion of convergence, it is notably harder to prove than the weaker result. Hence we consider only the special case of iid sequences, where the added difficulty is moderate. In this section convergence with probability one is considered and a strong law of large numbers is proved. The key new tools are the Borel–Cantelli lemma, which provides a condition ensuring convergence with probability one, and the Chernoff inequality, an improvement on the Tchebychev inequality.

Lemma 4.5 *If Y_n converges to Y with probability one, then it also converges in probability.*

Proof Given an $\epsilon > 0$, define the sequence of sets

$$\begin{aligned} G_n(\epsilon) &= \{\omega : |Y_n(\omega) - Y(\omega)| \geq \epsilon\} \\ F_n(\epsilon) &= \{\omega : |Y_m(\omega) - Y(\omega)| \geq \epsilon \text{ for some } m \geq n\} \\ &= \bigcup_{m \geq n} G_m(\epsilon). \end{aligned}$$

Thus $G_n(\epsilon) \subset F_n(\epsilon)$ and the $F_n(\epsilon)$ form a decreasing sequence of sets as n grows, that is, $F_n \subset F_{n-1}$ for all n. Thus from the continuity of probability,

$$\lim_{n\to\infty} \Pr(F_n(\epsilon)) = \Pr\left(\lim_{n\to\infty} F_n(\epsilon)\right) = \Pr\left(\bigcap_{n=1}^{\infty} F_n(\epsilon)\right)$$
$$= \Pr\left(\bigcap_{n=1}^{\infty} \bigcup_{m\geq n} G_m(\epsilon)\right). \qquad \square$$

If $\omega \in B = \bigcap_{n=1}^{\infty} \bigcup_{m\geq n} G_m(\epsilon)$, then for every value of n there must exist an $m \geq n$ for which $|Y_m(\omega) - Y(\omega)| \geq \epsilon$, which means that $Y_m(\omega)$ does not converge to $Y(\omega)$. Since Y_n is assumed to converge to Y with probability one, $\Pr(B) = 0$ and hence

$$\lim_{n\to\infty} \Pr(G_n(\epsilon)) \leq \lim_{n\to\infty} \Pr(F_n(\epsilon)) = \Pr(B) = 0.$$

Convergence in probability does not imply convergence with probability one; i.e., they are not equivalent. This can be shown by counterexample (Problem 4.36). There is, however, a test that can be applied to determine convergence with probability one. The result is one form of a result known as the first *Borel–Cantelli lemma.*

Lemma 4.6 *Y_n converges to Y with probability one if for any $\epsilon > 0$*

$$\sum_{n=1}^{\infty} \Pr(|Y_n - Y| > \epsilon) < \infty. \tag{4.108}$$

Proof Consider two collections of bad sequences. Let $F(\epsilon)$ be the set of all ω such that the corresponding sequence $Y_n(\omega)$ does *not* satisfy the convergence criterion, i.e.,

$$F(\epsilon) = \{\omega : |Y_n - Y| > \epsilon, \text{ for some } n \geq N, \text{ for any } N < \infty\}.$$

$F(\epsilon)$ is the set of points for which the sequence does not converge. Consider also the simpler sets where things look bad at a particular time:

$$F_n(\epsilon) = \{\omega : |Y_n - Y| > \epsilon\}.$$

The complicated collection of points with nonconvergent sequences can be written as a subset of the union of all of the simpler sets:

$$F(\epsilon) \subset \bigcup_{n\geq N}^{\infty} F_n(\epsilon) \triangleq G_N(\epsilon))$$

for any finite N. This in turn implies that

$$\Pr(F(\epsilon)) \leq \Pr(\bigcup_{n \geq N}^{\infty} F_n(\epsilon)).$$

From the union bound this implies that

$$\Pr(F(\epsilon)) \leq \sum_{n=N}^{\infty} \Pr(F_n(\epsilon)).$$

By assumption

$$\sum_{n=0}^{\infty} \Pr(F_n(\epsilon)) < \infty,$$

which implies that

$$\lim_{N \to \infty} \sum_{n=N}^{\infty} \Pr(F_n(\epsilon)) = 0$$

and hence $\Pr(F(\epsilon)) = 0$, proving the result. □

Convergence with probability one does not imply – nor is it implied by – convergence in mean square. This can be shown by counterexamples (Problem 4.36).

We now apply this result to sample averages to obtain a strong law of large numbers for an iid random process $\{X_n\}$. For simplicity we focus on a zero-mean Gaussian iid process and prove that with probability one

$$\lim_{n \to \infty} S_n = 0$$

where

$$S_n = \frac{1}{n} \sum_{k=0}^{n-1} X_k.$$

Assuming zero-mean does not lose any generality since if this result is true, the result for nonzero-mean m follows immediately by applying the zero-mean result to the zero-mean process $\{X_n - m\}$.

The approach is to use the Borel–Cantelli lemma with $Y_n = S_n$ and $Y = 0 = E[X_n]$ and hence the immediate problem is to bound $\Pr(|S_n| > \epsilon)$ in such a way that the sum over n will be finite. The Tchebychev inequality does not work here as it would yield a sum

$$\sigma_X^2 \sum_{n=1}^{\infty} \frac{1}{n},$$

which is *not* finite. A better upper bound than Tchebychev is needed. This is provided by a different application of the Markov inequality. Given a random variable Y, fix a $\lambda > 0$ and observe that $Y > y$ if and only if $e^{\lambda Y} > e^{\lambda y}$. Application of the Markov inequality then yields

$$\begin{aligned}\Pr(Y > y) &= \Pr(e^{\lambda Y} > e^{\lambda y}) = \Pr(e^{\lambda(Y-y)} > 1) \\ &\leq E[e^{\lambda(Y-y)}].\end{aligned} \tag{4.109}$$

This inequality is called the *Chernoff inequality* and it provides the needed bound.

Applying the Chernoff inequality yields for any $\lambda > 0$

$$\begin{aligned}\Pr(|S_n| > \epsilon) &= \Pr(S_n > \epsilon) + \Pr(S_n < -\epsilon) \\ &= \Pr(S_n > \epsilon) + \Pr(-S_n > \epsilon) \\ &\leq E[e^{\lambda(S_n-\epsilon)}] + E[e^{\lambda(-S_n-\epsilon)}] \\ &= e^{-\lambda\epsilon}\left(E[e^{\lambda S_n}] + E[e^{-\lambda S_n}]\right) \\ &= e^{-\lambda\epsilon}\left(M_{S_n}(\lambda) + M_{S_n}(-\lambda)\right).\end{aligned}$$

These moment-generating functions are easily found from Lemma 4.1 to be

$$E[e^{\gamma S_n}] = M_X^n\left(\frac{\gamma}{n}\right), \tag{4.110}$$

where $M_X(ju) = E[e^{juX}]$ is the common characteristic function of the iid X_i and $M_X(w)$ is the corresponding moment-generating function. Combining these steps yields the bound

$$\Pr(|S_n| > \epsilon) \leq e^{-\lambda\epsilon}\left(M_X^n\left(\frac{\lambda}{n}\right) + M_X^n\left(-\frac{\lambda}{n}\right)\right). \tag{4.111}$$

So far $\lambda > 0$ is completely arbitrary. We can choose a different λ for each n. Choosing $\lambda = n\epsilon/\sigma_X^2$ yields

$$\Pr(|S_n| > \epsilon) \leq e^{-n\frac{\epsilon^2}{\sigma_X^2}}\left(M_X^n\left(\frac{\epsilon}{\sigma_X^2}\right) + M_X^n\left(-\frac{\epsilon}{\sigma_X^2}\right)\right). \tag{4.112}$$

Plugging in the form for the Gaussian moment-generating function $M_X(w) = e^{w^2\sigma_X^2/2}$ yields

$$\Pr(|S_n| > \epsilon) \leq 2e^{-n\epsilon^2/\sigma_X^2}\left(e^{(\epsilon/\sigma_X^2)^2\sigma_X^2/2}\right)^n = 2\left(e^{-\epsilon^2/2\sigma_X^2}\right)^n \tag{4.113}$$

which has the form $\Pr(|S_n| > \epsilon) \leq 2\beta^n$ for $\beta < 1$. Hence summing a geometric progression yields

$$\sum_{n-1}^{\infty} \Pr(|S_n| > \epsilon) \leq 2\sum_{n=1}^{\infty} \beta^n = 2\frac{\beta}{1-\beta} < \infty, \tag{4.114}$$

which completes the proof for the iid Gaussian case.

The non-Gaussian case can be handled by combining the above approach with the approximation of (4.15). The bound for the Borel–Cantelli limit need only be demonstrated for small ϵ since if it is true for small ϵ it must also be true for large ϵ. For small ϵ, however, (4.15) implies that $M_X(\pm\epsilon/\sigma_X^2)$ in (4.112) can be written as $1 + \epsilon^2/2\sigma_X^2 + o(\epsilon^2/2\sigma_X^2)$ which is arbitrarily close to $e^{\epsilon^2/2\sigma_X^2}$ for sufficiently small ϵ, and the proof is completed as above.

The following theorem summarizes the results of this section.

Theorem 4.13 *Strong law of large numbers*
Given an iid process $\{X_n\}$ with finite mean $E[X]$ and variance, then

$$\lim_{n\to\infty} \frac{1}{n}\sum_{k=0}^{n-1} X_k = E[X] \text{ with probability one.} \tag{4.115}$$

4.18 Stationarity

Stationarity properties

In the development of the weak law of large numbers we made two assumptions on a random process $\{X_t;\ t \in \mathcal{Z}\}$: that the mean EX_t of the process did not depend on time and that the covariance function had the form $K_X(t,s) = \sigma_X^2 \delta_{t-s}$.

The assumption of a constant mean, independent of time, is an example of a *stationarity property* in the sense that it assumes that some property describing a random process does not vary with time (or is time-invariant). The process itself is not usually "stationary" in the usual literal sense of remaining still, but attributes of the process, such as the first moment in this case, can remain still in the sense of not changing with time. In the mean example we can also express this as

$$EX_t = EX_{t+\tau};\quad \text{all } t, \tau, \tag{4.116}$$

which can be interpreted as saying that the mean of a random variable at time t is not affected by a *shift* of any amount of time τ. Conditions on moments can be thought of as *weak* stationarity properties since they constrain only an expectation and not the distribution itself. Instead of simply constraining a moment, we could make the stronger assumption of constraining the marginal distribution. The assumption of a constant mean would follow, for example, if the marginal distribution of the process, the distribution of a single random variable X_t, did not depend on the sample time t. Thus a *sufficient* (but not necessary) condition for ensuring that a random process has a constant mean is that its marginal distribution P_{X_t} satisfies the condition

$$P_{X_t} = P_{X_{t+\tau}}; \quad \text{all } t, \tau. \tag{4.117}$$

This will be true, for example, if the same relation holds with the distribution replaced by cdf's, pdf's, or pmf's. If a process meets this condition, it is said to be *first-order stationary*. For example, an iid process is clearly first-order stationary. The word *stationary* refers to the fact that the first-order distribution (in this case) does not change with time, i.e., it is not affected by *shifting* the sample time by an amount τ.

Next consider the covariance used to prove the weak law of large numbers. It has a very special form in that it is the variance if the two sample times are the same, and zero otherwise. This class of constant mean, constant variance, and uncorrelated processes is admittedly an unusual case. A more general class of processes which will share many important properties with this very special case is formed by requiring a mean and variance that do not change with time, but easing the restriction on the covariance. We say that a random process is *weakly stationary* or *stationary in the weak sense* if EX_t does not depend on t, $\sigma^2_{X_t}$ does not depend on t, and the covariance $K_X(t,s)$ depends on t and s only through the difference $t-s$, that is, if

$$K_X(t,s) = K_X(t+\tau, s+\tau) \tag{4.118}$$

for all t, s, τ for which $s, s+\tau, t, t+\tau \in \mathcal{T}$. When this is true, it is often expressed by writing

$$K_X(t, t+\tau) = K_X(\tau). \tag{4.119}$$

for all t, τ such that $t, t+\tau \in \mathcal{T}$. A function of two variables of this type is said to be *Toeplitz* [36, 31]. Much of the theory of weakly stationary processes follows from the theory of Toeplitz forms.

If we form a covariance matrix by sampling such a covariance function, then the matrix (called a *Toeplitz matrix*) will have the property that all elements on any fixed diagonal of the matrix will be equal. For example, the (3,5) element will be the same as the (7,9) element since $5-3=9-7$. Thus, for example, if the sample times are $0,1,\ldots,n-1$, then the covariance matrix is $\{K_X(k,j)=K_X(j-k);\ k=0,1,\ldots,n-1, j=0,1,\ldots,n-1$ or

$$\begin{bmatrix} K_X(0) & K_X(1) & K_X(2) \ldots & K_X(n-1) \\ K_X(-1) & K_X(0) & K_X(1) & \\ K_X(-2) & K_X(-1) & K_X(0) & \vdots \\ \vdots & & \ddots & \\ K_X(-(n-1)) & & \ldots & K_X(0) \end{bmatrix}$$

As in the case of constant mean, the adjective *weakly* refers to the fact that the constraint is placed on the moments and not on the distributions. Mimicking the earlier discussion, we could make a stronger assumption that is sufficient to ensure weak stationarity. A process is said to be *second-order stationary* if the pairwise distributions are not affected by shifting; that is, if analogous to the moment condition (4.118) we make the stronger assumption that

$$P_{X_t,X_s}=P_{X_{t+\tau},X_{s+\tau}};\quad \text{all } t,s,\tau. \tag{4.120}$$

Observe that second-order stationarity implies first-order since the marginals can be computed from the joints. The class of iid processes is second-order stationary since the joint probabilities are products of the marginals, which do not depend on time.

There are a variety of such stationarity properties that can be defined, but weakly stationary is one of the two most important for two reasons. The first reason will be seen shortly – combining weak stationarity with an asymptotic version of uncorrelated gives a more general law of large numbers than the ones derived previously. The second reason will be seen in the next chapter: if a covariance depends only on a single argument (the difference of the sample times), then it will have an ordinary Fourier transform. Transforms of correlation and covariance functions provide useful analysis tools for stochastic systems.

Before proceeding it is useful to consider the other most important stationarity property: strict stationarity (sometimes the adjective "strict" is omitted). As the notion of weak stationary can be considered as a generalization of uncorrelated, the notion of strict stationary can be considered as a generalization of iid: if a process is iid, the probability distribution of a k-dimensional random vector $X_n, X_{n+1}, \ldots, X_{n+k-1}$ does not depend on the starting time of the collection of samples, i.e., for an iid process we have that

$$P_{X_n,X_{n+1},\ldots,X_{n+k-1}} = P_{X_{n+m},X_{n+m+1},\ldots,X_{n+m+k-1}}(\mathbf{x}), \text{ all } n,k,m. \quad (4.121)$$

This property can be interpreted as saying that the probability of any event involving a finite collection of samples of the random process does not depend on the starting time n of the samples and hence on the definition of time 0. Alternatively, these joint distributions are not affected by *shifting* the samples by a common amount m. In the simple Bernoulli process case this means things like

$$p_{X_n}(0) = p_{X_0}(0) = 1-p, \text{ all } n$$
$$p_{X_n,X_k}(0,1) = p_{X_0,X_{k-n}}(0,1) = p(1-p), \text{ all } n,k$$
$$p_{X_n,X_k,X_l}(0,1,0) = p_{X_0,X_{k-n},X_{l-n}}(0,1,0) = (1-p)^2 p, \text{ all } n,k,m,$$

and so on. Note that the *relative* sample times stay the same, that is, the differences between the sample times are preserved, but all of the samples together are *shifted* without changing the probabilities. A process need not be iid to possess the property of joint probabilities being unaffected by shifts, so we formalize this idea with a definition.

A discrete time random process $\{X_n\}$ is said to be *stationary* or *strictly stationary* or *stationary in the strict sense* if (4.121) holds. We have argued that a discrete alphabet iid process is an example of a stationary random process. This definition extends immediately to continuous alphabet discrete time processes by replacing the pmf's by pdf's. Both cases can be combined by using cdf's or the distributions. Hence we can make a more general definition for discrete time processes: a discrete time random process $\{X_n\}$ is said to be *stationary* if

$$P_{X_n,X_{n+1},\ldots,X_{n+k-1}} = P_{X_{n+m},X_{n+m+1},\ldots,X_{n+m+k-1}}, \text{ all } k,n,m. \quad (4.122)$$

This will hold if the corresponding formula holds for pmf's, pdf's, or cdf's. For example, any iid random process is stationary.

Generalizing the definition to include continuous time random processes requires only a little more work, much like that used to describe the Kolmogorov extension theorem. We would like all joint distributions involving a finite collection of samples to be independent of the starting time or, equivalently, to be unaffected by shifts. The following general definition does this. It reduces to the previous definition when the process is a discrete time process.

A random process $\{X_t;\ t \in \mathcal{T}\}$ is *stationary* if

$$P_{X_{t_0},X_{t_1},\ldots,X_{t_{k-1}}} = P_{X_{t_0-\tau},X_{t_1-\tau},\ldots,X_{t_{k-1}-\tau}},\ \text{all } k, t_0, t_1, \ldots, t_{k-1}, \tau. \quad (4.123)$$

The word "all" above must be interpreted with care; it means all choices of dimension k, sample times $t_0, \ldots, t_{k-1}$, and shift τ for which the equation makes sense, e.g., k must be a positive integer and $t_i \in \mathcal{T}$ and $t_i - \tau \in \mathcal{T}$ for $i = 0, \ldots, k-1$.

It should be obvious that strict stationarity implies weak stationarity since it implies that P_{X_t} does not depend on t, and hence the mean computed from this distribution does not depend on t, and it implies that $P_{X_t,X_s} = P_{X_{t-s},X_0}$ and hence $K_X(t,s) = K_X(t-s,0)$. The converse is generally not true – knowing that two moments are unaffected by shifts does not in general imply that all finite-dimensional distributions will be unaffected by shifts. This is why weak stationarity is indeed a "weaker" definition of stationarity. There is, however, one extremely important case where weak stationarity is sufficient to ensure strict stationarity – the case of Gaussian random processes. We shall not construct a careful proof of this fact because it is a notational mess that obscures the basic idea, which is actually rather easy to describe. A Gaussian process $\{X_t;\ t \in \mathcal{T}\}$ is completely characterized by knowledge of its mean function $\{m_t;\ t \in \mathcal{T}\}$ and its covariance function $\{K_X(t,s);\ t, s \in \mathcal{T}\}$. All joint pdf's for all possible finite collections of sample times are expressed in terms of these two functions. If the process is known to be weakly stationary, then $m_t = m$ for all t, and $K_X(t,s) = K_X(t-s,0)$ for all t, s. This implies that all of the joint pdf's will be unaffected by a time shift, since the mean vector stays the same and the covariance matrix depends only on the relative differences of the sample times, not on where they begin. Thus in this special case, knowing a process is weakly stationary is sufficient to conclude it is stationary. However, in general, stationarity can be difficult to prove, even for simple processes.

⋆*Strict stationarity*

In fact the above is not the definition of stationarity used in the mathematical and statistical literature, but it is equivalent to it. We pause for a moment to describe the more fundamental (but abstract) definition and its relation to the above definition. The reader should keep in mind that it is the above definition that is the important one in practice: it is the definition that is almost always used to verify that a process is stationary or not.

To state the alternative definition, recall that a random process $\{X_t;\ t \in \mathcal{T}\}$ can be considered to be a mapping from a probability space $(\Omega, \mathcal{F}, P)$ into a space of sequences or waveforms $\{x_t;\ t \in \mathcal{T}\}$ and that the inverse image formula implies a probability measure called a *process distribution*, say P_X, on this complicated space, i.e., $P_X(F) = P_X(\{\{x_t;\ t \in \mathcal{T}\} : \{x_t;\ t \in \mathcal{T}\} \in F\}) = P(\{\omega : \{X_t(\omega);\ t \in \mathcal{T}\} \in F\})$. The abstract definition of stationarity places a condition on the process distribution: a random process $\{X_t;\ t \in \mathcal{T}\}$ is stationary if the process distribution P_X is unchanged by shifting, that is, if

$$P_X(\{\{x_t;\ t \in \mathcal{T}\} : \{x_t;\ t \in \mathcal{T}\} \in F\}) \\ = P_X(\{\{x_t;\ t \in \mathcal{T}\} : \{x_{t+\tau};\ t \in \mathcal{T}\} \in F\});\quad \text{all } F, \tau. \tag{4.124}$$

The only difference between the left- and right-hand side is that the right-hand side takes every sample waveform and shifts it by a common amount τ. If the abstract definition is applied to finite-dimensional events, that is, events which actually depend only on a finite number of sample times, then this definition reduces to that of (4.123). Conversely, it turns out that having this property hold on all finite-dimensional events is enough to imply that the property holds for all possible events, even those depending on an infinite number of samples (such as the event one gets an infinite binary sequence with exactly p limiting relative frequency of heads). Thus the two definitions of strict stationarity are equivalent. The proof of this equivalence is one of the classic results of the theory of random processes and requires tools beyond those assumed for this book. The interested reader can find a thorough development in [32]. The underlying idea is that the collection of all finite-dimensional events generates the entire event space of the sample space of all waveforms or sequences. Given a consistent set of probabilities on a generating collection of events, it can be shown there must exist a probability measure on the full event space that agrees with the given probabilities.

Why is stationarity important? Are processes that are not stationary interesting? The answer to the first question is that this property leads to the most famous of the law of large numbers, which will be quoted without proof later. The answer to the second question is yes, nonstationary processes play an important role in theory and practice, as will be seen by example. In particular, some nonstationary processes will have a form of a law of large numbers, and others will have no such property, yet be quite useful in modeling real phenomena. Keep in mind that strict stationarity is stronger than weak stationarity. Thus if a process is not even weakly stationary then the process is also not strictly stationary. Two examples of nonstationary processes already encountered are the binomial counting process and the discrete time Wiener process. These processes have marginal distributions which change with time and hence the processes cannot be stationary. We shall see in Chapter 5 that these processes are also not weakly stationary.

4.19 Asymptotically uncorrelated processes

We close this chapter with a generalization of the mean ergodic theorem and the weak law of large numbers that demonstrates that weak stationarity plus an asymptotic form of uncorrelation is sufficient to yield a weak law of large numbers by a fairly modest variation of the earlier proof. The class of asymptotically uncorrelated processes is often encountered in practice. Only the result itself is important; the proof is a straightforward but tedious extension of the proof for the uncorrelated case.

An advantage of the more general result over the result for uncorrelated discrete time random processes is that it extends in a sensible way to continuous time processes.

A discrete time weakly stationary process $\{X_n;\ n \in \mathcal{Z}\}$ is said to be *asymptotically uncorrelated* if its covariance function is absolutely summable, that is, if

$$\sum_{k=-\infty}^{\infty} |K_X(k)| < \infty. \tag{4.125}$$

This condition implies that also

$$\lim_{k\to\infty} K_X(k) = 0, \tag{4.126}$$

and hence the property (4.125) can be considered as a weak form of uncorrelation, a generalization of the fact that a weakly stationary process is uncorrelated if $K_X(k) = 0$ when $k \neq 0$. If a process is uncorrelated, then X_n and X_{n+k} are uncorrelated random variables for all nonzero k; if it is asymptotically uncorrelated, the correlation between the two random variables decreases to zero as k grows. We use (4.125) rather than the weaker (4.126) as the definition as it also ensures the existence of a Fourier transform of K_X, which will be useful later. It also simplifies the proof of the resulting law of large numbers.

Theorem 4.14 *A mean ergodic theorem*

Let $\{X_n\}$ be a weakly stationary asymptotically uncorrelated discrete time random process such that $EX_n = \overline{X}$ is finite and $\sigma^2_{X_n} = \sigma^2_X < \infty$ for all n.

Then .

$$\underset{n\to\infty}{\text{l.i.m.}} \frac{1}{n}\sum_{i=0}^{n-1} X_i = \overline{X},$$

that is, $\frac{1}{n}\sum_{i=0}^{n-1} X_i \to \overline{X}$ *in mean square.*

Note that the theorem is indeed a generalization of the previous mean ergodic theorem since a weakly stationary uncorrelated process is trivially an asymptotically uncorrelated process. Note also that the Tchebychev inequality and this theorem immediately imply convergence in probability and hence establish a weak law of large numbers for weakly stationary asymptotically uncorrelated processes. Common examples of asymptotically uncorrelated processes are processes with exponentially decreasing covariance, i.e., of the form $K_X(k) = \sigma_X^2 \rho^{|k|}$ for $\rho < 1$.

⋆Proof for the asymptotically uncorrelated case

Proof Exactly as in the proof of Theorem 4.11 we have with $S_n = n^{-1}\sum_{i=0}^{n-1} X_i$ that

$$E[(S_n - \overline{X})^2] = E[(S_n - ES_n)^2] = \sigma_{S_n}^2.$$

From (4.101) we have that

$$\sigma_{S_n}^2 = n^{-2}\sum_{i=0}^{n-1}\sum_{j=0}^{n-1} K_X(i-j). \tag{4.127}$$

This sum can be rearranged as in Lemma B.1 of Appendix B as

$$\sigma_{S_n}^2 = \frac{1}{n}\sum_{k=-n+1}^{n-1}\left(1 - \frac{|k|}{n}\right) K_X(k). \tag{4.128}$$

From Lemma B.2

$$\lim_{n\to\infty}\sum_{k=-n+1}^{n-1}\left(1 - \frac{|k|}{n}\right) K_X(k) = \sum_{k=-\infty}^{\infty} K_X(k),$$

which is finite by assumption, hence dividing by n yields

$$\lim_{n\to\infty}\frac{1}{n}\sum_{k=-n+1}^{n-1}\left(1 - \frac{|k|}{n}\right) K_X(k) = 0$$

which completes the proof. □

A similar definition applies to continuous time processes. We say that $\{X(t);\ t \in \Re\}$ is *asymptotically uncorrelated* if its covariance function is absolutely integrable,

$$\int_{-\infty}^{\infty} |K_X(\tau)| < \infty, \tag{4.129}$$

which implies that

$$\lim_{\tau\to\infty} K_X(\tau) = 0. \tag{4.130}$$

No sensible continuous time random process can be uncorrelated (*why not?*), but many are asymptotically uncorrelated. A sample or time average can be defined for a continuous time process by replacing the sum operation by an integral, that is, by

$$S_T = \frac{1}{T}\int_0^T X(t)\,dt. \tag{4.131}$$

For the moment we ignore the technical difficulties that must be considered to assure that the integral exists in a suitable fashion. Suffice it to say that an integral can be considered as a limit of sums, and we have seen ways to make such limits of random variables precise. We will return to this issue with more care in Section 5.8. The definition of weakly stationary extends immediately to continuous time processes. The following result can be proved by extending the discrete time result to continuous time and integrals.

Theorem 4.15 *A mean ergodic theorem*
Let $\{X(t)\}$ be a weakly stationary asymptotically uncorrelated continuous time random process such that $EX(t) = \overline{X}$ is finite and $\sigma^2_{X(t)} = \sigma^2_X < \infty$ for all t. Then .

$$\underset{T\to\infty}{\text{l.i.m.}}\ \frac{1}{T}\int_0^T X(t)\,dt = \overline{X},$$

that is, $\frac{1}{T}\int_0^T X(t)\,dt \to \overline{X}$ *in mean square.*

As in the discrete time case, convergence in mean square immediately implies converges in probability, but much additional work is required to prove convergence with probability one (when such convergence indeed holds). Also as in the discrete case, we can define a limiting time average

$$< X(t) >= \lim_{T\to\infty} \frac{1}{T}\int_0^T X(t)\,dt \tag{4.132}$$

and interpret the law of large numbers as stating that the time average $< X(t) >$ exists in some sense and equals the expectation.

4.20 Problems

1. Wanda is going shopping for mystery books to take on a trip. She will spend X hours in the bookstore, where X is a discrete random variable equally likely to take on values of 1, 2, 3, or 4. She will buy N books, where N is a discrete random variable which depends on the amount of time she shops as described by a conditional probability mass function

$$p_{N|X}(n|k) = \frac{1}{k};\ n = 1, 2, \ldots, k.$$

   (a) Find the joint pmf of X and N.
   (b) Find the marginal pmf for N.
   (c) Find the conditional pmf of X given that $N = 2$.
   (d) Suppose you know that Wanda bought at least two but not more than three books. Find the conditional mean and variance of X given this information.
   (e) The cost of each book is a random variable (independent of all the other random variables in the problem) with mean 3. What is the expected value of Wanda's total expenditure?

2. The *Cauchy* pdf is defined by

$$f_X(x) = \frac{1}{\pi}\frac{1}{1 + x^2};\quad x \in \Re.$$

   Find EX. *Hint:* this is a trick question. Check the definition of Riemann integration over $(-\infty, \infty)$ before deciding on a final answer.

3. Suppose that Z is a discrete random variable with probability mass function

$$p_Z(k) = C\frac{a^k}{(1 + a)^{k+1}},\quad k = 0, 1, \ldots$$

   (This is sometimes called "Pascal's distribution.") Find the constant C and the mean, characteristic function, and variance of Z.

4. State and prove the fundamental theorem of expectation for the case where a discrete random variable X is defined on a probability space where the probability measure is described by a pdf f.

5. Given the setup of Problem 3.10:
   (a) Find $E(X)$ and $E(XY)$.
   (b) Find the conditional expectation $E(Y|X = x)$. For what value of x is this maximized?
   (c) Let A denote the event $X^2 \geq Y$. Find $P(A)$ and $E(XY|A)$.

6. Suppose that X is a random variable with pdf $f_X(\alpha)$ and characteristic function $M_X(ju) = E[e^{juX}]$. Define the new random variable $Y = aX + b$, where both a

and b are positive constants. Find the pdf f_Y and characteristic function $M_Y(ju)$ in terms of f_X and M_X, respectively.

7. X, Y and Z are iid Gaussian random variables with $\mathcal{N}(1,1)$ distributions. Define the random variables:

$$\begin{aligned} V &= 2X + Y \\ W &= 3X - 2Z + 5. \end{aligned}$$

(a) Find $E[VW]$.

(b) Find the two parameters that completely specify the random variable $V + W$.

(c) Find the characteristic function of the random vector $[V\ W]^t$, where t denotes "transpose."

(d) Find the linear estimator $\hat{V}(W)$ of V, given W.

(e) Find the optimal (smallest MSE) affine estimator $\hat{V}(W)$ of V, given W. In particular, find a and b such that $\hat{V}(W) = aW + b$ yields the minimum

$$\text{MSE} = E[(V - \hat{V}(W))^2]$$

(f) Is this an optimal estimator? Why?

(g) The zero-mean random variables $X - \bar{X}$, $Y - \bar{Y}$ and $Z - \bar{Z}$ are the inputs to a black box. There are two outputs, A and B. It is determined that the covariance matrix of the vector of its outputs $[A\ B]^t$ should be

$$\Lambda_{AB} = \begin{bmatrix} 3 & 2 \\ 2 & 5 \end{bmatrix}.$$

Find expressions for A and B in terms of the black box inputs so that this is in fact the case (design the black box). Your answer does not necessarily have to be unique.

(h) You are told that a *different* black box results in an output vector $[C\ D]^t$ with the following covariance matrix:

$$\Lambda_{CD} = \begin{bmatrix} 2 & 0 \\ 0 & 7 \end{bmatrix}.$$

How much information about output C does output D give you? Briefly but fully justify your answer.

8. Assume that $\{X_n\}$ is an iid process with Poisson marginal pmf

$$p_X(l) = \frac{\lambda^l e^{-\lambda}}{l!};\ l = 0, 1, 2, \ldots.$$

and define the process $\{N_k;\ k = 0, 1, 2, \ldots\}$

$$N_k = \begin{cases} 0 & k = 0 \\ \sum_{l=1}^{k} X_l & k = 1, 2, \ldots \end{cases}.$$

Define the process $\{Y_k\}$ by $Y_k = (-1)^{N_k}$ for $k = 0, 1, 2, \ldots$

(a) Find the mean $E[N_k]$, characteristic function $M_{N_k}(ju) = E[e^{juN_k}]$, and pmf $p_{N_k}(m)$.

(b) Find the mean $E[Y_k]$ and variance $\sigma^2_{Y_k}$.

(c) Find the conditional pmf's $p_{N_k|N_1,N_2,\ldots,N_{k-1}}(n_k|n_1,n_2,\ldots,n_{k-1})$ and $p_{N_k|N_{k-1}}(n_k|n_{k-1})$. Is $\{N_k\}$ a Markov process?

9. Let $\{X_n\}$ be an iid binary random process with equal probability of $+1$ or -1 occurring at any time n. Show that if Y_n is the standardized sum

$$Y_n = n^{-(1/2)} \sum_{k=0}^{n-1} X_k,$$

then

$$M_{Y_n}(ju) = e^{n \log \cos (u/\sqrt{n})}.$$

Find the limit of this expression as $n \to \infty$.

10. Suppose that a fair coin is flipped 1,000,000 times. Write an exact expression for the probability that between 400,000 and 500,000 heads occur. Next use the central limit theorem to find an approximation to this probability. Use tables to evaluate the resulting integral.

11. Using an expansion of the form of (4.99), show directly that the central limit theorem is satisfied for a sequence of iid random variables with pdf

$$p(x) = \frac{2}{\pi(1+x^2)^2}, \quad x \in \Re.$$

Try to use the same expansion for

$$p(x) = \frac{1}{\pi(1+x^2)}, \quad x \in \Re.$$

Explain your result.

12. If $\text{l.i.m.}_{n\to\infty} X_n = X$ and $\text{l.i.m.}_{n\to\infty} Y_n = Y$, show that for any constants a, b

$$\underset{n\to\infty}{\text{l.i.m.}}(aX_n + bY_n) = aX + bY. \tag{4.133}$$

13. Suppose that $\{X_n\}$ is a weakly stationary random process with a marginal pdf $f_X(\alpha) = 1$ for $0 < \alpha < 1$ and a covariance function

$$K_X(k) = \frac{1}{12}\rho^{|k|}$$

for all integer k ($\rho < 1$). What is

$$\underset{n\to\infty}{\text{l.i.m.}} \frac{1}{n} \sum_{k=1}^{n} X_k \; ?$$

What is

$$\underset{n\to\infty}{\text{l.i.m.}}\ \frac{1}{n^2}\sum_{k=1}^{n} X_k\ ?$$

14. If $\{X_n\}$ is an uncorrelated process with constant first and second moments, does it follow for an arbitrary function g that

$$\underset{n\to\infty}{\text{l.i.m.}}\ n^{-1}\sum_{i=0}^{n-1} g(X_i) = E[g(X)]?$$

($E[g(X)]$ denotes the unchanging value of $E[g(X_n)]$.) Show that it does follow if the process is iid.

15. Apply Problem 4.14 to indicator functions to prove that relative frequencies of order n converge to pmf's in mean square and in probability for iid random processes. That is, if $r_a^{(n)}$ is defined as in the chapter, then $r_a^{(n)} \to p_X(a)$ as $n \to \infty$ in both senses for any a in the range space of X.

16. Define the subsets of the real line

$$F_n = \left\{r : |r| > \frac{1}{n}\right\},\quad n = 1, 2, \ldots$$

and

$$F = \{0\}.$$

Show that

$$F^c = \bigcup_{n=1}^{\infty} F_n.$$

Use this fact, the Tchebychev inequality, and the continuity of probability to show that if a random variable X has variance 0, then $\Pr(|X - EX| \geq \epsilon|) \leq 0$ independent of $\in$ and hence $\Pr(X = EX) = 1$.

17. True or false? Given a nonnegative random variable X, for any $\epsilon > 0$ and $a > 0$:

$$\Pr(X \geq \epsilon) \leq \frac{E[e^{aX}]}{e^{a\epsilon}}.$$

18. Show that for a discrete random variable X,

$$|E(X)| \leq E(|X|).$$

Repeat for a continuous random variable.

19. This problem considers some useful properties of autocorrelation or covariance functions for real-valued random processes.
   (a) Use the fact that $E[(X_t - X_s)^2] \geq 0$ to prove that if $EX_t = EX_0$ for all t and $E(X_t^2) = R_X(t,t) = R_X(0,0)$ for all t – that is, if the mean and variance

do not depend on time – then

$$|R_X(t,s)| \le R_X(0,0)$$

and

$$|K_X(t,s)| \le K_X(0,0).$$

Thus both functions take on their maximum value when $t = s$. This can be interpreted as saying that no random variable can be more correlated with a given random variable than it is with itself.

(b) Show that autocorrelations and covariance functions are symmetric functions, e.g., $R_X(t,s) = R_X(s,t)$.

20. The Cauchy–Schwarz Inequality: given random variables X and Y, define $a = [E(X^2)]^{(1/2)}$ and $b = [E(Y^2)]^{(1/2)}$. By considering the quantity $E[(X/a \pm Y/b)^2]$ prove the following inequality:

$$|E(XY)| \le [E(X^2)]^{\frac{1}{2}}[E(Y^2)]^{\frac{1}{2}}.$$

21. Given two random processes $\{X_t;\ t \in \mathcal{T}\}$ and $\{Y_t;\ t \in \mathcal{T}\}$ defined on the same probability space, the *cross-correlation function* $R_{XY}(t,s);\ t,s \in \mathcal{T}$ is defined as

$$R_{XY}(t,s) = E(X_t Y_s).$$

since $R_X(t,s) = R_{XX}(t,s)$. Show that R_{XY} is not, in general, a symmetric function of its arguments. Use the Cauchy–Schwarz inequality of Problem 4.20 to find an upper bound to $|R_{XY}(t,s)|$ in terms of the autocorrelation functions R_X and R_Y.

22. Let Θ be a random variable described by a uniform pdf on $[-\pi,\pi]$ and let Y be a random variable with mean m and variance σ^2; assume that Θ and Y are independent. Define the random process $\{X(t);\ t \in \Re\}$ by $X(t) = Y\cos(2\pi f_0 t + \Theta)$, where f_0 is a fixed frequency in hertz. Find the mean and autocorrelation function of this process. Find the limiting time average

$$\lim_{T\to\infty} \frac{1}{T}\int_0^T X(t)dt.$$

(Only in trivial processes such as this can one find exactly such a limiting time average.)

23. Suppose that $\{X_n\}$ is an iid process with a uniform pdf on [0,1). Does $Y_n = X_1 X_2 \ldots X_n$ converge in mean square as $n \to \infty$? If so, to what?

24. Let $r^{(n)}(a)$ denote the relative frequency of the letter a in a sequence $x_0,\ldots,x_{n-1}$. Show that if we define $q(a) = r^{(n)}(a)$, then $q(a)$ is a valid pmf. (This pmf is called the "sample distribution," or "empirical distribution.")
One measure of the distance or difference between two pmf's p and q is

$$||p-q||_1 \triangleq \sum_a |p(a)-q(a)|.$$

Show that if the underlying process is iid with marginal pmf p, then the empirical pmf will converge to the true pmf in the sense that

$$\lim_{n\to\infty} ||p - r^{(n)}||_1 = 0$$

in probability.

25. Given two sequences of random variables $\{X_n;\, n = 1, 2, \ldots\}$ and $\{Y_n;\, n = 1, 2, \ldots\}$ and a random variable X, suppose that with probability one $|X_n - X| \le Y_n$ all n and that $EY_n \to 0$ as $n \to \infty$. Prove that $EX_n \to EX$ and that X_n converges to X in probability as $n \to \infty$.
26. This problem provides another example of the use of covariance functions. Say that we have a discrete time random process $\{X_n\}$ with a covariance function $K_X(t, s)$ and a mean function $m_n = EX_n$. We are told the value of the past sample, say $X_{n-1} = \alpha$, and are asked to make a good guess of the next sample on the basis of the old sample. Furthermore, we are required to make a *linear* guess or estimate, called a prediction, of the form

$$\widehat{X}_n(\alpha) = a\alpha + b,$$

for some constants a and b. Use ordinary calculus techniques to find the values of a and b that are "best" in the sense of minimizing the mean squared error

$$E[(X_n - \widehat{X}_n(X_{n-1}))^2].$$

Give your answer in term of the mean and covariance function. Generalize to a linear prediction of the form

$$\widehat{X}_n(X_{n-1}, X_{n-m}) = a_1 X_{n-1} + a_m X_{n-m} + b,$$

where m is an arbitrary integer, $m \ge 2$. When is $a_m = 0$?
27. We developed the mean and variance of the sample average S_n for the special case of uncorrelated random variables. Evaluate the mean and variance of S_n for the opposite extreme, where the X_i are highly correlated in the sense that $E[X_i X_k] = E[X_i^2]$ for all i, k.
28. Given n independent random variables X_i, $i = 1, 2, \ldots, n$ with variances σ_i^2 and means m_i, define the random variable

$$Y = \sum_{i=1}^{n} a_i X_i,$$

where the a_i are fixed real constants. Find the mean, variance, and characteristic function of Y.
Now let the mean be constant; i.e., $m_i = m$. Find the minimum variance of Y over the choice of the $\{a_i\}$ subject to the constraint that $EY = m$. The result is called the *minimum variance unbiased estimate of* m.
Now suppose that $\{X_i;\, i = 0, 1, \ldots\}$ is an iid random process and that N is a Poisson random variable with parameter λ and that N is independent of the

$\{X_i\}$. Define the random variable

$$Y = \sum_{i=1}^{N} \frac{X_i}{\sigma_X^2}.$$

Use iterated expectation to find the mean, variance, and characteristic function of Y.

29. The random process of Example [3.27] can be expressed as follows: let Θ be a continuous random variable with a pdf

$$f_\Theta(\theta) = \frac{1}{2\pi}; \quad \theta \in [-\pi, +\pi]$$

and define the process $\{X(t);\ t \in \Re\}$ by $X(t) = \cos(t + \Theta)$.
(a) Find the cdf $F_{X(0)}(x)$.
(b) Find $EX(t)$.
(c) Find the covariance function $K_X(t, s)$.

30. Let $\{X_n\}$ be a random process with mean m and autocorrelation function $R_X(n, k)$, and let $\{W_n\}$ be an iid random process with zero-mean and variance σ_W^2. Assume that the two processes are independent of each another; that is, any collection of the X_i is independent of any collection of the W_i. Form a new random process $Y_n = X_n + W_n$. *Note:* this is a common model for a communication system or measurement system with $\{X_n\}$ a "signal" process or "source," $\{W_n\}$ a "noise" process, and $\{Y_n\}$ the "received" process; see Problem 3.38 for example.
(a) Find the mean EY_n and covariance $K_Y(n, k)$ in terms of the given parameters.
(b) Find the cross-correlation function defined by

$$R_{XY}(k, j) = E[X_k Y_j].$$

(c) As in Problem 4.26, find the minimum mean squared error estimate of X_n of the form

$$\widehat{X}(Y_n) = aY_n + b.$$

The resulting estimate is called a *filtered* value of X_n.
(d) Extend to a linear filtered estimate that uses Y_n and Y_{n-1}.

31. Suppose that there are two independent data sources $\{W_i(n),\ i = 1, 2\}$. Each data source is modeled as a Bernoulli random process with parameter 1/2. The two sources are encoded for transmission as follows. First, three random processes $\{Y_i(n);\ i = 1, 2, 3\}$ are formed, where $Y_1 = W_1$, $Y_2 = W_2$, $Y_3 = W_1 + W_2$. The last sum is taken modulo 2 and is formed to provide redundancy for noise protection in transmission. These are time-multiplexed to form a random process $\{X(3n + i) = Y_i(n)\}$. Show that $\{X(n)\}$ has identically distributed components and is pairwise independent but is not iid.

32. Let $\{U_n;\ n = 0, 1, \ldots\}$ be an iid random process with marginal pdf $f_{U_n} = f_U$, the uniform pdf of Problem A.1. In other words, the joint pdf's can be written as

$$f_{U^n}(u^n) = f_{U_0, U_1, \ldots, U_{n-1}}(u_0, u_1, \ldots, u_{n-1}) = \prod_{i=0}^{n-1} f_U(u_i).$$

Find the mean $m_n = E[U_n]$ and covariance function $K_U(k, j) = E[(U_k - m_k)(U_j - m_j)]$ for the process and verify that the weak law of large numbers holds for this process.

33. Let $\{U_n\}$ be an iid process with a uniform marginal pdf on $[0, 1)$. Define a new process $\{W_n;\ n = 0, 1, \ldots\}$ by $W_0 = 2U_0$ and $W_n = U_n + U_{n-1}$ for $n = 1, 2, \ldots$ Find the mean $E[W_n]$ and covariance function $K_W(k, j)$. Does the weak law of large numbers hold for this process? Find the pdf f_{W_n}.

34. Show that if Y_n converges to Y in probability and also converges to W in probability, then $\Pr(Y = W) = 1$.

35. Show that the convergence of the average of the means in (4.100) to a constant and convergence of (4.101) to zero are sufficient for a mean ergodic theorem of the form of Theorem 4.11. In what sense if any does $\{S_n\}$ converge?

36. The purpose of this problem is to demonstrate the relationships among the four forms of convergence that we have presented. In each case $([0, 1], \mathcal{B}([0, 1]), P)$ is the underlying probability space, with probability measure described by the uniform pdf. For each of the following sequences of random variables, determine the pmf of $\{Y_n\}$, the senses in which the sequences converge, and the random variable and pmf to which the sequences converge.

    (a) $Y_n(\omega) = \begin{cases} 1 & \text{if } n \text{ is odd and } \omega < 1/2 \text{ or } n \text{ is even and } \omega > 1/2 \\ 0 & \text{otherwise} \end{cases}.$

    (b) $Y_n(\omega) = \begin{cases} 1 & \text{if } \omega < 1/n \\ 0 & \text{otherwise} \end{cases}.$

    (c) $Y_n(\omega) = \begin{cases} n & \text{if } \omega < 1/n \\ 0 & \text{otherwise} \end{cases}.$

    (d) Divide $[0, 1]$ into a sequence of intervals $\{F_n\} = \{[0, 1], [0, (1/2)), [(1/2), 1], [0, \frac{1}{3}), [\frac{1}{3}, 2/3), [2/3, 1], [0, \frac{1}{4}), \ldots\}$ Let

    $$Y_n(\omega) = \begin{cases} 1 & \text{if } \omega \in F_n \\ 0 & \text{otherwise} \end{cases}.$$

    (e) $Y_n(\omega) = \begin{cases} 1 & \text{if } \omega < 1/2 + 1/n \\ 0 & \text{otherwise} \end{cases}.$

37. Suppose that X is a random variable with mean m and variance σ^2. Let g_k be a deterministic periodic pulse train such that g_k is 1 whenever k is a multiple of a fixed positive integer N and g_k is 0 for all other k. Let U be a random variable

that is independent of X such that $p_U(u) = 1/N$ for $u = 0, 1, \ldots, N-1$. Define the random process Y_n by

$$Y_n = Xg_{U+n}$$

that is, Y_n looks like a periodic pulse train with a randomly selected amplitude and a randomly selected phase. Find the mean and covariance functions of the Y process. Find a random variable $\widehat{Y}$ such that

$$\lim_{n\to\infty} \frac{1}{n} \sum_{i=0}^{n-1} Y_i = \widehat{Y}$$

in the sense of convergence with probability one. (This is an example of a process that is simple enough for the limit to be evaluated explicitly.) Under what conditions on the distribution of X does the limit equal EY_0 (and hence the conclusion of the weak law of large numbers holds for this process with memory)?

38. Let $\{X_n\}$ be an iid zero-mean Gaussian random process with autocorrelation function $R_X(0) = \sigma^2$. Let $\{U_n\}$ be an iid random process with $\Pr(U_n = 1) = Pr(U_n = -1) = (1/2)$. Assume that the two processes are mutually independent of each other. Define a new random process $\{Y_n\}$ by

$$Y_n = U_n X_n.$$

(a) Find the autocorrelation function $R_Y(k, j)$.
(b) Find the characteristic function $M_{Y_n}(ju)$.
(c) Is $\{Y_n\}$ an iid process?
(d) Does the sample average

$$S_n = n^{-1} \sum_{i=0}^{n-1} Y_i$$

converge in mean square? If so, to what?

39. Assume that $\{X_n\}$ is an iid zero-mean Gaussian random process with $R_X(0) = \sigma^2$, that $\{U_n\}$ is an iid binary random process with $\Pr(U_n = 1) = 1 - \epsilon$ and $\Pr(U_n = 0) = \epsilon$ (in other words, $\{U_n\}$ is a Bernoulli process with parameter $1 - \epsilon$), and the processes $\{X_n\}$ and $\{U_n\}$ are mutually independent of each other. Define a new random process

$$V_n = X_n U_n.$$

(This is a model for the output of a communication channel that has the X process as an input but has "dropouts" in the sense that it occasionally sets an input symbol to zero.)
(a) Find the mean EV_n and characteristic function $M_{V_n}(ju) = Ee^{juV_n}$.
(b) Find the mean squared error $E[(X_n - V_n)^2]$.
(c) Find $\Pr(X_n \neq V_n)$.
(d) Find the covariance of V_n.

(e) Is the following true?

$$\left(\operatorname*{l.i.m.}_{n\to\infty} \frac{1}{n}\sum_{i=0}^{n-1} X_i\right)\left(\operatorname*{l.i.m.}_{n\to\infty} \frac{1}{n}\sum_{i=0}^{n-1} U_i\right) = \operatorname*{l.i.m.}_{n\to\infty} \frac{1}{n}\sum_{i=0}^{n-1} V_i.$$

40. Let $\{X_n\}$ be a finite-alphabet iid random process with marginal pmf p_x. The *entropy* of an iid random process is defined as

$$H(X) = -\sum_x p_x(x)\log p_X(x) = E(-\log p_X(X)),$$

where care must be taken to distinguish the use of the symbol X to mean the name of the random variable in $H(X)$ and p_X, and its use as the random variable itself in the argument of the right-hand expression. If the logarithm is base two then the units of entropy are called *bits*. Use the weak law of large numbers to show that $-(1/n)\sum_{i=0}^{n-1}\log p_X(X_i)$ converges to $H(X)$ in the sense of convergence in probability. Show that this implies that

$$\lim_{n\to\infty}\Pr(|p_{X_0,\ldots,X_{n-1}}(X_0,\ldots,X_{n-1}) - 2^{-nH(X)}| > \epsilon) = 0$$

for any $\epsilon > 0$. This result was first developed by Claude Shannon and is sometimes called the *asymptotic equipartition property* of information theory. It forms one of the fundamental results of the mathematical theory of communication. Roughly stated, with high probability an iid process with n-dimensional sample vector $X^n = (x_0, x_1, \ldots, x_{n-1})$ has nth order probability mass function evaluated at x^n approximately equal to $2^{-nH(X)}$ for large n. In other words, the process produces long vectors that appear to have an approximately uniform distribution over some collection of possible vectors.

41. Given $X^{n-1} = (X_0, X_1, \ldots, X_{n-2})$, the *conditional differential entropy* of X_{n-1} is defined by

$$h(X_{n-1}|X^{n-1}) = -\int f_{X_0,X_1,\ldots,X_{n-1}}(x_0, x_1, \ldots, x_{n-1}) \times \log f_{X_{n-1}|X_1,\ldots,X_{n-2}}(x_{n-1}|x_1,\ldots,x_{n-2})\, dx_0\, dx_1 \ldots dx_{n-1} \quad (4.134)$$

Show that

$$h(X^n) = h(X_n|X^{n-1}) + h(X^{n-1}). \quad (4.135)$$

Now suppose that $\{X_n\}$ is a stationary Gaussian random process with zero mean and covariance function K. Evaluate $h(X_n|X^{n-1})$.

42. Let $X \geq 0$ be an integer-valued random variable with $E(X) < \infty$.

(a) Prove that

$$E(X) = \sum_{k=1}^{\infty} P(X \geq k).$$

(b) Based on (a) argue that

$$\lim_{N\to\infty} P(X \geq N) = 0.$$

(c) Prove the stronger statement

$$P(X \geq N) \leq \frac{E(X)}{N}.$$

Hint: write an expression for the expectation $E(X)$ and break up the sum into two parts, a portion where the summation dummy variable is larger than N and a portion where it is smaller. A simple lower bound for each part gives the desired result.

(d) Let X be a geometric random variable with parameter p, $p \neq 0$. Calculate the quantity $P(X \geq N)$ and use this result to show that $\lim_{N\to\infty} P(X \geq N) = 0$.

(e) Based on the previous parts show that

$$(1-p)^{N-1} \leq \frac{1}{pN}$$

for any $0 < p \leq 1$ and for any integer N.

43. Suppose that $\{X_n\}$ is an iid random process with mean $E(X_n) = \bar{X}$ and variance $E[(X_n - \bar{X})^2] = \sigma_X^2$. A new process $\{Y_n\}$ is defined by the relation

$$Y_n = \sum_{k=0}^{\infty} r^k X_{n-k}$$

where $|r| < 1$. Find $E(Y_n)$ and the autocorrelation $R_Y(k,j)$ and the covariance $K_Y(k,j)$.

Define the sample average

$$S_n = \frac{1}{n}\sum_{i=0}^{n-1} Y_i.$$

Find the mean $E(S_n)$ and variance $\sigma_{S_n}^2$. Does $S_n \to 0$ in probability?

44. Let $\{U_n\}$ be an iid Gaussian random process with mean 0 and variance σ^2. Suppose that Z is a random variable having a uniform distribution on $[0,1]$. Suppose Z represents the value of a measurement taken by a remote sensor and that we wish to guess the value of Z based on a noisy sequence of measurements $Y_n = Z + U_n$, $n = 0, 1, 2, \ldots$; that is, we observe only Y_n and wish to estimate the underlying value of Z. To do this we form a sample average and define the estimate

$$\widehat{Z}_N = \frac{1}{N}\sum_{i=0}^{n-1} Y_i.$$

(a) Find a simple upper bound to the probability

$$\Pr(|\widehat{Z}_n - Z| > \epsilon)$$

that goes to zero as $n \to \infty$. (This means that our estimator is asymptotically good.)

Suppose next that we have a two-dimensional random process $\{U_n, W_n\}$ (i.e., the output at each time is a random pair or a two-dimensional random variable) with the following properties. Each pair (U_n, W_n) is independent of all past and future pairs (U_k, W_k) $k \neq n$. Each pair (U_n, W_n) has an identical joint cdf $F_{U,W}(u, w)$. For each n $E[U_n] = E[W_n] = 0$, $E[U_n^2] = E[W_n^2] = \sigma^2$, and $E[U_n W_n] = \rho\sigma^2$. (The quantity ρ is called the *correlation coefficient*.) Instead of just observing a noisy sequence $Y_n = Z + U_n$, we also observe a separate noisy measurement sequence $X_n = Z + W_n$ (the same Z, but different noises). Suppose further that we try to improve our estimate of Z by using both of these measurements to form an estimate

$$\tilde{Z} = a\frac{1}{n}\sum_{i=0}^{n-1} Y_i + (1-a)\frac{1}{n}\sum_{i=0}^{n-1} X_i.$$

for some a in $[0, 1]$.

(a) Show that $|\rho| \leq 1$. Find a simple upper bound to the probability

$$\Pr(|\tilde{Z}_n - Z| > \epsilon)$$

that goes to zero as $n \to \infty$. What value of a gives the smallest upper bound in part (b) and what is the resulting bound? (Note as a check that the bound should be no worse than part (a) since the estimator of part (a) is a special case of that of part (b).) In the special case where $\rho = -1$, what is the best a and what is the resulting bound?

45. Suppose that $\{X_n\}$ are iid random variables described by a common marginal distribution F. Suppose that the random variables

$$S_n = \frac{1}{n}\sum_{i=0}^{n-1} X_i$$

also have the distribution F for all positive integers n. Find the form of the distribution F. (This is an example of what is called a *stable distribution*.) Suppose that the $1/n$ in the definition of S_n is replaced by $1/\sqrt{n}$. What must F then be?

46. Consider the following *nonlinear* modulation scheme: define

$$W(t) = e^{j(2\pi f_0 t + cX(t) + \Theta)},$$

$\{X(t)\}$ is a zero-mean weakly stationary Gaussian random process with autocorrelation function $R_X(\tau)$, f_0 is a fixed frequency, Θ is a uniform random variable on $[0, 2\pi]$, Θ is independent of all of the $X(t)$, and c is a modulation constant. (This is a mathematical model for phase modulation.)

Define the expectation of a complex random variable in the natural way: if $Z =$

$\Re(Z) + j\Im(Z)$, then $E(Z) = E[\Re(Z)] + jE[\Im(Z)]$.) Define the autocorrelation of a complex -valued random process $W(t)$ by

$$R_W(t,s) = E(W(t)W(s)^*),$$

where $W(s)^*$ denotes the complex conjugate of $W(s)$.
Find the mean $E(W(t))$ and the autocorrelation function $R_W(t,s) = E[W(t)W(s)^*]$.
Hint: the autocorrelation is admittedly a trick question (but a very useful trick). Keep characteristic functions in mind when pondering the evaluation of the autocorrelation function.

47. Suppose that $\{X_n;\ n = 0, 1, \ldots\}$ is a discrete time iid random process with pmf

$$p_{X_n}(k) = (1/2);\ k = 0, 1.$$

Two other random processes are defined in terms of the X process:

$$Y_n = \sum_{i=0}^{n} X_i;\ n = 0, \ldots$$

$$W_n = (-1)^{Y_n}\ n = 0, 1, \ldots$$

and

$$V_n = X_n - X_{n-1};\ n = 1, \ldots$$

(a) Find the covariance functions for the X and Y processes.
(b) Find the mean and variance of the random variable W_n. Find the covariance function of the process W_n.
(c) Find the characteristic function of the random variable V_n.
(d) Which of the above four processes are weakly stationary? Which are not?
(e) Evaluate the limits $\text{l.i.m.}_{n\to\infty} \frac{Y_n}{n+1}$, $\text{l.i.m.}_{n\to\infty} \frac{Y_n}{n^2}$ and $\text{l.i.m.}_{n\to\infty} \frac{1}{n}\sum_{l=1}^{n} V_l$.
(f) For the showoffs: does the last limit above converge with probability one? (Only elementary arguments are needed.)

48. Suppose that $\{X_n\}$ is a discrete time iid random process with uniform marginal pdf's

$$f_{X_n}(\alpha) = \begin{cases} 1 & 0 \le \alpha < 1 \\ 0 & \text{otherwise} \end{cases}.$$

Define the following random variables:
- $U = X_0^2$
- $V = \max(X_1, X_2, X_3, X_4)$
- $W = \begin{cases} 1 & \text{if } X_1 \ge 2X_2 \\ 0 & \text{otherwise} \end{cases}.$

- For each integer n $Y_n = X_n + X_{n-1}$. Note that this defines a new random process $\{Y_n\}$.

(a) Find the expected values of the random variables U, V, and W.
(b) What are the mean $E(X_n)$ and covariance function $K_X(k,j)$ of $\{X_n\}$?
(c) What are the mean $E(Y_n)$ and covariance function $K_Y(k,j)$ of $\{Y_n\}$?
(d) Define the sample average

$$S_n = \frac{1}{n}\sum_{k=1}^{n} Y_k.$$

Find the mean $E(S_n)$ and variance $\sigma^2_{S_n}$ of S_n. Using only these results (and no results not yet covered), find $\text{l.i.m.}_{n\to\infty} S_n$.
(e) Does the sequence of random variables

$$Z_n = \prod_{i=1}^{n} X_i$$

converge in probability to 0?

49. A discrete time *martingale* $\{Y_n;\ n = 0, 1, \ldots\}$ is a process with the property that

$$E[Y_n|Y_0, Y_1, \ldots, Y_{n-1}] = Y_n.$$

In words, the conditional expectation of the current value is the previous value. Suppose that $\{X_n\}$ is iid. Is

$$Y_n = \sum_{n=0}^{n-1} X_n$$

a martingale?

50. Let $\{Y_n\}$ be the one-dimensional *random walk* of Chapter 3.
(a) Find the pmf p_{Y_n} for $n = 0, 1, 2$.
(b) Find the mean $E[Y_n]$ and variance $\sigma^2_{Y_n}$.
(c) Does Y_n/n converge as n gets large?
(d) Find the conditional pmf's $p_{Y_n|Y_0,Y_1,\ldots,Y_{n-1}}(y_n|y_0, y_1, \ldots, y_{n-1})$ and $p_{Y_n|Y_{n-1}}(y_n|y_{n-1})$. Is this process Markov?
(e) What is the minimum MSE estimate of Y_n given Y_{n-1}? What is the probability that Y_n actually equals its minimum MSE estimate?

51. Let $\{X_n\}$ be a binary iid process with $p_X(\pm 1) = 0.5$. Define a new process $\{W_n;\ n = 0, 1, \ldots\}$ by

$$W_n = X_n + X_{n-1}.$$

This is an example of a *moving-average* process, so called because it computes a short term average of the input process. Find the mean, variance, and covariance function of $\{W_n\}$. Prove a weak law of large numbers for W_n.

52. How does one generate a random process? It is often of interest to do so in order

to simulate a physical system in order to test an algorithm before it is applied to genuine data. Using genuine physical data may be too expensive, dangerous, or politically risky. One might connect a sensor to a resistor and heat it up to produce thermal noise, or flip a coin a few million times. One solution requires uncommon hardware and the other physical effort. The usual solution is to use a computer to generate a sequence that is not actually random, but *pseudo-random* in that it can produce a long sequence of numbers that *appear* to be random and will satisfy several tests for randomness, provided that the tests are not too stringent. An example is the `rand` command used in Matlab™. It uses the linear congruential method which starts with a "seed" X_0 and then recursively defines the sequence

$$X_n = (7^7 X_{n-1}) \bmod (2^{31} - 1). \tag{4.136}$$

This produces a sequence of integers in the range from 0 to $2^{31} - 1$. Dividing by 2^{31} (which is just a question of shifting in binary arithmetic) produces a number in the range $[0, 1)$. Find a computer with Matlab or program this algorithm yourself and try it out with different starting sequences. Find the sample average S_n of a sequence of 100, 1000, and 10,000 samples and compare them to the expected value of the uniform pdf random variable considered in this chapter. How might you determine whether or not the sequence being viewed was indeed random or not if you did not know how it was generated?

53. Suppose that U is a random variable with pdf $f_U(u) = 1$ for $u \in [0, 1)$. Describe a function $q : [0, 1) \to A$, where $A = \{0, 1, \ldots, K - 1\}$, so that the random variable $X = q(U)$ is discrete with pmf

$$p_X(k) = \frac{1}{K};\ k = 0, 1, \ldots, K - 1.$$

You have produced a uniform discrete random variable from a uniform continuous random variable.

(a) What is the minimum mean squared error estimator of U given $X = k$? Call this estimator $\hat{U}(k)$. Write an expression for the resulting MSE

$$E[(U - \hat{U}(q(U))]^2.$$

(b) Show that the estimator $\hat{U}$ found in the previous part minimizes the MSE $E[(U - \hat{U}(q(U))^2]$ between the original input and the final output (assuming that q is fixed). You have just demonstrated one of the key properties of a Lloyd–Max quantizer (see, e.g., [26]).

(c) Find the pmf for the random variable $\hat{U} = \hat{U}(q(U))$. Find $E[\hat{U}]$ and $\sigma^2_{\hat{U}}$. How do the mean and variance of the $\hat{U}$ compare with those of U? (equal, bigger, smaller?)

54. Modify the development in the text for the minimum mean squared error estimator to work for discrete random variables. What is the minimum MSE estimator for Y_n given Y_{n-1} for the binary Markov process developed in the chapter?

Which do you think makes more sense for guessing the next outcome for a binary Markov process, the minimum probability of error classifier or the minimum MSE estimator? Explain.

55. Let $\{Y_n;\ n = 0, 1, \ldots\}$ be the binary Markov process developed in the chapter. Find a new process $\{W_n;\ n = 1, 2, \ldots\}$ defined by $W_n = Y_n \oplus Y_{n-1}$. Describe the process W_n.

Problems 56–62 courtesy of the EE Department of the Technion.

56. Let X be a Gaussian random variable with zero-mean and variance σ^2.
    (a) Find $E[\cos(nX)]$, $n = 1, 2, \ldots$
    (b) Find $E[X^n]$, $n = 1, 2, \ldots$
    (c) Let N be a Poisson random variable with parameter λ and assume that X and N are independent. Find $E[X^N]$.
    Hint: use characteristic functions and iterated expectation.
57. Let X be a random variable with uniform pdf on $[-1, 1]$. Define a new random variable Y by

$$Y = \begin{cases} X & X \leq 0 \\ 1 & X > 0 \end{cases}.$$

    (a) Find the cdf $F_Y(y)$ and plot it.
    (b) Find the pdf $f_Y(y)$.
    (c) Find $E(Y)$ and σ_Y^2.
    (d) Find $E(X|Y)$.
    (e) Find $E[(X - E(X|Y))^2]$.
58. Let $X_1, X_2, \ldots, X_n$ be zero-mean statistically independent random variables. Define

$$Y_n = \sum_{i=1}^{n} X_i.$$

    Find $E(Y_7|Y_1, Y_2, Y_3)$.
59. Let U denote a binary random variable with pmf $p_U(u) = 0.5$ for $u = \pm 1$. Let $Y = U + X$, where X is $\mathcal{N}(0, \sigma^2)$ and where U and X are independent. Find $E(U|Y)$.
60. Let Y, N_1, N_2 be zero-mean, unit-variance, mutually independent random variables. Define

$$X_1 = Y + N_1 + \sqrt{\alpha} N_1$$
$$X_2 = Y + 3N_1 + \sqrt{\alpha} N_1.$$

    (a) Find the linear MMSE estimator of Y given X_1 and X_2.
    (b) Find the resulting MSE.
    (c) For what value of $\alpha \in [0, \infty)$ does the mean squared error become zero? Provide an intuitive explanation.

61. Let $\{X_n;\ n = 1, 2, \ldots\}$ be an iid sequence of $\mathcal{N}(m, \sigma^2)$ random variables. Define for any positive integer N

$$S_N = \sum_{n=1}^{N} X_n.$$

(a) For $K < N$ find the pdf $f_{S_N|S_K}(\alpha|\beta)$.

(b) Find the MMSE estimator of S_K given S_N, $E(S_K|S_N)$. Define $V_K = \sum_{n=1}^{K} X_n^2$. Find the linear MMSE of V_K given V_N.

62. Let $X_i = S + W_i$, $i = 1, 2, \ldots, N$, where S and the W_i are mutually independent with zero-mean. The variance of S is σ_S and the variances of all the W_i are σ_W^2.

(a) Find the linear MMSE of S given the observations X_i, $i = 1, 2, \ldots, N$.

(b) Find the resulting MSE.

63. Suppose that $\{X_n;\ n = 0, 1, \ldots\}$ is a Gaussian iid process with zero-mean and variance σ_X^2. Let $\{Y_n;\ n = 0, 1, 2, \ldots\}$ be an iid Bernoulli process with parameter p. Define the random variables

$$S_n = \frac{1}{n} \sum_{i=0}^{n-1} X_i;\ n = 1, 2, 3, \ldots$$

and

$$N_n = \sum_{i=0}^{n-1} Y_i;\ n = 1, 2, 3, \ldots$$

(a) Find the mean, variance, transform, and pdf for S_n.

(b) Find the mean, variance, transform, and pmf for N_n.

(c) Find the mean, variance, and transform for S_{N_n}.

(d) Does S_n converge in mean square as $n \to \infty$? If so, to what?

(e) Does S_{N_n} converge in probability as $n \to \infty$? If so, to what?

5

Second-order theory

In Chapter 4 we saw that the second-order moments of a random process – the mean and covariance or, equivalently, the autocorrelation – play a fundamental role in describing the relation of limiting sample averages and expectations. We also saw, for example in Section 4.6.1 and Problem 4.26, that these moments also play a key role in signal processing applications of random processes, especially in linear least squares estimation. Because of the fundamental importance of these particular moments, this chapter considers their properties in greater depth and their evaluation for several important examples. A primary focus is on a second-order moment analog of a derived distribution problem. Suppose we are given the second-order moments of one random process and this process is then used as an input to a linear system. What are the resulting second-order moments of the output random process? These results are collectively known as second-order moment input/output or I/O relations for linear systems.

Linear systems may seem to be a very special case. As we will see, their most obvious attribute is that they are easier to handle analytically, which leads to more complete, useful, and stronger results than can be obtained for the class of all systems. This special case, however, plays a central role and is by far the most important class of systems. The design of engineering systems frequently involves the determination of an optimum system – perhaps the optimum signal detector for a signal in noise, the filter that provides the highest signal-to-noise ratio, the optimum receiver, etc. Surprisingly enough, the optimum is frequently a linear system. Even when it is not, often a linear system is a good enough approximation to be used for the sake of economical design. For these reasons it is of interest to study the properties of the output random process from a linear system that is driven by a specified input random process. In this chapter we consider only second-order moments. In the next chapter we consider examples in which one can develop a more complete probabilistic description of the output process. As one might suspect the less complete second-order descriptions are possible under far more general conditions.

When a linear system is driven by a given random process, the second-order properties of the output process provide the fundamental tools for analysis and optimization. As an example of such analysis, the chapter closes with an application of second-order moment theory to the design of systems for linear least squares estimation.

Because the primary engineering application of these systems is to noise discrimination, we will group them together under the name "linear filters." This designation denotes the suppression or "filtering out" of noise from the combination of signal and noise. The methods of analysis are not limited to this application, of course.

As usual, we emphasize discrete time in the development, with the obvious extensions to continuous time provided by integrals. Furthermore, we restrict attention in the basic development to linear time-invariant filters. The extension to time-varying systems is obvious but cluttered with obfuscating notation. Time-varying systems will be encountered briefly when considering recursive estimation.

We initially follow a somewhat cavalier attitude in terms of how infinite sums and integrals are defined for random processes. We subsequently explore the use of second-order moments to provide rigorous definitions of these ideas using mean square convergence of random variables. This provides a more careful treatment of integrals and hence also of linear filtering and linear operations such as differentiation and integration of continuous time random processes.

5.1 Linear filtering of random processes

Suppose that a random process $\{X(t);\, t \in \mathcal{T}\}$ (or $\{X_t;\, t \in \mathcal{T}\}$) is used as an input to a linear time-invariant system described by a δ-response h. Hence the output process, say $\{Y(t)\}$ or $\{Y_t\}$ is described by the convolution integral (A.35) in the continuous time case and by the convolution sum (A.42) in the discrete time case. To be precise, we have to be careful about how the integral or sum is defined; integrals and infinite sums of random processes are really limits of random variables, and those limits can converge in a variety of ways, such as in quadratic mean or with probability one. For the moment we will assume that the convergence is with probability one, i.e., that each realization or sample function of the output is related to the corresponding realization of the input via (A.35) or (A.42). That is, we take

$$Y(t) = \int_{s:\, t-s \in \mathcal{T}} X(t-s)h(s)\, ds \tag{5.1}$$

or

$$Y_n = \sum_{k: n-k \in \mathcal{T}} X_{n-k} h_k \tag{5.2}$$

to mean actual equality for a collection of elementary events ω on the underlying probability space Ω collectively having probability one. More precisely, with probability one

$$Y(t,\omega) = \int_{s:\, t-s \in \mathcal{T}} X(t-s,\omega)h(s)\, ds$$

or

$$Y_n(\omega) = \sum_{k: n-k \in \mathcal{T}} X_{n-k}(\omega) h_k,$$

respectively. As has been discussed earlier, it can be extremely difficult to demonstrate that such forms of convergence actually hold for a particular random process and δ-response, but the results obtained by such formal manipulations can be demonstrated rigorously under suitable technical assumptions either through advanced mathematical methods (which are not considered here) or by using weaker notions of convergence of random variables – in particular by using mean square convergence to define both infinite sums and integrals (which we will do later in this chapter).

Rigorous consideration of conditions under which the various limits exist is straightforward for the discrete time case. It is obvious that the limits exist for the so-called finite impulse response (FIR) discrete time filters where only a finite number of the h_k are nonzero and hence the sum has only a finite number of terms. It is also possible to show mean square convergence for the general discrete time convolution if the input process has finite mean and variance and if the filter is stable in the sense of (A.43). In particular, for a one-sided input process, (5.2) converges in quadratic mean; i.e.,

$$\operatorname*{l.i.m.}_{n \to \infty} \sum_{k=0}^{n-1} X_{n-k} h_k$$

converges. Initially we ignore these technical difficulties and just assume that the sums and integrals are well defined.

Fourier analysis plays a fundamental role in the study of linear systems, but unfortunately (A.37) and (A.43) are not satisfied in general for sample functions of interesting random processes and hence one cannot simply take Fourier transforms of both sides of (5.1) and (5.2) and obtain a useful spectral relation. Even if one could, the Fourier transform of a random process would be a random variable for each value of frequency! Because of this, the frequency domain theory for random processes is quite different from that for deterministic processes. Relations such as (A.39) may on occasion be useful for intuition, but they must be used with extreme care.

With the foregoing notation and preliminary considerations, we now turn to the analysis of discrete time linear filters with random process inputs.

5.2 Linear systems I/O relations

Discrete time systems

Ideally one would like to have a complete specification of the output of a linear system as a function of the specification of the input random process. Usually this is a difficult proposition because of the complexity of the computations required. However, it is a relatively easy task to determine the mean and covariance function at the output. As we will show, the output mean and covariance function depend only on the input mean and covariance function and on no other properties of the input random process. Furthermore, in many, if not most, applications, the mean and covariance functions of the output are all that are needed to solve the problem at hand. As an important example: if the random process is Gaussian, then the mean and covariance functions provide a complete description of the process.

Linear filter input/output (I/O) relations are most easily developed using the convolution representation of a linear system. Let $\{X_n\}$ be a discrete time random process with mean function $m_n = EX_n$ and covariance function $K_X(n,k) = E[(X_n - m_n)(X_k - m_k)]$. Let $\{h_k\}$ be the Kronecker δ-response of a discrete time linear filter. For notational convenience we assume that the δ-response is causal. The noncausal case simply involves a change of the limits of summation. Next we will find the mean and covariance functions for the output process $\{Y_n\}$ that is given in the convolution equation of (5.2).

From (5.2) the mean of the output process is found using the linearity of expectation as

$$EY_n = \sum_k h_k EX_{n-k} = \sum_k h_k m_{n-k}, \tag{5.3}$$

assuming, of course, that the sum exists. The sum does exist if the filter is stable and the input mean is bounded. That is, if there is a constant $m < \infty$, such that $|m_n| \le |m|$ for all n and if the filter is stable in the sense of equation (A.43), then

$$|EY_n| = \left|\sum_k h_k m_{m-k}\right| \leq \max_k |m_{n-k}| \sum_k |h_k| \leq |m| \sum_k |h_k| < \infty.$$

If the input process $\{X_n\}$ is weakly stationary, then the input mean function equals the constant, m, and

$$EY_n = m \sum_k h_k, \tag{5.4}$$

which is the dc response of the filter times the mean. For reference we specify the precise limits for the two-sided random process where $\mathcal{T} = \mathcal{Z}$ and for the one-sided input random process where $\mathcal{T} = \mathcal{Z}_+$:

$$EY_n = m \sum_{k=0}^{\infty} h_k, \quad \mathcal{T} = \mathcal{Z} \tag{5.5}$$

$$EY_n = m \sum_{k=0}^{n} h_k, \quad \mathcal{T} = \mathcal{Z}_+. \tag{5.6}$$

Thus, if the input random process is weakly stationary, then the output mean exists if the input mean is finite and the filter is stable. In addition, it can be seen that for two-sided weakly stationary random processes, the expected value of the output process does not depend on the time index n since the limits of the summation do not depend on n. For one-sided weakly stationary random processes, however, the output mean is not constant with time but approaches a constant value as $n \to \infty$ if the filter is stable. Note that this means that if a one-sided stationary process is put into a linear filter, the output is in general not stationary!

If the filter is not stable, the magnitude of the output mean is unbounded with time. For example, if we set $h_k = 1$ for all k in (5.6) then $EY_n = (n+1)m$, which very strongly depends on the time index n and which is unbounded.

Turning to the calculation of the output covariance function, we use equations (5.2) and (5.3) to evaluate the covariance with some bookkeeping as

$$\begin{aligned} K_Y(k,j) &= E[(Y_k - EY_k)(Y_j - EY_j)] \\ &= E\left[\left(\sum_n h_n(X_{k-n} - m_{k-n})\right)\left(\sum_m h_m(X_{j-m} - m_{j-m})\right)\right] \\ &= \sum_n \sum_m h_n h_m E[(X_{k-n} - m_{k-n})(X_{j-m} - m_{j-m})] \\ &= \sum_n \sum_m h_n h_m K_X(k-n, j-m). \end{aligned} \tag{5.7}$$

A careful reader might note the similarity between (5.7) and the corresponding matrix equation (4.28) derived during the consideration of Gaussian vectors (but true generally for covariance matrices of linear functions of random vectors).

As before, the range of the sums depends on the index set used. Since we have specified causal filters, the sums run from 0 to ∞ for two-sided processes and from 0 to k and 0 to j for one-sided random processes.

It can be shown that the sum of (5.7) converges if the filter is stable in the sense of (A.43) and if the input process has bounded variance; i.e., there is a constant $\sigma^2 < \infty$ such that $|K_X(n,n)| < \sigma^2$ for all n (Problem 5.20).

If the input process is weakly stationary, then K_X depends only on the difference of its arguments. This is made explicit by replacing $K_X(m,n)$ by $K_X(m-n)$. Then (5.7) becomes

$$K_Y(k,j) = \sum_n \sum_m h_n h_m K_X((k-j) - (n-m)). \tag{5.8}$$

Specifying the limits of the summation for the one-sided and two-sided cases, we have that

$$K_Y(k,j) = \sum_{n=0}^{\infty} \sum_{m=0}^{\infty} h_n h_m K_X((k-j) - (n-m)) \ ; \ \mathcal{T} = \mathcal{Z}. \tag{5.9}$$

and

$$K_Y(k,j) = \sum_{n=0}^{k} \sum_{m=0}^{j} h_n h_m K_X((k-j) - (n-m)) \ ; \ \mathcal{T} = \mathcal{Z}_+. \tag{5.10}$$

If the sum of (5.9) converges (e.g., if the filter is stable and $K_X(n,n) = K_X(0) < \infty$), then two interesting facts follow. First, if the input random process is weakly stationary and if the processes are two-sided, then the covariance of the output process depends only on the time lag; i.e., $K_Y(k,j)$ can be replaced by $K_Y(k-j)$. Note that this is not the case for a one-sided process, even if the input process is stationary and the filter stable! This fact, together with our earlier result regarding the mean, can be summarized as follows.

Given a two-sided random process as input to a linear filter, if the input process is weakly stationary and the filter is stable, the output random process is also weakly stationary. The output mean and covariance functions are given by

$$EY_n = m \sum_{k=0}^{\infty} h_k \tag{5.11}$$

$$K_Y(k) = \sum_{n=0}^{\infty} \sum_{m=0}^{\infty} h_n h_m K_X(k - (n - m)). \tag{5.12}$$

The second observation is that (5.8), (5.9), (5.10) or (5.12) is a double discrete convolution! The direct evaluation of (5.8), (5.9), and (5.10), while straightforward in concept, can be an exceedingly involved computation in practice. As in other linear systems applications, the evaluations of convolutions can often be greatly simplified by resort to transform techniques, as be considered shortly.

Continuous time systems

For each of the discrete time filter results there is an analogous continuous time result. For simplicity, however, we consider only the simpler case of two-sided processes. Let $\{X(t)\}$ be a two-sided continuous time input random process to a linear time-invariant filter with impulse response $h(t)$.

We can evaluate the mean and covariance functions of the output process in terms of the mean and covariance functions of the input random process by using the same development as was used for discrete time random processes. This time we have integrals instead of sums. Let $m(t)$ and $K_X(t, s)$ be the respective mean and covariance functions of the input process. Then the mean function of the output process is

$$EY(t) = \int E[X(t - s)]h(s)\, ds = \int m(t - s)h(s)\, ds. \tag{5.13}$$

The covariance function of the output random process is obtained by computations analogous to (5.7) as

$$K_Y(t, s) = \int d\alpha \int d\beta K_X(t - \alpha, s - \beta)h(\alpha)h(\beta). \tag{5.14}$$

Thus if $\{X(t)\}$ is weakly stationary with mean $m = m(t)$ and covariance function $K_X(\tau)$, then

$$EY(t) = m \int h(t)\, dt \tag{5.15}$$

and

$$K_Y(t,s) = \int d\alpha \int d\beta K_X((t-s) - (\alpha - \beta))h(\alpha)h(\beta). \tag{5.16}$$

In analogy to the discrete time result, the output mean is constant for a two-sided random process, and the covariance function depends on only the time difference. Thus a weakly stationary two-sided process into a stable linear time-invariant filter yields a weakly stationary output process in both discrete and continuous time. We leave it to the reader to develop conclusions that are parallel to the discrete time results for one-sided processes.

Transform I/O relations

In both discrete and continuous time, the covariance function of the output can be found by first convolving the input autocorrelation with the δ-response h_k or $h(t)$ and then convolving the result with the reflected δ-response h_{-k} or $h(-t)$. A way of avoiding the double convolution is found in Fourier transforms. Taking the Fourier transform (continuous or discrete time) of the double convolution yields the transform of the covariance function, which can be used to arrive at the output covariance function – essentially the same result with (in many cases) less overall work.

We shall show the development for discrete time. A similar sequence of steps provides the proof for continuous time by replacing the sums by integrals. Using (5.12),

$$\begin{aligned}
&\mathcal{F}_f(K_Y) \\
&= \sum_k \left(\sum_n \sum_m h_n h_m K_X(k - (n-m)) \right) e^{-j2\pi fk} \\
&= \sum_n \sum_m h_n h_m \left(\sum_k K_X(k-(n-m)) e^{-j2\pi f(k-(n-m))} \right) e^{-j2\pi f(n-m)} \\
&= \left(\sum_n h_n e^{-j2\pi fn} \right) \left(\sum_m h_m e^{+j2\pi fm} \right) \mathcal{F}(K_X) \\
&= \mathcal{F}_f(K_X)\mathcal{F}_f(h)\mathcal{F}_f(h^*),
\end{aligned} \tag{5.17}$$

where the asterix denotes complex conjugate. If we define $H(f) = \mathcal{F}_f(h)$, the transfer function of the filter, then the result can be abbreviated for both continuous and discrete time as

$$\mathcal{F}_f(K_Y) = |H(f)|^2 \mathcal{F}_f(K_X). \tag{5.18}$$

We can also conveniently describe the mean and autocorrelation functions in the frequency domain. From (5.5) and (5.15) the mean m_Y of the output is related to the mean m_Y of the input simply as

$$m_Y = H(0) m_X. \tag{5.19}$$

Since $K_X(k) = R_X(k) - |m_X|^2$ and $K_Y(k) = R_Y(k) - |m_Y|^2$, (5.18) implies that

$$\mathcal{F}_f(R_Y - |m_Y|^2) = |H(f)|^2 \mathcal{F}_f(R_X - |m_X|^2)$$

or

$$\begin{aligned}
\mathcal{F}_f(R_Y) - |m_Y|^2 \delta(f) &= |H(f)|^2 \left(\mathcal{F}_f(R_X) - |m_X|^2 \delta(f)\right) \\
&= |H(f)|^2 \mathcal{F}_f(R_X) - |H(f)|^2 |m_X|^2 \delta(f) \\
&= |H(f)|^2 \mathcal{F}_f(R_X) - |H(0)|^2 |m_X|^2 \delta(f),
\end{aligned}$$

where we have used the property of Dirac deltas that $g(f)\delta(f) = g(0)\delta(f)$ (provided $g(f)$ has no jumps at $f = 0$). Thus the autocorrelation function satisfies the same transform relation as the covariance function. This result is abbreviated by giving a special notation to the transform of an autocorrelation function: given a weakly stationary process $\{X(t)\}$ with autocorrelation function R_X, the *power spectral density* of the process is defined by

$$S_X(f) = \mathcal{F}_f(R_X) = \begin{cases} \sum_k R_X(k) e^{-j2\pi f k} & \text{discrete time} \\ \int R_X(\tau) e^{-j2\pi f \tau} \, d\tau & \text{continuous time} \end{cases}, \tag{5.20}$$

the Fourier transform of the autocorrelation function. The reason for the name will be given in the next section and discussed at further length later in the chapter. Given the definition we have now proved the following result.

If a weakly stationary process $\{X(t)\}$ with power spectral density $S_X(f)$ is the input to a linear time-invariant filter with transfer function H, then the output process $\{Y(t)\}$ is also weakly stationary and has mean

$$m_Y = H(0) m_X \tag{5.21}$$

and power spectral density

$$S_Y(f) = |H(f)|^2 S_X(f). \tag{5.22}$$

This result is true for both discrete and continuous time.

5.3 Power spectral densities

Under suitable technical conditions the Fourier transform can be inverted to obtain the autocorrelation function from the power spectral density. Thus the reader can verify from the definitions (5.20) that

$$R_X(\tau) = \begin{cases} \int_{-1/2}^{1/2} S_X(f) e^{j2\pi f\tau}\, df & \text{discrete time} \\ \int_{-\infty}^{\infty} S_X(f) e^{j2\pi f\tau}\, d\tau & \text{continuous time} \end{cases}. \tag{5.23}$$

The limits of $-1/2$ to $+1/2$ for the discrete time integral correspond to the fact that time is measured in units; e.g., adjacent outputs are one second or one minute or one year apart. Sometimes, however, the discrete time process is formed by sampling a continuous time process at every, say, T seconds, and it is desired to retain seconds as the unit of measurement. Then it is more convenient to incorporate the scale factor T into the time units and scale (5.20) and the limits of (5.23) accordingly – i.e., kT replaces k in (5.20), and the limits become $-1/2T$ to $1/2T$.

Power spectral densities inherit the property of symmetry from autocorrelation functions. As seen from the definition in Chapter 4, covariance and autocorrelation functions are symmetric ($R_X(t,s) = R_X(s,t)$). Therefore $R_X(\tau)$ is an even function. From (5.20) it can be seen with a little juggling that $S_X(f)$ is also even; that is, $S_X(-f) = S_X(f)$ for all f.

The name "power spectral density" comes from observing how the average power of a random process is distributed in the frequency domain. The autocorrelation function evaluated at 0 lag, $P_X = R_X(0) = E(|X(t)|^2)$, can be interpreted as the average power dissipated in a unit resistor by a voltage $X(t)$. Since the autocorrelation is the inverse Fourier transform of the power spectral density, this means that

$$P_X = \int S_X(f)\, df, \tag{5.24}$$

that is, the total average power in the process can be found by integrating $S_X(f)$. Thus if S_X were nonnegative, it could be considered as a density of power analogous to integrating a probability or mass density to find total probability or mass. For the probability and mass analogues, however, we know that integrating over any reasonable set will give the probability or mass of that set, i.e., we do not wish to confine interest to integrating over all possible frequencies. The analogous consideration for power is to look at the total average power within an arbitrary frequency band, which we do next. The fact that power spectral densities are nonnegative can be derived from the fact that the autocorrelation function is nonnegative definite (which can be shown in the same way it was shown for covariance functions) – a result known as Bochner's theorem. We shall prove nonnegativity of the power spectral density as part of the development.

Suppose that we wish to find the power of a process, say $\{X_t\}$ in some frequency band $f \in F$. Then a physically natural way to accomplish this would be to pass the given process through a bandpass filter with transfer function $H(f)$ equal to 1 for $f \in F$ and 0 otherwise and then to measure the output power. This is depicted in Figure 5.1 for the special case of a frequency interval $F = \{f : f_0 \leq |f| < f_0 + \Delta f\}$. Calling the output process

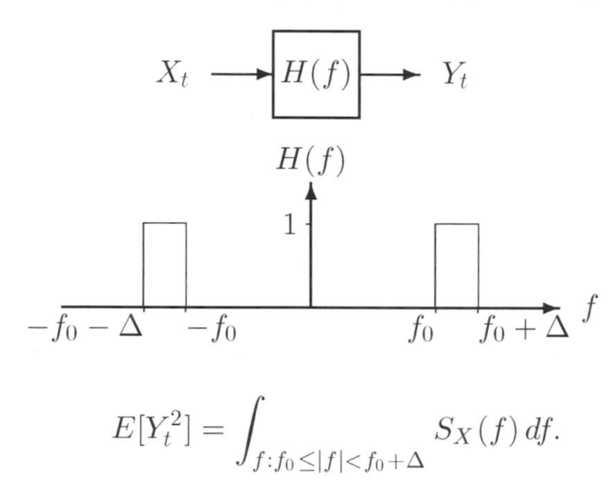

Figure 5.1 Power spectral density

$\{Y_t\}$, we have from (5.24) that the output power is

$$R_Y(0) = \int S_Y(f)\, df = \int |H(f)|^2 S_X(f)\, df = \int_F S_X(f)\, df. \tag{5.25}$$

Thus to find the average power contained in any frequency band we integrate the power spectral density over the frequency band. Because the average power must be nonnegative for any choice of f_0 and Δf, it follows that any power spectral density must be nonnegative, i.e.,

$$S_X(f) \geq 0, \text{ all } f. \tag{5.26}$$

To elaborate further, suppose that this is not true; i.e., suppose that $S_X(f)$ is negative for some range of frequencies. If we put $\{X_t\}$ through a filter that passes only those frequencies, the filter output power would have to be negative – clearly an impossibility.

From the foregoing considerations it can be deduced that the name *power spectral density* derives from the fact that $S_X(f)$ is a nonnegative function that is integrated to get power; that is, a "spectral" (meaning frequency content) density of power. Keep in mind the analogy to evaluating probability by integrating a probability density.

5.4 Linearly filtered uncorrelated processes

If the input process $\{X_n\}$ to a discrete time linear filter with δ-response $\{h_k\}$ is a weakly stationary uncorrelated process with mean m and variance σ^2 (for example, if it is iid), then $K_X(k) = \sigma^2\delta_k$ and $R_X(k) = \sigma^2\delta_k + m^2$. In this case the power spectral density is easily found to be

$$S_X(f) = \sum_k \sigma^2\delta_k e^{-j2\pi fk} + m^2\delta(f) = \sigma^2 + m^2\delta(f); \text{ all } f,$$

since the only nonzero term in the sum is the $k = 0$ term. The presence of the delta is due to the nonzero mean. When the mean is zero, this simplifies to

$$S_X(f) = \sigma^2; \text{ all } f. \tag{5.27}$$

Because the power spectral density is flat in this case, in analogy to the flat electromagnetic spectrum of white light, such a process (a discrete time, weakly stationary, zero mean, uncorrelated process) is said to be *white* or *white noise.* The inverse Fourier transform of the white noise spectral density is found from (5.23) (or simply by uniqueness) to be $R_X(k) = \sigma^2\delta_k$. Thus a discrete time random process is white if and only if it is weakly stationary, zero mean, and uncorrelated.

For the two-sided case we have from (5.12) that the output covariance is

$$K_Y(k) = \sigma^2 \sum_{n=0}^{\infty} h_n h_{n-k} = \sigma^2 \sum_{n=k}^{\infty} h_n h_{n-k} \ ; \ \mathcal{T} = \mathcal{Z}, \tag{5.28}$$

where the lower limit of the sum follows from the causality of the filter. If we assume for simplicity that $m = 0$, the power spectral density in this case reduces to

$$S_Y(f) = \sigma^2|H(f)|^2. \tag{5.29}$$

For a one-sided process, (5.10) yields

$$K_Y(k, j) = \sigma^2 \sum_{n=0}^{k} h_n h_{n-(k-j)}; \ \mathcal{T} = \mathcal{Z}_+. \tag{5.30}$$

Note that if $k > j$, then the sum can be taken over the limits $n = k - j$ to k since causality of the filter implies that the first few terms are 0. If $k < j$, then all of the terms in the sum may be needed. The covariance for the one-sided case appears to be asymmetric, but recalling that h_l is 0 for negative l, we can write the terms of the sum of (5.30) in descending order to obtain

$$\sigma^2(h_k h_j + h_{k-1}h_{j-1} + \cdots + h_0 h_{j-k})$$

if $j \geq k$ and

$$\sigma^2(h_k h_j + h_{k-1}h_{j-1} + \cdots + h_{k-j}h_0)$$

if $j \leq k$. By defining the function $\min(k, j)$ to be the smaller of k and j, we can rewrite (5.30) in two symmetric forms:

$$K_Y(k, j) = \sigma^2 \sum_{n=0}^{\min(k.j)} h_{k-n}h_{j-n}; \ \mathcal{T} = \mathcal{Z}_+ \tag{5.31}$$

and

$$K_Y(k, j) = \sigma^2 \sum_{n=0}^{\min(k.j)} h_n h_{n+|k-j|}. \tag{5.32}$$

The one-sided process is not weakly stationary because of the distinct presence of k and j in the sum, so the power spectral density is not defined.

In the two-sided case, the expression (5.28) for the output covariance function is the convolution of the unit δ-response with its *reflection* h_{-k}. Such a convolution between a waveform or sequence and its own reflection is also called a *sample autocorrelation*.

We next consider specific examples of this computation. These examples point out how two processes – one one-sided and the other two-sided – can be apparently similar and yet have quite different properties.

[5.1] Suppose that an uncorrelated discrete time two-sided random process $\{X_n\}$ with mean m and variance σ^2 is put into a linear filter with causal δ-response $h_k = r^k$, $k \geq 0$, with $|r| < 1$. Let $\{Y_n\}$ denote the output process, i.e.,

$$Y_n = \sum_{k=0}^{\infty} r^k X_{n-k}. \tag{5.33}$$

Find the output mean and covariance.

From the geometric series summation formula,

$$\sum_{k=0}^{\infty} |r|^k = \frac{1}{1-|r|} < \infty,$$

and hence the filter is stable. From (5.4), (5.5), and (5.6)

$$EY_n = m\sum_{k=0}^{\infty} r^k = \frac{m}{1-r}; \quad n \in \mathcal{Z}.$$

From (5.28), the output covariance for nonnegative k is

$$\begin{aligned} K_Y(k) &= \sigma^2 \sum_{n=k}^{\infty} r^n r^{n-k} \\ &= \sigma^2 r^{-k} \sum_{n=k}^{\infty} (r^2)^n = \sigma^2 \frac{r^k}{1-r^2} \end{aligned}$$

using the geometric series formula. Repeating the development for negative k (or appealing to symmetry) we find in general the covariance function is

$$K_Y(k) = \sigma^2 \frac{r^{|k|}}{1-r^2}; \quad k \in \mathcal{Z}.$$

Observe in particular that the output variance is

$$\sigma_Y^2 = K_y(0) = \frac{\sigma^2}{1-r^2}.$$

As $|r| \to 1$ the output variance grows without bound. However, as long as $|r| < 1$, the variance is defined and the process is clearly weakly stationary.

The previous example has an alternative construction that demonstrates how two models that appear quite different can lead to the same thing. From (5.33) we have

$$\begin{aligned} Y_n - rY_{n-1} &= \sum_{k=0}^{\infty} r^k X_{n-k} - r\sum_{k=0}^{\infty} r^k X_{n-1-k} \\ &= X_n + \sum_{k=1}^{\infty} r^k X_{n-k} - r\sum_{k=0}^{\infty} r^k X_{n-1-k} \\ &= X_n, \end{aligned}$$

since the two sums are equal. This yields a difference equation relating the two processes, expressing the output process Y_n in a recursive form:

$$Y_n = X_n + rY_{n-1}. \tag{5.34}$$

Thus the new Y_n is formed by adding the new X_n to the previous Y_n. This representation shows that in a sense the X_n process represents the "new information" in the Y_n process. We will see in the next chapter that if X_n is actually iid and not just uncorrelated, this representation leads to a complete probabilistic description of the output process. The representation (5.34) is called a first-order *autoregressive* model for the process, in contrast to the ordinary convolution representation of (5.33), which is often called a *moving-average* model.

The output spectral density can be found directly by taking the Fourier transform of the output covariance as

$$S_Y(f) = \sum_{k=-\infty}^{\infty} \frac{\sigma^2 r^{|k|}}{1-r^2} e^{-j2\pi fk},$$

a summation that can be evaluated using the geometric series formula – first from 1 to ∞ and then from 0 to $-\infty$ – and then summing the two complex terms. The reader should perform this calculation as an exercise. It is easier, however, to find the output spectral density through the linear system I/O relation. The transfer function of the filter is evaluated by a single application of the geometric series formula as

$$H(f) = \sum_{k=0}^{\infty} r^k e^{-j2\pi fk} = \frac{1}{1-re^{-j2\pi f}}.$$

Therefore the output spectral density from (5.22) is

$$S_Y(f) = \frac{\sigma^2}{|1-re^{-2\pi f}|^2} = \frac{\sigma^2}{1+r^2-2r\cos(2\pi f)}.$$

By a quick table lookup the reader can verify that the inverse transform of the output spectral density agrees with the covariance function previously found.

[5.2] Suppose that a one-sided uncorrelated process $\{X_n\}$ with mean m and variance σ^2 is put into a one-sided filter with δ-response as in Example [5.1]. Let $\{Y_n\}$ be the resulting one-sided output process. Find the output mean and covariance.

This time (5.6) yields

$$EY_n = m\sum_{k=0}^{n} r^k = m\frac{1-r^{n+1}}{1-r}$$

from the geometric series formula. From (5.32) the covariance is

$$K_Y(k,j) = \sigma^2 \sum_{n=0}^{\min(k,j)} r^{2n+|k-j|} = \sigma^2 r^{|k-j|}\frac{1-r^{2(\min(k,j)+1)}}{1-r^2}.$$

Observe that since $|r| < 1$, if we let $n \to \infty$, then the mean of this example goes to the mean of the preceding example in the limit. Similarly, if one fixes the lag $|k - j|$ and lets k (and hence j) go to ∞, then in the limit the one-sided covariance looks like the two-sided example. This simple example points out a typical form of nonstationarity: a linearly filtered uncorrelated process is not stationary by any definition, but as one gets farther and farther from the origin, the parameters look more and more stationary. This can be considered as a form of asymptotic stationarity. In fact, a process is defined as being asymptotically weakly stationary if the mean and covariance converge in the sense just given. One can view such processes as having transients that die out with time. It is not difficult to show that if a process is asymptotically weakly stationary and if the limiting mean and covariance meet the conditions of the ergodic theorem, then the process itself will satisfy the ergodic theorem. Intuitively stated, transients do not affect the long-term sample averages.

[5.3] Next consider the one-sided process of Example [5.2], but now choose the δ-response with $r = 1$; that is, $h_k = 1$ for all $k \geq 0$. Find the output mean and covariance. (Note that this filter is *not* stable.) Applying (5.4)–(5.6) and (5.28), (5.30), and (5.31) yields

$$EY_n = m \sum_{k=0}^{n} h_k = m(n+1)$$

and

$$K_Y(k,j) = \sigma^2(\min(k,j) + 1) = \sigma^2 \min(k+1, j+1).$$

Observe that like Example [5.2], the process of Example [5.3] is not weakly stationary. Unlike [5.2], however, it does not behave asymptotically like a weakly stationary process – even for large time, the moments very much depend on the time origin. Thus the nonstationarities of this process are not only transients – they last forever! In a sense, this process is much more nonstationary than the previous one and, in fact, does not have a mean ergodic theorem. If the input process is Gaussian with zero mean, then we shall see in Chapter 6 that the output process $\{Y_n\}$ is also Gaussian. Such a Gaussian process with zero mean and with the covariance function of this example is called the discrete time *Wiener process*.

[5.4] *A binary Markov process*

The linear filtering ideas can be applied when other forms of arithmetic than real arithmetic is used. Rather than try to be general we illustrate the approach by an example, a process formed by linear filtering using binary (modulo 2) arithmetic on an iid sequence of coin flips.

Given a known input process and a filter (a modulo 2 linear recursion in the present case), find the covariance function of the output. General formulas will be derived later in the book; here a direct approach to the problem at hand is taken.

First observe that $K_Y(k,j) = E[(Y_k - E(Y_k))(Y_j - E(Y_j))]$ is easily evaluated for the case $k = j$ because the marginal for Y_k is equiprobable:

$$E[Y_k] = \sum_y y p_Y(y) = \frac{1}{2}(0+1) = \frac{1}{2}$$

$$K_Y(k,k) = \sigma_Y^2 = E\left[\left(Y_k - \frac{1}{2}\right)^2\right]$$

$$= \sum_y \left(y - \frac{1}{2}\right)^2 p_Y(y) = \frac{1}{2}\left(\frac{1}{4} + \frac{1}{4}\right) = \frac{1}{4}.$$

As we have seen, a covariance function is *symmetric* in the sense that

$$\begin{aligned} K_Y(k,j) &= E[(Y_k - E(Y_k))(Y_j - E(Y_j))] \\ &= E[(Y_j - E(Y_j))(Y_k - E(Y_k))] \\ &= K_Y(j,k) \end{aligned}$$

so that we will be done if we evaluate $K_Y(k,j)$ for the special case where $k = j + l$ for $l \geq 1$. Let p be the probability of a 1 and consider therefore

$$\begin{aligned} K_Y(j+l,j) &= E\left[\left(Y_{j+l} - \frac{1}{2}\right)\left(Y_j - \frac{1}{2}\right)\right] \\ &= E\left[\left(X_{j+l} \oplus Y_{j+l-1} - \frac{1}{2}\right)\left(Y_j - \frac{1}{2}\right)\right] \\ &= \sum_{x,y,z} \left(x \oplus y - \frac{1}{2}\right)\left(z - \frac{1}{2}\right) p_{X_{j+l},Y_{j+l-1},Y_j}(x,y,z) \\ &= \sum_{x,y,z} \left(x \oplus y - \frac{1}{2}\right)\left(z - \frac{1}{2}\right) p_{X_{j+l}}(x) p_{Y_{j+l-1},Y_j}(y,z) \\ &= \sum_{y,z} \left(\left(0 \oplus y - \frac{1}{2}\right)\left(z - \frac{1}{2}\right)(1-p) p_{Y_{j+l-1},Y_j}(y,z)\right. \\ &\qquad \left. + \left(1 \oplus y - \frac{1}{2}\right)\left(z - \frac{1}{2}\right) p\, p_{Y_{j+l-1},Y_j}(y,z)\right). \end{aligned}$$

Since $0 \oplus y = y$ and $1 \oplus y = 1 - y$, this becomes

$$\begin{aligned} K_Y(j+l,j) &= (1-p)\sum_{y,z}\left(y-\frac{1}{2}\right)\left(z-\frac{1}{2}\right)p_{Y_{j+l-1},Y_j}(y,z) \\ &\quad +p\sum_{y,z}\left(1-y-\frac{1}{2}\right)\left(z-\frac{1}{2}\right)p_{Y_{j+l-1},Y_j}(y,z) \\ &= (1-2p)K_Y(j+l-1,j);\ l=1,2,\ldots \end{aligned}$$

This is a simple linear difference equation with initial condition $K_Y(j,j)$ and hence the solution is

$$K_Y(j+l,j) = (1-2p)^l K_Y(j,j) = \frac{1}{4}(1-2p)^l;\ l=1,2,\ldots \tag{5.35}$$

(Just plug it into the difference equation to verify that it is indeed a solution.) Invoking the symmetry property the covariance function is given by

$$K_Y(k,j) = \frac{1}{4}(1-2p)^{|k-j|} = K_Y(k-j). \tag{5.36}$$

Note that $K_Y(k)$ is absolutely summable (use the geometric progression) so that the weak law of large numbers holds for the process.

5.5 Linear modulation

In this section we consider a different form of linear system: a linear modulator. Unlike the filters considered thus far, these systems are generally time-varying and contain random parameters. They are simpler than the general linear filters, however, in that the output depends on the input in an instantaneous fashion; that is, the output at time t depends only on the input at time t and not on previous inputs.

In general, the word *modulation* means the methodical altering of one waveform by another. The waveform being altered is often called a carrier, and the waveform or sequence doing the altering, which we will model as a random process, is called the signal. Physically, such modulation is usually done to transform an information-bearing signal into a process suitable for communication over a particular medium; e.g., simple amplitude modulation of a carrier sinusoid by a signal in order to take advantage of the fact that the resulting high-frequency signals will better propagate through the atmosphere than will audio frequencies.

The emphasis will be on continuous time random processes since most communication systems involve at some point such a continuous time link.

Several of the techniques, however, work virtually without change in a discrete environment.

The prime example of linear modulation is the ubiquitous amplitude modulation or AM used for much of commercial broadcasting. If $\{X(t)\}$ is a continuous time weakly stationary random process with zero mean and covariance function $K_X(\tau)$, then the output process

$$Y(t) = (a_0 + a_1 X(t)) \cos(2\pi f t + \theta) \tag{5.37}$$

is called amplitude modulation of the cosine by the original process. The parameters a_0 and a_1 are called modulation constants. Amplitude modulation is depicted in a block diagram in Figure 5.2 Observe that linear modula-

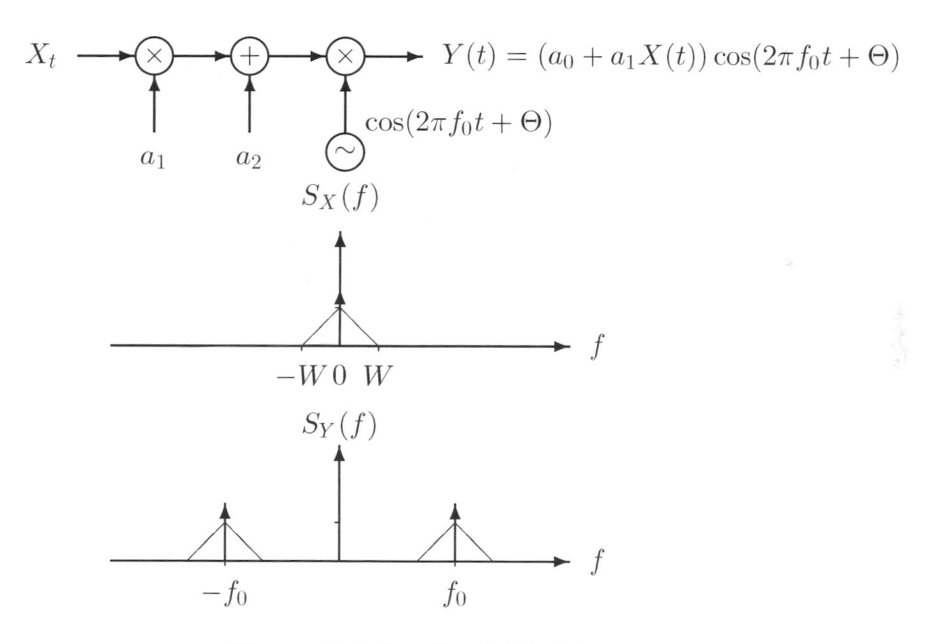

Figure 5.2 Amplitude Modulation

tion is not a linear operation in the normal linear systems sense unless the constant a_0 is 0. (It is, however, an *affine* operation – linear in the sense that straight lines in the two-dimensional $x - y$ space are said to be linear.) Nonetheless, as is commonly done, we will refer to this operation as linear modulation.

The phase term θ may be a fixed constant or a random variable, say Θ.

(We point out a subtle source of confusion here: if Θ is a random variable, then the system is affine or linear for the input process only when the actual sample value, say θ, of Θ is known.) We usually assume for convenience that Θ is a random variable, independent of the X process and uniformly distributed on $[0, 2\pi]$ – one complete rotation of the carrier phaser in the complex plane. This is a mathematical convenience, that, as we will see, makes $Y(t)$ weakly stationary. Physically it corresponds to the simple notion that we are modeling the modulated waveform as seen by a receiver. Such a receiver will not know a priori the phase of the transmitter oscillator producing the sinusoid. Furthermore, although the transmitted phase could be monitored and related to the signal as part of the transmission process, this is never done with AM. Hence, so far as the receiver is concerned, the phase is equally likely to be anything; that is, it has a uniform distribution independent of the signal.

If $a_0 = 0$, the modulated process is called double sideband suppressed carrier (DSB or DSB-SC). The a_0 term clearly wastes power, but it makes the recovery or demodulation of the original process easier and cheaper, as explained in any text on elementary communication theory. Our goal here is only to look at the second-order properties of the AM process.

Observe that for any fixed phase angle, say $\Theta = 0$ for convenience, a system taking a waveform and producing the DSB modulated waveform is indeed linear in the usual linear systems sense. It is actually simpler than the output of a general linear filter since the output at a given time depends only on the input at that time.

Since Θ and the X process are independent, the mean of the output is

$$EY(t) = (a_0 + a_1 EX(t))E\cos(2\pi ft + \Theta).$$

But Θ is uniformly distributed. Thus for any fixed time and frequency,

$$\begin{aligned} E\cos(2\pi ft + \Theta) &= \int_0^{2\pi} \cos(2\pi ft + \theta)\frac{d\theta}{2\pi} \\ &= \frac{1}{2\pi}\int_0^{2\pi} \cos(2\pi ft + \theta)\, d\theta = 0 \end{aligned} \tag{5.38}$$

since the integral of a sinusoid over a period is zero; hence $EY(t) = 0$ whether or not the original signal has zero mean.

The covariance function of the output is given by the following expansion of the product $Y(t)Y(s)$ using (5.37):

$$\begin{aligned}K_Y(t,s) &= a_0^2 E[\cos(2\pi f t+\Theta)\cos(2\pi f s+\Theta)] \\ &\quad + a_0a_1(EX(t)E[\cos(2\pi f t+\Theta)\cos(2\pi f s+\Theta)] \\ &\qquad + a_0a_1 EX(s)E[\cos(2\pi f t+\Theta)\cos(2\pi f s+\Theta)]) \\ &\quad + a_1^2 K_X(t,s)E[(\cos 2\pi f t+\Theta)(\cos 2\pi f s+\Theta)].\end{aligned}$$

Using the fact that the original process has zero mean eliminates the middle lines in the preceding. Combining the remaining two terms and using the cosine identity

$$\cos x\cos y = \frac{1}{2}\cos(x+y) + \frac{1}{2}\cos(x-y) \tag{5.39}$$

yields

$$\begin{aligned}K_Y(t,s) &= (a_0^2 + a_1^2 K_X(t,s)) \\ &\quad \times \left(\frac{1}{2}E\cos(2\pi f(t+s)+2\Theta) + \frac{1}{2}E\cos(2\pi f(t-s))\right).\end{aligned}$$

Exactly as in the mean computation of (5.38), the expectation of the term with the Θ in it is zero, leaving

$$K_Y(\tau) = \frac{1}{2}(a_0^2 + a_1^2 K_X(\tau))\cos(2\pi f\tau).$$

Thus we have demonstrated that amplitude modulation of a carrier by a weakly stationary random process results in an output that is weakly stationary.

The power spectral density of the AM process that we considered in the section on linear modulation can be found directly by transforming the covariance function or by using standard Fourier techniques: the transform of a covariance function times a cosine is the convolution of the original power spectral density with the generalized Fourier transform of the cosine – that is, a pair of impulses. This yields a pair of replicas of the original power spectral densities, centered at plus and minus the carrier frequency f_0 and symmetric about f_0, as depicted in Figure 5.2.

If further filtering is desired, e.g., to remove one of the symmetric halves of the power spectral density to form single sideband modulation, then the usual linear techniques can be applied, as indicated by (5.22).

5.6 White noise

Let $\{X_n\}$ be an uncorrelated, weakly stationary, discrete time random process with zero mean. We have seen that for such a process the covariance function is a pulse at the origin; that is,

$$K_X(\tau) = \sigma_X^2 \delta_\tau,$$

where δ_τ is a Kronecker delta function. As noted earlier, taking the Fourier transform results in the spectral density

$$S_X(f) = \sigma_X^2; \text{ all } f,$$

that is, the power spectral density of such a process is flat over the entire frequency range. We remarked that a process with such a flat spectrum is said to be white. We now make this definition formally for both discrete and continuous time processes:

A random process $\{X_t\}$ is said to be *white* if its power spectral density is a constant for all f. (A white process is also almost always assumed to have a zero mean, an assumption that we will make.)

The concept of white noise is clearly well defined and free of analytical difficulties in the discrete time case. In the continuous time case, however, there is a problem if white noise is defined as a process with constant power spectral density for all frequencies. Recall from (5.24) that the average power in a process is the integral of the power spectral density. In the discrete time case, integrating a constant over a finite range causes no problem. In the continuous time case, we find from (5.24) that a white noise process has infinite average power. In other words, if such a process existed, it would blow up the universe! A quick perusal of the stochastic systems literature shows, however, that this problem has not prevented models of continuous time white noise processes from being popular and useful. The resolution of the apparent paradox is fairly simple: indeed, white noise is a physically impossible process. But there do exist noise sources that have a flat power spectral density over a range of frequencies that is much larger than the bandwidths of subsequent filters of measurement devices. In fact, this is exactly the case with the thermal noise process caused by heat in resistors in amplifier circuits. A derivation based on the physics of such a process (see Chapter 6) yields a covariance function of the form $K_X(\tau) = kTR\alpha e^{-\alpha|\tau|}$, where k is Boltzman's constant, T is the absolute temperature, and R and α are parameters of the physical medium. The application of (5.20) results in the power spectral density

$$S_X(f) = kTR\frac{2\alpha^2}{\alpha^2 + (2\pi f)^2}.$$

As $\alpha \to \infty$, the power spectral density tends toward the value $2kTR$ for all f; that is, the process looks like white noise over a large bandwidth. Thus, for example, the total noise power in a bandwidth $(-B, B)$ is approximately $2kTR \times 2B$, a fact that has been verified closely by experiment.

If such a process is put into a filter having a transfer function whose magnitude become negligible long before the power spectral density of the input process decreases much, then the output process power spectral density $S_Y(f) = |H(f)|^2 S_X(f)$ will be approximately the same at the output as it would have been if $S_X(f)$ were flat forever since $S_X(f)$ is flat for all values of f where $|H(f)|$ is nonnegligible. Thus, so far as the output process is concerned the input process can be either the physically impossible white noise model or a more realistic model with finite power. However, since the input white noise model is much simpler to work with analytically, it is usually adopted.

In summary, continuous time white noise is often a useful model for the input to a filter when we are trying to study the output. Commonly the input random process is represented as being white with flat spectral density equal to $N_0/2$. The factor of 2 is included because of the "two-sided" nature of filter transfer functions; a low pass filter with cutoff frequency B applied to the white noise input will have output power equal to $N_0 B$ in accordance with (5.25). Such a white noise process makes mathematical sense only if seen through a filter. The process itself is not rigorously defined. Its covariance function, however, can be represented in terms of a Dirac delta function for the purposes of analytical manipulations. Note that in (5.23) the generalized Fourier transform of the flat spectrum results in a Dirac delta function of unit impulse. In particular, if the continuous time, white noise random process has power spectral density

$$S_X(f) = \frac{N_0}{2},$$

then it will have a covariance or autocorrelation function

$$K_X(\tau) = \frac{N_0}{2}\,\delta(\tau).$$

Thus adjacent samples of the random process are uncorrelated (and hence also independent if the process is Gaussian) *no matter how close together in time the samples are.* At the same time, the variance of a single sample is infinite. Clearly such behavior is physically impossible. It is reasonable, however, to state qualitatively that adjacent samples are uncorrelated at all times greater than the shortest time delays in subsequent filtering.

Perhaps the nicest attribute of white noise processes is the simple form of the output power spectral density of a linear filter driven by white noise. If a discrete or continuous time random process has power spectral density $S_X(f) = N_0/2$ for all f and it is put into a linear filter with transfer function $H(f)$, then from (5.22) the output process $\{Y_t\}$ has power spectral density

$$S_Y(f) = |H(f)|^2 \frac{N_0}{2}. \tag{5.40}$$

The result given in (5.40) is of more importance than first appearances indicate. A basic result of the theory of weakly stationary random processes, called the spectral factorization theorem, states that if a random process $\{Y_t\}$ has a spectral density $S_Y(f)$ such that

$$\int \ln S_Y(f)\, df > -\infty \quad \text{(discrete time)} \tag{5.41}$$

or

$$\int \frac{\ln S_Y(f)}{1+f^2}\, df > -\infty \quad \text{(continuous time)}, \tag{5.42}$$

then the power spectral density has the form of (5.40) for some *causal* linear stable time-invariant filter. That is, the second-order properties of any random process satisfying these conditions can be *modeled* as the output of a causal linear filter driven by white noise. Such random processes are said to be *physically realizable* and comprise most random processes seen in practice. The conditions (5.41)–(5.42) are referred to as the *Paley–Wiener criteria*[70]. This result is of extreme importance in estimation, detection, prediction, and system identification. We note in passing that in such models the white noise driving process is called the *innovations* process of the output process if the filter has a causal and stable inverse.

As a word of caution, there do exist processes which are not "physically realizable" in the above sense of violating the Paley–Wiener criteria (5.41 - 5.42), yet which are still "physically realizable" in the sense that simple models describe the processes. Consider the following example suggested to the authors by A. V. Balakrishnan: let X be a zero-mean Gaussian random variable with variance 1 and let Θ be a random variable with a uniform distribution on $[-\pi, \pi)$ which is independent of X. Define the random process $Y(t) = \cos(Xt - \Theta)$. Then analogous to the development of the autocorrelation function for linear modulation, we have that

$$\begin{aligned} E[Y(t)] &= E[\cos(Xt - \Theta)] = 0 \\ R_Y(\tau) &= E[\cos(Xt - \Theta)\cos(X(t-\tau) - \Theta)] \\ &= \frac{1}{2}E[\cos(X\tau)] = \frac{1}{4}E[e^{j\tau X} + e^{-j\tau X}] \\ &= \frac{1}{4}(M_X(j\tau) + M_X(-j\tau)) = \frac{1}{2}e^{-\tau^2} \end{aligned}$$

so that the power spectral density is

$$S_Y(f) = \frac{1}{2}e^{-f^2}, \tag{5.43}$$

which fails to meet the Paley–Wiener criterion for a continuous time random process.

5.6.1 Low pass and narrow band noise

Suppose that white noise is put through a filter with transfer function $H(f)$ which has magnitude 1 for $f \in [-W, W]$ and is zero elsewhere. This will produce a process with a flat power spectral density in a low frequency range, a process sometimes referred to as *low band noise* or *baseband noise.* If the original input is Gaussian, than the output low band noise process will also be Gaussian. Now suppose that this process, say $n_c(t)$, is put into an AM system. Then by an obvious modification of Figure 5.2 the resulting process will be flat within a band around $\pm f_0$ and 0 elsewhere, an example of *narrow band noise.*

5.7 ⋆Time averages

Recall the definitions of mean, autocorrelation, and covariance as expectations of samples of a weakly stationary random process $\{X_n;\ n \in \mathcal{Z}\}$:

$$\begin{aligned} m &= E[X_n] \\ R_X(k) &= E[X_n X_{n-k}^*] \\ K_X(k) &= E[(X_n - m)(X_{n-k} - m)^*] = R_X(k) - |m|^2. \end{aligned}$$

These are collectively considered as the second-order moments of the process. The corresponding time-average moments can be described if the limits are assumed to exist in some suitable sense:

$$\mathcal{M} = < X_n >= \lim_{N\to\infty} \frac{1}{N}\sum_{n=0}^{N-1} X_n$$

$$\mathcal{R}_X(k) = < X_n X_{n-k}^* >= \lim_{N\to\infty} \frac{1}{N}\sum_{n=0}^{N-1} X_n X_{n-k}^*$$

$$\begin{aligned}\mathcal{K}_X(k) &= < (X_n - m)(X_{n-k} - m)^* > \\ &= \lim_{N\to\infty} \frac{1}{N}\sum_{n=0}^{N-1} (X_n - m)(X_{n-k} - m)^*.\end{aligned}$$

Keep in mind that these quantities, if they exist at all, are random variables. For example, if we actually view a sample function $\{X_n(\omega);\ n \in \mathcal{Z}\}$, then the sample autocorrelation is

$$\mathcal{R}_X(k) = \lim_{N\to\infty} \frac{1}{N}\sum_{n=0}^{N-1} X_n(\omega) X_{n-k}^*(\omega),$$

also a function of the sample point ω and hence a random variable. Of particular interest is the autocorrelation for 0 lag:

$$\mathcal{P}_X = \mathcal{R}_X(0) = \lim_{N\to\infty} \frac{1}{N}\sum_{n=0}^{N-1} |X_n|^2,$$

which can be considered as the sample- or time-average *power* of the sample function in the sense that it is the average power dissipated in a unit resistance if X_n corresponds to a voltage.

Analogous to the expectations, the time-average autocorrelation function and the time-average covariance function are related by

$$\mathcal{K}_X(k) = \mathcal{R}_X(k) - |\mathcal{M}|^2. \tag{5.44}$$

In fact, subject to suitable technical conditions as described in the laws of large numbers, the time averages should be the same as the expectations, that is, under suitable conditions a weakly stationary random process $\{X_n\}$ should have the properties that

$$\begin{aligned} m &= \mathcal{M} \\ R_X(k) &= \mathcal{R}_X(k) \\ K_X(k) &= \mathcal{K}_X(k), \end{aligned}$$

which provides a suggestion of how the expectations can be estimated in practice. Typically the actual moments are not known a priori, but the random process is observed over a finite time N and the results used to estimate the moments, e.g., the sample mean

$$\mathcal{M}_N = \frac{1}{N}\sum_{n=0}^{N-1} X_n$$

and the sample autocorrelation function

$$\mathcal{R}_N(k) = \frac{1}{N}\sum_{n=0}^{N-1} X_n X_{n-k}^*.$$

These provide intuitive estimates of the actual moments which should converge to the true moments as $N \to \infty$.

There are in fact many ways to estimate second-order moments and there is a wide literature on the subject. For example, the observed samples may be weighted or "windowed" so as to diminish the impact of samples in the distant past or near the borders of separate blocks of data which are handled separately. The literature on estimating correlations and covariances is particularly rich in the speech processing area.

If the process meets the conditions of the law of large numbers, then its sample-average power $\mathcal{P}_X$ will be $R_X(0)$, which is typically some nonzero positive number. But if the limit $\lim_{N\to\infty}(1/N)\sum_{n=0}^{N-1}|X_n|^2$ is not zero, then observe that necessarily the limit

$$\lim_{N\to\infty}\sum_{n=0}^{N-1}|X_n|^2 = \sum_{n=0}^{\infty}|X_n|^2$$

must blow up since it lacks the normalizing N in the denominator. In other words, a sample function with nonzero average power will have infinite energy. The point of this observation is that a sample function from a perfectly reasonable random process will not meet the conditions for the existence of a Fourier transform, which suggests we might not be able to apply the considerable theory of Fourier analysis when considering the behavior of random processes in linear systems. Happily this is not the case, but Fourier analysis of random processes will be somewhat different from the Fourier analysis of deterministic finite energy signals and of deterministic periodic signals.

To motivate a possible remedy, first "window" the sample signal $\{X_n;\ n \in \mathcal{Z}\}$ to form a new signal $\{X_n^{(N)};\ n \in \mathcal{Z}\}$ defined by

$$X_n^{(N)} = \begin{cases} X_n & \text{if } n \le N-1 \\ 0 & \text{otherwise} \end{cases}. \tag{5.45}$$

The new random process $\{X_n^{(N)}\}$ clearly has finite energy (and is absolutely summable) so it has a Fourier transform in the usual sense, which can be defined as

$$\mathcal{X}_N(f) = \sum_{n=0}^{\infty} X_n^{(N)} e^{-j2\pi fn} = \sum_{n=0}^{N-1} X_n e^{-j2\pi fn},$$

which is the *Fourier transform* or *spectrum* of the truncated sample signal. Keep in mind that this is a random variable. It depends on the underlying sample point ω through the sample waveform selected. From Parceval's (or Plancherel's) theorem, the energy in the truncated signal can be evaluated from the spectrum as

$$\mathcal{E}_N = \sum_{n=0}^{N-1} |X_n|^2 = \int_{-1/2}^{1/2} |\mathcal{X}_N(f)|^2 \, df. \tag{5.46}$$

The average power is obtained by normalizing the average energy by the time duration N:

$$\mathcal{P}_N = \frac{1}{N} \sum_{n=0}^{N-1} |X_n|^2 = \frac{1}{N} \int_{-1/2}^{1/2} |\mathcal{X}_N(f)|^2 \, df. \tag{5.47}$$

Because of this formula $|\mathcal{X}_N(f)|^2/N$ can be considered as the power spectral density of the truncated waveform because, analogous to a probability density or a mass density, it is a nonnegative function which when integrated gives the power. Unfortunately it gives only the power spectral density for a particular truncated sample function. What is really desired is a notion of power spectral density for the entire random process. An alternative definition of power spectral density resolves these two issues by taking the expectation to get rid of the randomness, and using the limit to look at the entire signal, that is, to define the average power spectral density as the limit (if it exists)

$$\lim_{N\to\infty} \frac{E(|\mathcal{X}_N(f)|^2)}{N}.$$

To evaluate this limit consider

$$\begin{aligned}
\lim_{N\to\infty}\frac{E(|\mathcal{X}_N(f)|^2)}{N} &= \lim_{N\to\infty}\frac{1}{N}E\left(|\sum_{k=0}^{N-1}X_k e^{-i2\pi fk}|^2\right) \\
&= \lim_{N\to\infty}\frac{1}{N}E\left(\sum_{k=0}^{N-1}\sum_{l=0}^{N-1}X_k e^{-i2\pi fk}X_l^* e^{+i2\pi fl}\right) \\
&= \lim_{N\to\infty}\frac{1}{N}\sum_{k=0}^{N-1}\sum_{l=0}^{N-1}E[X_kX_l^*]e^{-i2\pi f(k-l)} \\
&= \lim_{N\to\infty}\frac{1}{N}\sum_{k=0}^{N-1}\sum_{l=0}^{N-1}R_X(k-l)e^{-i2\pi f(k-l)} \\
&= \lim_{N\to\infty}\sum_{k=-(N-1)}^{N-1}(1-\frac{|k|}{N})R_X(k)e^{-i2\pi fk},
\end{aligned}$$

where the last term involves reordering terms using Lemma B.1 (analogous to what was done to prove the law of large numbers for asymptotically uncorrelated weakly stationary processes). If the autocorrelation function is absolutely summable, i.e., if

$$\sum_{k=-\infty}^{\infty}|R_X(k)|<\infty, \tag{5.48}$$

then Lemma B.2 implies that

$$\lim_{N\to\infty}\frac{E(|\mathcal{X}_N(f)|^2)}{N}=\sum_{k=-\infty}^{\infty}R_X(k)e^{-i2\pi fk}=S_X(f), \tag{5.49}$$

the power spectral density as earlier defined.

5.8 ⋆Mean square calculus

The goal of this section is to develop in some detail the ideas that permit a rigorous approach to several of the results developed earlier in this chapter. We consider only one approach to rigor, that of mean square calculus. There are others, but the mean square calculus approach is popular for engineering applications because the details involve techniques already considered – in particular the idea of mean square convergence. Although the correctness of the informally derived results is usually simply assumed by engineering practitioners, it can prove useful when developing new models to be aware of the kinds of conditions necessary for the good behavior and correctness of many common tools, including limits, interchanging limits and expectations, integration, and differentiation. In other words, we look deeper into one approach to calculus for random processes. As with ordinary calculus, arguably the key underlying idea is that of a limit, and in this section the limit used is the limit in mean square. This section is strongly influenced by unpublished notes of Wilbur B. Davenport, Jr., from MIT course 6.573 in fall 1965, and on the first author's notes taken in that course. Material has been added to put mean square convergence into the context of metric spaces, linear normed spaces, Hilbert spaces, Banach spaces, and L_2 spaces to provide an explicit connection to the mathematical literature, especially functional analysis.

5.8.1 Mean square convergence revisited

Recall that a sequence of real random variables Y_n is said to converge in mean square to a random variable Y if $E[Y_n^2] < \infty$ all n and

$$\lim_{n\to\infty} E(|Y_n - Y|^2) = 0$$

in which case we write

$$Y = \operatorname*{l.i.m.}_{n\to\infty} Y_n.$$

Mean square convergence of random variables can be viewed as simply an engineering interpretation of the mathematical topics in analysis and functional analysis of L_2 convergence. A brief consideration of the mathematical point of view can provide some useful geometric interpretations of both the convergence and of some of its properties and implications. The reader is referred to the literature, e.g., [1] or [32] section 2.5, for extensive treatments of such spaces.

Since random variables are functions on probability spaces, convergence of random variables is just an example of convergence of functions. Suppose we have a probability space $(\Omega, \mathcal{F}, P)$ and we define L_2 to be the space of all complex-valued random variables with finite second moment, that is, if $X \in L_2$, then

$$E[|X|^2] < \infty,$$

where the expectation is with respect to the probability measure P. We consider two random variables (functions) to be the same if they are equal with probability one. The L_2 space is also referred to as the space of square-integrable functions.

We can treat this space of random variables or functions like ordinary Euclidean space in many ways. In Euclidean space we can define a *metric* $|x|$ which denotes the "size" of a member of the space (the Euclidean distance to the origin) and a *distance* between any two members of the space x and y by the Euclidean distance

$$d(x, y) = |x - y|.$$

The notions of a metric and distance lead to many useful ideas in geometry and calculus. For example, to say that a sequence x_n converges to x can be defined by

$$\lim_{n\to\infty} d(x_n, x) = \lim_{n\to\infty} |x_n - x| = 0.$$

These notions in the real line or in k-dimensional Euclidean vector space can be extended to any space of objects for which we can define a distance or norm.

In general, a space A is called a *metric space* if for every pair of points $a, b \in A$ there is an associated nonnegative real number $d(a, b)$ called a *distance* or *metric* such that

$$d(a, b) = 0 \text{ if and only if } a = b, \tag{5.50}$$

$$d(a, b) = d(b, a) \text{ (symmetry)}, \tag{5.51}$$

$$d(a, b) \leq d(a, c) + d(c, b), \text{ all } c \in A \text{ (triangle inequality)}. \tag{5.52}$$

If A is the real line, then the magnitude of the difference between two points trivially meets these conditions and is therefore a metric.

An example of a general metric space that is also inspired by Euclidean space is a *normed linear space*. A normed linear vector space is a space A with norm $|.|$ and distance $d(a, b) = |a - b|$. A vector space or linear space A is a space consisting of points called vectors and two operations – one called *addition* which associates with each pair of vectors a, b a new vector $a + b \in A$ and one called *multiplication* which associates with each vector a and each number $r \in \Re$ (called a scalar) a vector $ra \in A$ – for which the usual rules of arithmetic hold, that is, if $a, b, c \in A$ and $r, s \in \Re$, then

$$\begin{aligned} a+b &= b+a \text{ (commutative law)} \\ (a+b)+c &= a+(b+c) \text{ (associative law)} \\ r(a+b) &= ra+rb \text{ (left distributive law)} \\ (r+s)a &= ra+sa \text{ (right distributive law)} \\ (rs)a &= r(sa) \text{ (associative law for multiplication)} \\ 1a &= a. \end{aligned}$$

In addition, it is assumed that there is a zero vector, also denoted 0, for which

$$a + 0 = a$$

$$0a = 0.$$

We also define the vector $-a = (-1)a$. A normed linear space is a linear vector space A together with a function $||a||$ called a norm defined for each a such that for all $a, b \in A$ and $r \in \Re$, $||a||$ is nonnegative and

$$\|a\| = 0 \text{ if and only if } a = 0, \tag{5.53}$$

$$\|a+b\| \leq \|a\| + \|b\| \text{ (triangle inequality).} \tag{5.54}$$

$$\|ra\| = |r|\|a\|. \tag{5.55}$$

Returning to the space L_2 it is easy to see that it is a linear space since linear combinations of random variables with finite second moment have finite second moments (just use the Cauchy–Schwarz inequality)(Problem 4.20), that

$$\|X - Y\| = \sqrt{E[|X - Y|^2]} \tag{5.56}$$

satisfies the conditions for a norm (remembering that two random variables are considered equal in this space if they are equal with probability one), and that

$$d(X, Y) = \|X - Y\| \tag{5.57}$$

satisfies the requirements for a distance.

It can now be recognized that mean square convergence of Y_n to Y is almost identical to L_2 convergence of Y_n to Y in the sense that $d(Y_n, Y) \to 0$ or $\|Y_n - Y\| \to 0$. The only difference is that in the definition of mean square convergence we have not required that the limit random variable Y is itself in L_2, that is, it might not have a finite second or even first moment. These observations bring into play the entire theory of normed spaces and metric spaces when interpreting or applying mean square convergence.

As yet a third mathematical formulation note that we can define an *inner product* in L_2 by

$$(X, Y) \triangleq E[XY^*] \tag{5.58}$$

and use the inner product to define a norm and a distance

$$\|X\| = (X, X) \tag{5.59}$$

$$d(X, Y) = \|X - Y\|. \tag{5.60}$$

In general, an inner product space (or pre-Hilbert space) A is a linear space such that for each pair $a, b \in A$ there is a real number (a, b) called an *inner product* such that for $a, b, c \in A$, $r \in \Re$

$$\begin{aligned}
(a, b) &= (b, a) \\
(a + b, c) &= (a, c) + (b, c) \\
(ra, b) &= r(a, b) \\
(a, a) &\geq 0 \text{ and } (a, a) = 0 \text{ if and only if } a = 0.
\end{aligned}$$

An inner product $(\cdot, \cdot)$ then determines a distance $d(a, b) = ||a - b||$, where $||a|| = (a, a)^{1/2}$ is a norm.

By mimicking well-known properties of the real line, we have seen that there is a similar structure on the space L_2 of all random variables with a finite second moment and that mean square convergence is simply an example of convergence in a metric space, a normed linear space, and a pre-Hilbert space. We consider one more technical property of the real line which is less familiar, but often useful in proofs involving mean square convergence.

Suppose A is a metric space (e.g., the real line, L_2, or any other normed linear space or pre-Hilbert space). If a sequence $x_n \in A$ has the property that the double limit

$$\lim_{n.m \to \infty} \lim d(x_n, x_m) = 0,$$

then the sequence is called a *Cauchy sequence*. In other words, a sequence $\{x_n\}$ is Cauchy if for every $\epsilon > 0$ there is an integer N such that $d(x_n, x_m) < \epsilon$ if $n \geq N$ and $m \geq N$. A metric space is *complete* if every Cauchy sequence converges; that is, if $\{x_n\}$ is a Cauchy sequence, then there is an x in A for which $x = \lim_{n \to \infty} x_n$. In the words of Simmons [62], p. 71, a complete metric space is one wherein "every sequence that tries to converge is successful." A standard result of elementary real analysis is that Euclidean space (with the Euclidean distance) is complete (see Rudin [61], p. 46). A complete inner product space is called a *Hilbert space.* A complete normed space is called a *Banach space.*

At this point the reader may be somewhat overwhelmed by the profuse family connections of the space of random variables with finite second moments, but the primary point is to demonstrate that with all these connections the theory of mean square convergence can draw upon a wealth of literature and results and that this explains the common occurrence of these names in the signal processing literature. Those who pursue advanced topics in signal processing will almost certainly run into various branches of these topics. Here we collect some of the standard properties for future use.

Lemma 5.1 *Let L_2 be the space of all random variables with finite second moments. All random variables X_n and Y_n for all n are assumed to be in L_2.*

1. Completeness *L_2 is complete, that is, if X_n is a Cauchy sequence, then there is a random variable $X \in L_2$ such that* $\text{l.i.m.}_{n\to\infty} X_n = X$.
2. Linearity *Limits in the mean combine in a linear fashion (like ordinary limits), that is, if* $\text{l.i.m.}_{n\to\infty} X_n = X$ *and* $\text{l.i.m.}_{n\to\infty} Y_n = Y$, *then for any constants a, b*
$$\underset{n\to\infty}{\text{l.i.m.}}(aX_n + bY_n) = aX + bY. \tag{5.61}$$
3. *If $X \in L_2$, then*
$$|E(X)| \leq \|X\|. \tag{5.62}$$
4. Convergence of expectation *Suppose that* $Y = \text{l.i.m.}_{n\to\infty} Y_n$, *then*
$$\lim_{n\to\infty} E(Y_n) = E(\underset{n\to\infty}{\text{l.i.m.}}\, Y_n). \tag{5.63}$$
The result shows that the common engineering practice of interchanging limits and expectation works for random variables when the limit inside of the expectation is a limit in mean square.
5. Convergence of correlation *Suppose that X and Y have finite second moments and that* $\text{l.i.m.}_{n\to\infty} X_n = X$ *and* $\text{l.i.m.}_{n\to\infty} Y_n = Y$, *then*
$$\lim_{n,m\to\infty} E[X_n Y_m^*] = E[XY^*], \tag{5.64}$$
that is, one can interchange expectation and limits when computing correlations in the sense that
$$\lim_{n,m\to\infty} E[X_n Y_m^*] = E[\underset{n,m\to\infty}{\text{l.i.m.}} X_n Y_m^*]. \tag{5.65}$$

Proof

1. This is a fundamental result of functional analysis and will not be proved here. Note that it is easy to see that if $\text{l.i.m.}_{n\to\infty} X_n = X$, then X must be a Cauchy sequence. This follows from the triangle inequality: given ϵ pick N so large that $\|X - X_n\| \le \epsilon/2$ and $\|X - X_m\| \le \epsilon/2$ and then observe that $\|X_n - X_m\| \le \|X_n - X\| + \|X_m - X\| \le \epsilon$ so that the sequence is Cauchy.
2. The linearity follows easily using the triangle inequality and is left as an exercise.
3. Equation (5.62) follows from the Cauchy–Schwarz inequality since

$$|E(X)| = |E(X \times 1)| \le \|X\| \times \|1\| = \|X\|.$$

4. $0 \le |E(Y_n) - E(Y)|^2 = |E(Y - Y_n)|^2 \le E(|Y - Y_n|^2)$ where the last inequality uses the Cauchy–Schwarz inequality (see Problem 4.20) Since the rightmost term goes to zero with n by assumption, the claim is proved.
5. Use the Cauchy–Schwarz inequality:

$$\begin{aligned} |E[X_n Y_m^* - XY^*]| &= |E[X_n Y_m^* - XY_m^* + XY_m^* - XY^*]| \\ &= |E[(X_n - X)Y_m^*] + E[X(Y_m - Y)^*]| \\ &\le |E[(X_n - X)Y_m^*]| + |E[X(Y_m - Y)^*]| \\ &\le \|X_n - X\| \times \|Y_m\| + \|X\| \times \|Y_m - Y\| \\ &= \|X_n - X\| \times \|Y_m - Y + Y\| + \|X\| \times \|Y_m - Y\| \\ &\le \|X_n - X\| \times (\|Y_m - Y\| + \|Y\|) + \|X\| \times \|Y_m - Y\| \end{aligned}$$

which tends to 0 as n and m go to infinity. □

Throughout this section we will be approximating a random process at time t, say $X(t)$, by a sequence of processes, say $X_N(t)$ as $N \to \infty$. The final lemma of this subsection builds on the previous lemma to demonstrate the crucial role played by second-order moments – autocorrelations in particular – in studying mean square convergence.

Lemma 5.2 *Given a sequence of random processes $\{X_N(t); t \in \mathcal{T}\}$ for which*

$$E[|X_N(t)|^2] < \infty \text{ for all } t \in \mathcal{T}, N = 1, 2, \ldots,$$

then

$$X(t) = \underset{N\to\infty}{\text{l.i.m.}}\, X_N(t) \tag{5.66}$$

where

$$E[|X(t)|^2] < \infty \tag{5.67}$$

iff

$$\lim_{N,M\to\infty} E[X_N(t)X_M^*(t)] = R_X(t,t), \tag{5.68}$$

in which case also

$$\lim_{N\to\infty} R_{X_N}(t,s,) = R_X(t,s). \tag{5.69}$$

Proof The final part of the preceeding lemma with $X_N = X_N(t)$, $X = X(t)$, $Y_M = X_M(s)$, $Y = X(s)$ yields (5.69) and with $t = s$ yields the "only if" statement. For the converse suppose (5.68) holds. Then

$$\begin{aligned} E[|X_N(t) - X_M(t)|^2] =& E[X_N(t)X_N^*(t)] - E[X_N(t)X_M^*(t)] \\ &- E[X_M(t)X_N^*(t)] + E[X_M(t)X_M^*(t)], \end{aligned}$$

where all of the terms on the right go to 0 as $N, M \to \infty$, so $X_N(t)$ is a Cauchy sequence and must have a limit in the mean with a finite second moment from the previous lemma. □

5.8.2 Integrating random processes

It has been noted that some care is required to extend the notion of integrals over time from waveforms to random processes. One way of accomplishing this is to extend the usual definition of Riemann integrals as a limit of sums of samples to a corresponding limit of sums of random variables, where the limit is in the sense of mean squared convergence. In this subsection we develop this idea further and use it to provide conditions under which the integrals exist. Suppose that $x(t)$ has an ordinary Riemann integral $I = \int_a^b x(t)\,dt$. This will be the case if the sums

$$I_N = \sum_{n=1}^{N} x(t_n)\Delta t_n$$

where $a = t_0 < t_1 < t_2 < \cdots t_{N-1} < t_N = b$, and $\Delta t_n = (t_n - t_{n-1})/(b-a)$ (we have suppressed the dependence of the sample times t_n on N) will have a unique limit as $N \to \infty$ regardless of how the sample times are chosen provided that the differences satisfy

$$\lim_{N\to\infty} \max_{n:0\leq n\leq N} \Delta t_n = 0.$$

For example, a bounded function $x(t)$ on a bounded interval $[a, b]$ will have a Riemann integral if and only if it is continuous almost everywhere in the interval. In this case the integral is given by the limit of sums

$$I = \lim_{N\to\infty} I_N.$$

Now suppose that $\{X(t)\}$ is a continuous time random process with a finite second moment $E[|X(t)|^2] < \infty$ for all t and we again choose sample times with vanishing differences as above. Define the sum

$$I_N = \sum_{n=1}^{N} X(t_n)\Delta t_n,$$

where now the sum is itself a random variable. Since each $X(t_n)$ has finite second moment, the triangle inequality implies that I_N also has finite second moment since

$$\|I_N\| = \left\| \sum_{n=1}^{N} X(t_n)\Delta t_n \right\| \leq \sum_{n=1}^{N} \|X(t_n)\|\Delta t_n.$$

If there exists a random variable I such that

$$I = \underset{N\to\infty}{\text{l.i.m.}}\, I_N = \underset{N\to\infty}{\text{l.i.m.}} \sum_{n=1}^{N} X(t_n)\Delta t_n,$$

then I is the *mean square integral* of the given random process over $[a, b]$ and we write

$$I = \int_a^b X(t)\, dt \triangleq \underset{N\to\infty}{\text{l.i.m.}} \sum_{n=1}^{N} X(t_n)\Delta t_n.$$

Note that even though we use the familar Riemann notation for an integral, this is *not* a Riemann integral. In particular, it is a random variable!

To explore further, suppose that the mean square integral I exists. Lemma 5.1 implies that if I itself has a finite second moment, then the integral will behave in the expected sense that one can interchange expectation and integration so that the expected value of the integral satisfies

$$\begin{aligned}
E[I] &= E[\underset{N\to\infty}{\text{l.i.m.}}\, I_N] = \lim_{N\to\infty} E[I_N] \\
&= \lim_{N\to\infty} E[\sum_{n=1}^{N} X(t_n)\,\Delta t_n] = \lim_{N\to\infty} \sum_{n=1}^{N} E[X(t_n)]\,\Delta t_n,
\end{aligned}$$

where in the final step we have used the linearity of expectation. The right-most expression, however, is the Riemann integral of the mean function $E[X(t)]$. This Riemann integral must be finite since it is equal to $E[I]$, which is finite from Lemma 5.1 since I is assumed to be in L_2. Thus the existence of the mean squared integral with a finite second moment ensures that the mean function of the original random process has a finite Riemann integral.

Next consider the second-order moment. Again assume that the mean square integral I exists and has a finite second moment,

$$E[|I|^2] = E[\left|\int_a^b X(t)\,dt\right|^2].$$

Analogous to the corresponding result for Riemann integration, it is easy to show that for the mean square integral

$$\begin{aligned}\left|\int_a^b X(t)\,dt\right|^2 &= \left(\int_a^b X(t)\,dt\right) \times \left(\int_a^b X^*(s)\,ds\right) \\ &= \int_a^b \int_a^b X(t)X^*(s)\,dt\,ds \qquad (5.70)\end{aligned}$$

so that interchanging expectation and integration using Lemma 5.1

$$\begin{aligned}E[|I|^2] &= E\left[\int_a^b \int_a^b X(t)X^*(s)\,dt\,ds\right] \\ &= \int_a^b \int_a^b E[X(t)X(s)]\,dt\,ds \\ &= \int_a^b \int_a^b R_X(t,s)\,dt\,ds.\end{aligned}$$

Since I has a finite second moment, the double Riemann integral of the correlation function exists and is finite. We have now proved half of the following result, which shows that the converse is true as well and thus gives a necessary and sufficient condition for the existence of the mean square integral of a continuous time random process with a finite second moment.

Lemma 5.3 *A continuous time random process $\{X(t)\}$ has a mean square integral over $[a,b]$ with a finite second absolute moment if and only if*

$$\int_a^b \int_a^b R_X(t,s)\,dt\,ds < \infty. \qquad (5.71)$$

that is, the integral exists and is finite.

Proof We have already proved the "only if" part. Suppose for the converse that (5.71) holds and consider

$$\begin{aligned} E[I_N - I_M|^2] &= E[(I_N - I_M)(I_N^* - I_M^*)] \\ &= E[I_N I_N^*] - E[I_N I_M^*] - E[I_M I_N^*] + E[I_M I_M^*]. \quad (5.72) \end{aligned}$$

Consider for example the term

$$\begin{aligned} E[I_N I_M^*] &= E\left[\left(\sum_{i=1}^{N} X(t_i)\Delta t_i\right)\left(\sum_{j=1}^{M} X^*(t_j)\Delta t_j\right)\right] \\ &= \sum_{i=1}^{N}\sum_{j=1}^{M} X(t_i)X^*(t_j)\Delta t_i \Delta t_j = \sum_{i=1}^{N}\sum_{j=1}^{M} R(t_i, t_j)\Delta t_i \Delta t_j. \end{aligned}$$

By assumption the double Riemann integral of the autocorrelation exists and is finite, so

$$\lim_{N,M\to\infty} E[I_N I_M^*] = \int_a^b \int_a^b R_X(t,s)\,dt\,ds < \infty.$$

Every one of the four terms in (5.72) converges in the same way to the same thing and hence the sum is 0 since two are positive and two are negative. Thus

$$\lim_{N,M\to\infty} \|I_N - I_M\| = 0$$

and I_N is a Cauchy sequence, which from Lemma 5.1 must converge to a random variable I with a finite second moment. □

Observe that if either condition of the lemma is met, then we have shown that we can exchange the expectation with the mean square integral:

$$\int_a^b E[X(t)]\,dt = E\left[\int_a^b X(t)\,dt\right] \quad (5.73)$$

$$\int_a^b \int_a^b R_X(t,s)\,dt\,ds = E\left[\int_a^b \int_a^b X(t)X(s)\,dt\,ds\right]. \quad (5.74)$$

5.8.3 Linear filtering

As an application of mean square integration, we can now define a linear filter operating on a continuous time random process $\{X(t);\, t \in \mathcal{T}\}$, where for convenience we consider $\mathcal{T} = (-\infty, \infty)$. Define the output process $Y(t)$ of a linear filter described by a weighting function $h(t, \tau)$ as a mean square integral (if it exists)

$$Y(t) = \int_a^b X(\tau)h(t,\tau)\,d\tau, \tag{5.75}$$

i.e.,

$$Y(t) = \underset{N\to\infty}{\text{l.i.m.}}\, Y_N(t)$$

where

$$Y_N(t) \triangleq \sum_{n=1}^{N} X(\tau_n)h(t,\tau_i)\Delta\tau_n.$$

From Lemma 5.3 this integral will be well defined and have a finite second moment if and only if

$$\int_a^b \int_a^b h(t,\tau)h(t,\tau')R_X(\tau,\tau')\,d\tau\,d\tau' < \infty$$

in which case

$$\int_a^b E[Y(t)]\,dt = E\left[\int_a^b X(\tau)h(t,\tau)\,d\tau\right] \tag{5.76}$$

and

$$\begin{aligned}
&\int_a^b \int_a^b h(t,\tau)R_X(\tau,\tau')h^*(s,\tau')\,d\tau\,d\tau' \\
&= \int_a^b \int_a^b E[h(t,\tau)X(\tau)X^*(\tau')h^*(s,\tau')]\,d\tau\,d\tau' \\
&= E\left[\int_a^b \int_a^b X(\tau)h(t,\tau)X^*(\tau')h^*(s,\tau')\,d\tau d\tau'\right] \\
&= E[Y(t)Y^*(s)] < \infty.
\end{aligned} \tag{5.77}$$

This provides an alternative derivation of the output autocorrelation function and shows that its finiteness ensures the existence and finite second absolute moment of the limit in the mean defining the output process provided the output autocorrelation is finite.

Note that if the original process is Gaussian, then the sums $Y_N(t)$ are Gaussian. Since $Y_N(t)$ converges to $Y(t)$ in mean square, it also converges in distribution so that $Y(t)$ is also Gaussian for each N. This is not enough to prove that the mean square integral $Y(t)$ is a Gaussian process, but the idea can be extended to vectors of samples to provide a careful proof that the process put out by a linear filter with a Gaussian process input is also Gaussian.

Extra work is needed to extend these ideas to integrals over infinite limits.

5.8.4 Differentiating random processes

We have said that linear systems can often be described by means other than convolution integrals, e.g., difference equations for discrete time and differential equations in continuous time. Here we explore the I/O relations for a simple continuous time differentiator in order to demonstrate some of the techniques involved for handling such systems. In addition, the results developed will provide another interpretation of white noise.

Suppose now that we have a continuous time random process $\{X(t); t > 0\}$ and we form a new random process $\{Y(t)\}$ by differentiating; that is,

$$Y(t) = \frac{d}{dt} X(t).$$

Here we assume $\{X(t)\}$ to be a zero-mean random process for simplicity. Results for nonzero-mean processes are found by noting that $X(t)$ can be written as the sum of a zero-mean random process plus the mean function $m(t)$. That is, we can write $X(t) = X_0(t) + m(t)$ where $X_0(t) = X(t) - m(t)$. Then, the derivative of $X(t)$ is the derivative of a zero-mean random process plus the derivative of the mean function. The derivative of the mean function is a derivative in the usual sense and hence provides no special problems.

As with integration, there is a problem in interpreting what the derivative means when the thing being differentiated is a random process. A derivative is defined as a limit, and as we found in Chapter 4, there are several notions of limits of sequences of random variables. Care is required because the limit may exist in one sense but not necessarily in another. In particular, two natural definitions for the derivative of a random process correspond to convergence with probability one and convergence in mean square. As a first possibility we could assume that each sample function of $Y(t)$ is obtained by differentiating each sample function of $X(t)$; that is, we could use ordinary differentiation on the sample functions. This gives us a definition of the form

$$Y(t, \omega) = \frac{d}{dt} X(t, \omega) = \lim_{\Delta t \to 0} \frac{X(t + \Delta t, \omega) - X(t, \omega)}{\Delta t}.$$

If $P(\{\omega : \text{the limit exists}\}) = 1$, then the definition of differentiation corresponds to convergence with probability one. Alternatively, we could define $Y(t)$ as a limit in quadratic mean of the random variables $(X(t + \Delta t) -$

$X(t))/\Delta t$ as Δt goes to zero (which does not require that the derivative exist with probability one on sample functions). With this definition we obtain

$$Y(t) = \operatorname*{l.i.m.}_{\Delta t \to 0} \frac{X(t+\Delta t) - X(t)}{\Delta t}.$$

Clearly a choice of definition of derivative must be made in order to develop a theory for this simple problem and, more generally, for linear systems described by differential equations. Consistent with the approach of this section, we focus on mean square convergence. First, however, we sketch a heuristic development with the assumption that all of the derivatives exist as required. We will blithely ignore careful specification of conditions under which the formulas make sense during the initial pass. (Mathematicians sometimes refer to such derivations as formal developments: techniques are used as if they are applicable to see what happens. This often provides the answer to a problem, which, once known, can then be proved rigorously to be correct.) After the nonrigorous development we shall return to provide some rigor to the key points using mean square calculus.

Define a process $\{Y_{\Delta t}(t)\}$ for a fixed Δt as the following difference, which approximates the derivative of $X(t)$:

$$Y_{\Delta t}(t) = \frac{X(t+\Delta t) - X(t)}{\Delta t}.$$

This difference process is perfectly well defined for any fixed $\Delta t > 0$ and in some sense it should converge to the desired $Y(t)$ as $\Delta t \to 0$. We can easily find the following correlation:

$$\begin{aligned} &E[Y_{\Delta t}(t) Y_{\Delta s}(s)] \\ &= E\left[\frac{X(t+\Delta t) - X(t)}{\Delta t} \frac{X(s+\Delta s) - X(s)}{\Delta s}\right] \\ &= \frac{R_X(t+\Delta t, s+\Delta s) - R_X(t+\Delta t, s) - R_X(t, s+\Delta s) + R_X(t,s)}{\Delta t \Delta s}. \end{aligned}$$

If we (formally) let Δt and Δs go to zero, then, if the various limits exist, this formula becomes

$$R_Y(t,s) = \frac{\partial}{\partial t \partial s} R_X(t,s). \tag{5.78}$$

As previously remarked, we will not try to specify complete conditions under which this sleight of hand can be made rigorous. Suffice it to say that if the conditions on the X process are sufficiently strong, the formula is valid. Intuitively, since differentiation and expectation are linear operations, the formula follows from the assumption that the linear operations commute, as they usually do. There are, however, serious issues of existence involved in making the proof precise.

One obvious regularity condition to apply is that the double derivative of (5.78) exists. If it does and the processes are weakly stationary, then we can transform (5.78) by using the property of Fourier transforms that differentiation in the time domain corresponds to multiplication by $-j2\pi f$ in the frequency domain. Then for the double derivative to exist we obtain the requirement that for the spectral density of $\{Y(t)\}$ to have finite second moments, i.e., if $S_Y(f) = (2\pi f)^2 S_X(f)$, then

$$\int_{-\infty}^{\infty} S_Y(f)\, df < \infty. \tag{5.79}$$

As a rather paradoxical application of (5.78), suppose that we have a one-sided continuous time Gaussian random process $\{X(t);\, t \geq 0\}$ that has zero mean and a covariance function that is the continuous time analog of Example [5.3]; that is, $K_X(t,s) = \sigma^2 \min(t,s)$. (The Kolmogorov construction guarantees that there is such a random process; that is, that it is well defined.) This process is known as the continuous time Wiener process, a process that we will encounter again in the next chapter. Strictly speaking, the double derivative of this function does not exist because of the discontinuity of the function at $t = s$. From engineering intuition, however, the derivative of such a step discontinuity is an impulse, suggesting that

$$R_Y(t,s) = \sigma^2 \delta(t-s),$$

the covariance function for Gaussian white noise! Because of this formal relation, Gaussian white noise is sometimes described as the formal derivative of a Wiener process. We have to play loose with mathematics to find this result, and the sloppiness cannot be removed in a straightforward manner. In fact, it is known from the theory of Wiener processes that they have the strange attribute of producing with probability one sample waveforms that are continuous but *nowhere differentiable.* Thus we are considering white noise as the derivative of a process that is not differentiable. Nonetheless, this is a useful intuition that is consistent with the extremely pathological behavior of sample waveforms of white noise – an idealized concept of a process that cannot really exist.

Now return to the issue of rigor in defining the derivative of a random process. As with mean square integration, the idea is to look at Cauchy sequences. The proof is very similar. The following lemma summarizes the results. We continue to consider zero-mean real-valued random processes for simplicity.

Lemma 5.4 *Given a zero-mean real-valued random process $\{X(t)\}$ with finite second moments $E[|X(t)|^2] < \infty$, then the mean square derivative*

$$Y(t) = \underset{\Delta \to 0}{\text{l.i.m.}} \frac{X(t+\Delta) - X(t)}{\Delta} \tag{5.80}$$

will exist and have a finite second moment at t if and only if the correlation function

$$\frac{\partial^2 R_X(\tau, \tau')}{\partial \tau \partial \tau'}$$

is well defined at $\tau = \tau' = t$, in which case

$$R_Y(t,s) = \frac{\partial^2 R_X(t,s)}{\partial t \partial s}. \tag{5.81}$$

When the mean square derivative exists, it is often denoted $X'(t)$.

Proof Because of the similarity of the proof to that of the mean square integration result, we only present a brief sketch. Define as before $Y_\Delta(t) = (X(t+\Delta) - X(t)/\Delta$ and consider

$$\begin{aligned}
&\|Y_\Delta - Y_{\Delta'}\|^2 \\
&= E\left[\left|\frac{X(t+\Delta) - X(t)}{\Delta} - \frac{X(t+\Delta') - X(t)}{\Delta'}\right|^2\right] \\
&= E\left[\left|\frac{X(t+\Delta) - X(t)}{\Delta}\right|^2\right] - 2E\left[\left|\frac{X(t+\Delta) - X(t)}{\Delta}\right|\left|\frac{X(t+\Delta') - X(t)}{\Delta'}\right|\right] \\
&\quad + E\left[\left|\frac{X(t+\Delta') - X(t)}{\Delta'}\right|^2\right].
\end{aligned}$$

The leftmost term in the final formula is

$$\begin{aligned}
E\left[\left|\frac{X(t+\Delta) - X(t)}{\Delta}\right|^2\right] &= E\left[\frac{X(t+\Delta)^2 - 2X(t+\Delta)X(t) + X(t)^2}{\Delta^2}\right] \\
&= \frac{R_X(t+\Delta, t+\Delta) - 2R_X(t+\Delta, t) + R_X(t,t)}{\Delta^2}.
\end{aligned}$$

This second-order difference will converge to the second-order partial derivative of $R_X(\tau, \tau')$ evaluated at $\tau = \tau' = t$ if and only if the said derivative is well defined. Thus $Y_\Delta(t)$ will be a Cauchy sequence in L_2 if and only if the said derivative is well defined, in which case $Y_\Delta(t)$ converges in mean square.

The remainder of the lemma follows by interchanging the expectation and limits.

□

5.8.5 Fourier series

As another application of mean square convergence we consider Fourier series for periodic random processes. From basic Fourier analysis recall that if $\{x(t);\ t \in (-\infty, \infty)\}$ is a periodic function with period T, i.e., if $x(t+T) = x(t)$, then subject to suitable techincal conditions we can expand $x(t)$ in a Fourier series

$$x(t) = \sum_{n=-\infty}^{\infty} \mathcal{X}(n) e^{+j2\pi nt/T}, \text{ where } \mathcal{X}(n) \triangleq \frac{1}{T} \int_0^T x(t) e^{-j2\pi nt/T}\, dt \quad (5.82)$$

is the Fourier transform of one period of the function $x(t)$. Note also that a Fourier series can be used to represent a function $\{x(t);\ t \in [0, T)\}$ defined *only* on the interval $[0, T)$.

Consider the extension of this idea to a random process. Suppose that $\{X(t);\ t \in (-\infty, \infty)\}$ is a weakly stationary zero-mean process with autocorrelation function $R_X(\tau)$. We further assume that the process is periodic with period T in the sense that

$$R_X(\tau + T) = R_X(\tau); \quad \text{all } \tau. \quad (5.83)$$

If a random process is periodic in this sense, then for any t expanding the magnitude square gives

$$E[|X(t) - X(t+\tau)|^2] = 2R_X(0) - R_X(\tau) - R_X(-\tau) = 0$$

which from the Markov inequality implies that $\Pr(|X(t) - X(t+\tau)| > 0) = 0$, so that with probability one $X(t+T) = X(t)$.

Define the Fourier coefficients of the random process by the mean square integrals

$$\mathcal{X}(n) = \frac{1}{T} \int_0^T X(t) e^{-(j2\pi nt/T)(j2\pi nt/T)}\, dt. \quad (5.84)$$

From Lemma 5.3 this will exist iff

$$\int_0^T \int_0^T R_X(t, s) e^{-j2\pi nt/T} e^{+j2\pi ns/T}\, dt\, ds$$
$$= \int_0^T \int_0^T R_X(t - s) e^{-j2\pi n(t-s)/T}\, dt\, ds < \infty.$$

In order to ensure the existence, we assume that R_X satisfies this equation.

Note, however, two things. First, since both R_X and $e^{-j2\pi nt/T}$ are periodic with period T, taking the integral over T results in the same value regardless of s so that the constraint becomes

$$\int_0^T R_X(t)e^{-j2\pi nt/T}\,dt < \infty. \tag{5.85}$$

Second, note that since $R_X(t)$ is periodic with period T, assuming it is well behaved it has a Fourier series

$$R_X(\tau) = \sum_{n=-\infty}^{\infty} b_n e^{+j2\pi nt/T} \tag{5.86}$$

where

$$b_n = \frac{1}{T}\int_0^T R_X(\tau)e^{-j2\pi nt/T}\,dt \tag{5.87}$$

so that our constraint for existence of the mean square integral is simply that the Fourier coefficients of R_X be finite. Assuming this to be the case, then the correlation of the Fourier coefficients of $X(t)$ can be evaluated as

$$\begin{aligned}
&E[\mathcal{X}(n)\mathcal{X}^*(m)] \\
&= E\left[\left(\frac{1}{T}\int_0^T X(t)e^{-j2\pi nt/T}\,dt\right)\left(\frac{1}{T}\int_0^T X(s)e^{-j2\pi ms/T}\,ds\right)\right] \\
&= \frac{1}{T^2}\int_0^T\int_0^T R_X(t-s)e^{-j2\pi(nt-ms)/T}\,dt\,ds \\
&= \frac{1}{T^2}\int_0^T e^{-j2\pi(n-m)s/T}\int_0^T R_X(t-s)e^{-j2\pi n(t-s)/T}\,dt\,ds \\
&= b_n\int_0^1 e^{-j2\pi(n-m)s}\,ds = b_n\delta_{n-m},
\end{aligned}$$

so that the Fourier coefficients are orthogonal random variables.

We now have all the needed machinery to prove the following lemma characterizing the existence of Fourier series of random processes.

Lemma 5.5 *Suppose that $\{X(t);\ t \in \Re\}$ is a weakly stationary random process for which the $R_X(\tau)$ is periodic and has a Fourier series. Then the mean square integrals*

$$\mathcal{X}(n) = \frac{1}{T}\int_0^T X(t)e^{-j2\pi nt/T}\,dt \tag{5.88}$$

exist and have finite second moment,

$$E[\mathcal{X}(n)\mathcal{X}^*(m)] = E[|\mathcal{X}_n|^2\delta_{n-m}], \tag{5.89}$$

and

$$X(t) = \underset{n \to \infty}{\text{l.i.m.}} \sum_{n=-N}^{N} \mathcal{X}(n) e^{+j2\pi nt/T}. \tag{5.90}$$

Conversely, if $\{X(t);\ t \in \Re\}$ is a random process for which (5.88) exists for all n and (5.89)–(5.90) hold, then $X(t)$ is weakly stationary and periodic.

Proof We have already shown that if $X(t)$ is weakly stationary, periodic, and its autocorrelation $R_X(\tau)$ has a Fourier series, then (5.88)–(5.89) hold. To prove (5.90) define $X_N(t) = \sum_{n=-N}^{N} \mathcal{X}(n) e^{+j2\pi nt/T}$ and we have

$$\begin{aligned}
&E[|X(t) - X_N(t)|^*] \\
&= E[X(t)X^*(t)] - E[X(t)X_N^*(t)] - E[X_N(t)X^*(t)] + E[X_N(t)X_N^*(t)] \\
&= R_X(0) - E\left[X(t) \sum_{n=-N}^{N} \mathcal{X}^*(n) e^{-j2\pi nt/T}\right] - E\left[X^*(t) \sum_{n=-N}^{N} \mathcal{X}(n) e^{+j2\pi nt/T}\right] \\
&\quad + E\left[\sum_{n=-N}^{N} \mathcal{X}(n) e^{+j2\pi nt/T} \sum_{n=-N}^{N} \mathcal{X}^*(n) e^{-j2\pi nt/T}\right] \\
&= R_X(0) - \sum_{n=-N}^{N} E\left[X(t) \frac{1}{T} \int_0^T X^*(s) e^{+j2\pi ns/T}\, ds\right] e^{-j2\pi nt/T} \\
&\quad - \sum_{n=-N}^{N} E\left[X^*(t) \frac{1}{T} \int_0^T X(s) e^{-j2\pi ns/T}\, ds\right] e^{+j2\pi nt/T} \\
&\quad + \sum_{n=-N}^{N} \sum_{m=-N}^{N} E[\mathcal{X}(n)\mathcal{X}^*(m)] \\
&= R_X(0) - \sum_{n=-N}^{N} \frac{1}{T} \int_0^T R_X(t-s) e^{-j2\pi n(t-s)/T}\, ds \\
&\quad - \sum_{n=-N}^{N} \frac{1}{T} \int_0^T R_X(s-t) e^{-j2\pi n(s-t)/T}\, ds + \sum_{n=-N}^{N} b_n \\
&= R_X(0) - \sum_{n=-N}^{N} b_n.
\end{aligned}$$

From Parceval's theorem for Fourier series,

$$R_X(0) = \sum_{n=-\infty}^{\infty} b_n$$

and hence

$$\lim_{N\to\infty} E[|X(t) - X_N(t)|^*] = \lim_{N\to\infty}\left(\sum_{n=-\infty}^{\infty} b_n - \sum_{n=-N}^{N} b_n\right) = 0,$$

proving (5.90).

For the converse, assume (5.88)–(5.90) hold. From Lemma 5.2 we now that

$$R_X(t,s) = \lim_{N\to\infty} R_{X_N}(t,s) \tag{5.91}$$

and we evaluate

$$\begin{aligned} R_{X_N}(t,s) &= E[X_N(t)X_N^*(s)] \\ &= E[\sum_{n=-N}^{N} \mathcal{X}(n)e^{+j2\pi nt/T} \sum_{m=-N}^{N} \mathcal{X}^*(m)e^{-j2\pi ms/T}] \\ &= \sum_{n=-N}^{N}\sum_{m=-N}^{N} E[\mathcal{X}(n)\mathcal{X}^*(m)]e^{+j2\pi(nt-ms)/T} \\ &= \sum_{n=-N}^{N} E[|\mathcal{X}(n)|^2]e^{+j2\pi n(t-s)/T}, \end{aligned}$$

which is weakly stationary and periodic with period T for all N, hence the limit also be weakly stationary and periodic with period T. □

Thus a zero-mean continuous time random process can be expanded in a mean square Fourier series with uncorrelated coefficients if and only if it is weakly stationary and periodic.

5.8.6 Sampling

A key step in converting an analog signal processing problem into a digital signal processing problem is sampling a continuous time process $X(t)$ to produce a discrete time process $X_n = X(nT) = X(n/f_0)$ using a sampling interval of T seconds or a sampling frequency of $W = 1/T$ samples per second. It is a standard result of Fourier analysis and basic linear systems theory that if the Fourier transform of a deterministic signal $x(t)$ is zero outside some frequency range $[-W_0, W_0]$, then provided the sampling frequency W is fast enough, $W > W_0$, or the sampling interval or period $T = 1/2W$ is short enough, $T \leq 1/2W_0$, then the original signal can in theory be recovered from its samples using the *sampling expansion*

$$x(t) = \sum_{n=-\infty}^{\infty} x(nT) \frac{\sin(\pi(t - nT)/T)}{\pi(t - nT)/T}. \tag{5.92}$$

The sampling expansion is sometimes interpreted as *pulse amplitude modulation* since a sequence of "pulses" of the form $\sin(\pi(t - nT)/T)/\pi(t - nT)/T$ are multiplied or modulated by the samples $x(nT)$ and the resulting waveform allows the reconstruction of the original time function. In this subsection we use the ideas of mean square convergence to derive a sampling theorem for continuous time random processes. Here we state the result and devote the remainder of the subsection to deriving it.

Before proceeding to the case of random processes, it will be useful to provide a simple proof of the deterministic case using Fourier series. By assumption $x(t)$ has a Fourier transform

$$\mathcal{X}(f) = \int_{\infty}^{\infty} x(t) e^{-j2\pi ft}\, dt$$

and the original signal $x(t)$ can be recovered by inverting the Fourier transform,

$$x(t) = \int_{-\infty}^{\infty} \mathcal{X}(f) e^{+j2\pi ft}\, df \tag{5.93}$$

which is 0 for $f \notin (-W, W)$. Thus we can expand $\mathcal{X}$ on the interval $[-W, W]$ in a Fourier series as

$$\begin{aligned}
\mathcal{X}(f) &= \sum_{n=-\infty}^{\infty} a_n e^{-j2\pi nf/2W} \\
a_n &= \frac{1}{2W} \int_{-W}^{W} \mathcal{X}(f) e^{+j2\pi nf/2W}\, df \\
&= \frac{1}{2W} \int_{-\infty}^{\infty} \mathcal{X}(f) e^{+j2\pi nf/2W}\, df = \frac{1}{2W} x\left(\frac{n}{2W}\right)
\end{aligned}$$

using (5.93). Thus the transform of $x(t)$ can be expressed in terms of the samples of $x(nT)$ by

$$\mathcal{X}(f) = \sum_{n=-\infty}^{\infty} T x(nT) e^{-j2\pi nf/2W}. \tag{5.94}$$

Inverting the transform then yields the sampling expansion:

$$\begin{aligned} x(t) &= \int_{-W}^{W} \mathcal{X}(f) e^{+j2\pi ft}\, df \\ &= \int_{-W}^{W} \sum_{n=-\infty}^{\infty} \frac{1}{2W} x\left(\frac{n}{2W}\right) e^{-j2\pi nf/2W} e^{+j2\pi ft}\, df \\ &= \sum_{n=-\infty}^{\infty} \frac{1}{2W} x\left(\frac{n}{2W}\right) \int_{-W}^{W} e^{+j2\pi f(t-n/2W)}\, df \\ &= \sum_{n=-\infty}^{\infty} x(nT) \frac{\sin(\pi(t-nT)/T)}{\pi(t-nT)/T} \end{aligned}$$

as claimed.

As a special case that will be useful later, observe that $x(t) = \sin(\pi(t-s)/T))/\pi(t-s)/T)$ has a transform that meets the constraint – its Fourier transform is a box function at the origin running from $-W$ to W. Thus applying the ordinary sampling theorem to this function yields the identity

$$\frac{\sin(\pi(t-s)/T))}{\pi(t-s)/T)} = \sum_{n=-\infty}^{\infty} \frac{\sin(\pi(nT-s)/T)}{\pi(nT-s)/T} \frac{\sin(\pi(t-nT)/T)}{\pi(t-nT)/T}. \tag{5.95}$$

Theorem 5.1 *The mean square sampling theorem*
Given a continuous time weakly stationary random process with zero mean and a power spectral density $S_X(f)$ with the property that

$$S_X(f) = 0 \text{ for } |f| \geq W, \tag{5.96}$$

then for any $T \leq 1/2W$

$$X(t) = \underset{N\to\infty}{\text{l.i.m.}} \sum_{n=-N}^{N} X(nT) \frac{\sin(\pi(t-nT)/T)}{\pi(t-nT)/T}. \tag{5.97}$$

Proof Define

$$X_N(t) = \sum_{n=-N}^{N} X(nT) \frac{\sin(\pi(t-nT)/T)}{\pi(t-nT)/T}$$

and consider

$$\begin{aligned}
&E[|X(t) - X_N(t)|^2] \\
&= E[|X(t) - \sum_{n=-N}^{N} X(nT)\frac{\sin(\pi(t-nT)/T)}{\pi(t-nT)/T}|^2] \\
&= E[X(t)X^*(t)] - \sum_{n=-N}^{N} E[X(t)X^*(nT)]\frac{\sin(\pi(t-nT)/T)}{\pi(t-nT)/T} \\
&\quad - \sum_{n=-N}^{N} E[X^*(t)X(nT)]\frac{\sin(\pi(t-nT)/T)}{\pi(t-nT)/T} \\
&\quad + \sum_{n=-N}^{N}\sum_{m=-N}^{N} E[X(nT)X^*(mT)]\frac{\sin(\pi(t-nT)/T)}{\pi(t-nT)/T}\frac{\sin(\pi(t-mT)/T)}{\pi(t-mT)/T}
\end{aligned}$$

so that

$$\begin{aligned}
&E[|X(t) - X_N(t)|^2] \\
&= R_X(0) - \sum_{n=-N}^{N} R_X(t-nT)\frac{\sin(\pi(t-nT)/T)}{\pi(t-nT)/T} && (5.98) \\
&\quad - \sum_{n=-N}^{N} R_X(nT-t)\frac{\sin(\pi(t-nT)/T)}{\pi(t-nT)/T} && (5.99) \\
&\quad + \sum_{n=-N}^{N}\sum_{m=-N}^{N} R_X((n-m)T) \\
&\qquad \times \frac{\sin(\pi(t-nT)/T)}{\pi(t-nT)/T}\frac{\sin(\pi(t-mT)/T)}{\pi(t-mT)/T}. && (5.100)
\end{aligned}$$

The second and third terms above resemble a sampling expansion of the autocorrelation R_X, which is reasonable since its Fourier transform is assumed to be band-limited. To be more precise, consider the function $R_X(t-\tau)$ as a function of t. From Fourier transform theory its transform is $S_X(f)e^{j2\pi f\tau}$, which is band-limited. Hence we have the sampling expansion

$$R_X(t-\tau) = \sum_{n=-\infty}^{\infty} R_X(nT-\tau)\frac{\sin(\pi(t-nT)/T)}{\pi(t-nT)/T}. \tag{5.101}$$

Setting $\tau = t$ results in

$$R_X(0) = \sum_{n=-\infty}^{\infty} R_X(nT - t)\frac{\sin(\pi(t - nT)/T)}{\pi(t - nT)/T}.$$

This shows that the third term (5.99) converges to $R_X(0)$ and the same argument holds for the second term (5.100). This leaves the fourth term (5.100). Rewrite the term as

$$\sum_{n=-N}^{N} \sum_{m=-N}^{N} R_X((n-m)T)\frac{\sin(\pi(t - nT)/T)}{\pi(t - nT)/T}\frac{\sin(\pi(t - mT)/T)}{\pi(t - mT)/T}$$
$$= \sum_{k=-2N}^{2N} R_X(kT)$$
$$\times \sum\sum_{n,m:n-m=k,-N\leq m,n\leq N} \frac{\sin(\pi(t - nT)/T)}{\pi(t - nT)/T}\frac{\sin(\pi(t - mT)/T)}{\pi(t - mT)/T}.$$

As $N \to \infty$ the kth term in the summand converges to

$$\sum\sum_{n,m:n-m=k} \frac{\sin(\pi(t - nT)/T)}{\pi(t - nT)/T}\frac{\sin(\pi(t - mT)/T)}{\pi(t - mT)/T}$$
$$= \sum_{n=-\infty}^{\infty} \frac{\sin(\pi(t - nT)/T)}{\pi(t - nT)/T}\frac{\sin(\pi(t + kT - nT)/T)}{\pi(t + kT - nT)T} = \frac{\sin(\pi k)}{\pi k},$$

where (5.95) was used in the last step. Thus

$$\lim_{N\to\infty} \sum_{n=-N}^{N} \sum_{m=-N}^{N} R_X((n-m)T)\frac{\sin(\pi(t - nT)/T)}{\pi(t - nT)/T}\frac{\sin(\pi(t - mT)/T)}{\pi(t - mT)/T}$$
$$= \sum_{-\infty}^{\infty} R_X(k)\frac{\sin(\pi k)}{\pi k} = R_X(0)$$

and the proof is complete. □

5.8.7 Karhunen–Loeve expansion

Let $\{X(t);\ t \in \Re\}$ be a zero-mean random process with finite second moment and autocorrelation function $R_X(t, s)$. Consider a generalization of a Fourier series where a random process is represented as a linear combination of functions with certain specified properties. Specifically we seek a representation of the form

$$X(t) = \underset{n\to\infty}{\text{l.i.m.}}\ X_N(t) \tag{5.102}$$

for $t \in [a, b]$, where

$$X_N(t) \triangleq \sum_{n=1}^{N} X_n \phi_n(t) \tag{5.103}$$

$$X_n = \int_a^b X(t)\phi_n^*(t)\,dt \tag{5.104}$$

$$\int_a^b \phi_n(t)\phi_m^*(t)\,dt = \delta_{n-m} \tag{5.105}$$

$$E[X_m X_n^*] = \lambda_n \delta_{n-m} \tag{5.106}$$

We refer to such an expansion (if it exists) as a *Karhunen–Loeve expansion* of the random process.

If (5.102) holds, then from Lemma 5.2

$$R_X(t, s) = \lim_{N\to\infty} R_{X_N}(t, s)$$

where we use the assumed properties to compute

$$\begin{aligned} R_{X_N}(t, s) = E[X_N(t)X_N^*(s)] &= E\left[\sum_{n=1}^{N} X_n\phi_n(t) \sum_{m=1}^{N} X_m^*\phi_m^*(s)\right] \\ &= \sum_{n=1}^{N}\sum_{m=1}^{N} E[X_n X_m^*]\phi_n(t)\phi_m^*(s)] = \sum_{n=1}^{N} \lambda_N \phi_n(t)\phi_n^*(s) \end{aligned}$$

so that

$$R_X(t, s) = \lim_{N\to\infty} R_{X_N}(t, s) = \sum_{n=1}^{\infty} \lambda_n \phi_n(t)\phi_n^*(s).$$

Multiplying by a ϕ-function and integrating then yields

$$\begin{aligned} \int_a^b R_X(t, s)\phi_m(s)\,ds &= \sum_{n=1}^{\infty} \lambda_n\phi_n(t) \int_a^b \phi_m(s)\phi_n^*(s)\,ds \\ &= \lambda_m\phi_m(t). \end{aligned}$$

Thus if we wish to have a representation satisfying (5.102)–(5.105), then the ϕ-functions and energies λ must satisfy

$$\int_a^b R_X(t, s)\phi(s)\,ds = \lambda\phi(t), \tag{5.107}$$

that is, the ϕs must be *eigenfunctions* of the above integral equation with *eigenvalue* λ. In fact, this is an "if and only if" statement, that is, ϕ_n are the eigenfunctions and λ_n the eigenvalues satisfying (5.107) if and only if

(5.102)–(5.105) hold. To prove this, we next show that if ϕ_n and λ_n satisfy (5.107), then indeed (5.102)–(5.105) hold. We normalize eigenfunctions to have unit energy,

$$\int_a^b |\phi(t)|^2 \, dt = 1$$

and absorb the normalization in the eigenvalue λ. First suppose that ϕ_n and ϕ_m are distinct eigenfunctions of (5.107) with eigenvalues λ_n and λ_m. We show that this implies that ϕ_n and ϕ_m are orthogonal in that (5.105) holds and that λ_n is real and nonnegative, so it indeed is a valid energy for a random variable. By assumption

$$\lambda_n \phi_n(t) = \int_a^b R_X(t,s)\phi_n(s) \, ds$$

so that multiplying both sides by $\phi_m(t)$ for *any* m (including n) and integrating we have

$$\begin{aligned}
\lambda_n \int_a^b \phi_m^*(t)\phi_n(t) \, dt &= \int_a^b \int_a^b \phi_m^*(t) R_X(t,s)\phi_n(s) \, ds \, dt \\
&= \int_a^b \phi_n(s) \left(\int_a^b R_X(t,s)\phi_m^*(t) \, dt \right) ds \\
&= \int_a^b \phi_n(s) \left(\int_a^b R_X(s,t)\phi_m(t) \, dt \right)^* ds \\
&= \lambda_m^* \int_a^b \phi_n(s)\phi_m^*(s) \, ds
\end{aligned}$$

so that replacing the dummy variable s by t yields

$$(\lambda_n - \lambda_m^*) \int_a^b \phi_n(t)\phi_m^*(t) \, dt = 0.$$

If $n = m$ this implies that

$$(\lambda_n - \lambda_n^*) \int_a^b |\phi_n(t)|^2 \, dt = (\lambda_n - \lambda_n^*) = 0,$$

which forces λ_n to be real. For $n \neq m$ we have

$$(\lambda_n - \lambda_m) \int_a^b \phi_n(t)\phi_m^*(t) \, dt = 0.$$

If $\lambda_n \neq \lambda_m$, then ϕ_n and ϕ_m must be orthogonal. If the two eigenvalues are equal, then we have to work harder and quote material beyond the mathematical scope of this book to prove the desired result. It turns out (see, e.g.,

Appendix 2 of [14]) that there are at most a countably infinite number of distinct eigenfunctions. Each eigenfunction can have associated with it at most a finite number of linearly independent eigenfunctions and these eigenfunctions can be transformed by linear operations into a set of an equal number of orthogonal unit energy eigenfunctions using a Gram–Schmidt procedure. Thus any distinct eigenvalues automatically yield orthogonal eigenfunctions, while equal eigenvalues can correspond to several eigenfunctions which can be made orthogonal by linear operations.

Given now the collection ϕ_n of orthogonal and normalized eigenfunctions, we have

$$\begin{aligned} E[X_n X_m^*] &= E\left[\left(\int_a^b X(t)\phi_n^*(t)\,dt\right)\left(\int_a^b X^*(s)\phi_m(s)\,ds\right)\right] \\ &= \int_a^b \int_a^b E[X(t)X^*(s)]\phi_n^*(t)\phi_m(s)\,ds\,dt \\ &= \int_a^b \left(\int_a^b R_X(t,s)\phi_m(s)\,ds\right)\phi_n^*(t)\,dt \\ &= \int_a^b \lambda_m \phi_m(t)\phi_n^*(t)\,dt = \lambda_n \delta_{n-m}, \end{aligned}$$

so that the X_n are indeed uncorrelated. To prove mean square convergence consider as usual

$$\begin{aligned} &E[|X(t) - X_N(t)|^2] = \\ &\quad E[X(t)X(t)^*] - E[X(t)X_N(t)^*] - E[X_N(t)X(t)^*] + E[X_N(t)X_N(t)^*]. \end{aligned}$$

The first term is $R_X(t,s)$. The second term is

$$\begin{aligned} E[X(t)X_N(t)^*] &= E\left[X(t)\sum_{n=1}^N X_n^*\phi_n^*(t)\right] = \sum_{n=1}^N E[X(t)X_n^*]\phi_n^*(t)] \\ &= \sum_{n=1}^N E\left[X(t)\int_a^b X^*(s)\phi_n(s)\,ds]\phi_n^*(t)\right] \\ &= \sum_{n=1}^N \phi_n^*(t)\int_a^b E[X(t)X^*(s)]\phi_n(s)\,ds \\ &= \sum_{n=1}^N \phi_n^*(t)\int_a^b R_X(t,s)\phi_n(s)\,ds = \sum_{n=1}^N \lambda_n\phi_n^*(t)\phi_n^*(t). \end{aligned}$$

A similar evaluation shows that

$$E[X_N(t)X(t)^*] = \sum_{n=1}^{N} \lambda_n \phi_n^*(t)\phi_n^*(t),$$

and

$$\begin{aligned} E[X_N(t)X_N(t)^*] &= E\left[\sum_{n=1}^{N} X_n\phi_n(t) \sum_{m=1}^{N} X_m^*\phi_m^*(t)\right] \\ &= \sum_{n=1}^{N}\sum_{m=1}^{N} E[X_n X_m^*]\phi_n(t)\phi_m^*(t) \\ &= \sum_{n=1}^{N} \lambda_n\phi_n(t)\phi_n^*(t) \end{aligned}$$

so that

$$\lim_{n\to\infty} E[|X(t) - X_N(t)|^2] = R_X(t,t) - \sum_{n=1}^{\infty} \lambda_n\phi_n(t)\phi_n^*(t). \tag{5.108}$$

Once again we invoke a result from the theory of integral equations: Mercer's theorem (see, e.g., Appendix 2 of [14]) states that under suitable conditions – including the case where $R_X(t,s)$ is continuous in t and s over $[a,b]$, the right-hand side above is 0 and mean square convergence is proved.

As a first example, suppose that $\{X(t)\}$ is weakly stationary and periodic with period T and hence that $R_X(\tau)$ is periodic. If R_X has a Fourier series expansion

$$R_X(\tau) = \sum_{n=-\infty}^{\infty} b_n e^{j2\pi n\tau/T}$$

then the integral equation (5.107) to be solved for the Karhunen–Loeve expansion is

$$\lambda\phi(t) = \int_0^T \sum_{n=-\infty}^{\infty} b_n e^{j2\pi n(t-s)/T}\phi(s)\,ds = \sum_{n=-\infty}^{\infty} b_n \int_0^T e^{j2\pi n(t-s)/T}\phi(s)\,ds.$$

Guessing a solution $\phi_m(t) = c_n e^{j2\pi mt/T}$ where c_m is a normalizing constant, then

$$\begin{aligned} \int_0^T c_n e^{j2\pi nt/T} c_m^* e^{-j2\pi mt/T}\,dt &= c_n c_m^* \int_0^t e^{j2\pi(n-m)t/T}\,dt \\ &= |c_n|^2 T\delta_{n-m} \end{aligned}$$

so that $c_n = 1/\sqrt{T}$. Then with

$$\phi_m(t) = \frac{e^{j2\pi mt/T}}{\sqrt{T}}$$

the integral equation becomes

$$\begin{aligned}\lambda \frac{e^{j2\pi mt/T}}{\sqrt{T}} &= \sum_{n=-\infty}^{\infty} b_n \int_0^T e^{j2\pi n(t-s)/T} \frac{e^{j2\pi ms/T}}{\sqrt{T}}\, ds \\ &= \sum_{n=-\infty}^{\infty} b_n \frac{e^{j2\pi nt/T}}{\sqrt{T}} \int_0^T e^{\frac{j2\pi(m-n)s}{T}}\, ds \\ &= b_m T \frac{e^{j2\pi mt/T}}{\sqrt{T}} = b_m T \phi_m(t),\end{aligned}$$

which we already knew was a solution from the development for Fourier series.

As another example, suppose that $X(t)$ is white noise with $S_X(f) = N_0/2$ for all f and hence $R_X(\tau) = (N_0/2)\delta(\tau)$, a Dirac delta. In this case

$$\begin{aligned}\int_a^b R_X(t-s)\phi(s)\, ds &= \frac{N_0}{2} \int_a^b \delta(t-s)\phi(s)\, ds \\ &= \frac{N_0}{2}\phi(t) = \lambda\phi(t),\end{aligned}$$

yielding the strange behavior that *all* eigenvalues are identical so that all terms in the Karhunen–Loeve representation have the same second moment.

5.9 ⋆Linear estimation and filtering

In this section we give another application of second-order moments in linear systems by showing how they arise in one of the basic problems of communication; estimating the outcomes of one random process based on observations of another process using a linear filter. The initial results can be viewed as process variations on the vector results of Section 4.11, but we develop them independently here in the process and linear filtering context for completeness. We will obtain the classical orthogonality principle and the Wiener–Hopf equation and consider solutions for various simple cases. This section provides additional practice in manipulating second-order moments of random processes and provides more evidence for their importance.

We will focus on discrete time for the usual reasons, but the continuous time analogs are found, also as usual, by replacing the sums by integrals.

5.9.1 Discrete time

Suppose that we are given a record of observations of values of one random process; e.g., we are told the values of $\{Y_i;\ N < i < M\}$, and we are asked to form the best estimate of a particular sample, say X_n, of another, related random process $\{X_k;\ k \in \mathcal{T}\}$. We refer to the collection of indices of observed samples as $\mathcal{K} = (N, M)$. We permit N and M to take on infinite values. For convenience we assume throughout this section that both processes have zero means for all time. We place the strong constraint on the estimate that it must be linear; that is, the estimate $\widehat{X}_n$ of X_n must have the form

$$\widehat{X}_n = \sum_{k:\, n-k\in\mathcal{K}} h_k Y_{n-k} = \sum_{k\in\mathcal{K}} h_{n-k} Y_k$$

for some δ-response h. We wish to find the "best" possible filter h, perhaps under additional constraints such as causality. One possible notion of best is to define the error

$$\epsilon_n = X_n - \widehat{X}_n$$

and define that filter to be best within some class if it minimizes the mean squared error $E(\epsilon_n^2)$; that is, a filter satisfying some constraints will be considered optimum if no other filter yields a smaller expected squared error. The filter accomplishing this goal is often called a linear least squared error (LLSE) filter.

Many constraints on the filter or observation times are possible. Typical constraints on the filter and on the observations are the following:

1. We have a noncausal filter that can "see" into the infinite future and a two-sided infinite observation $\{Y_k;\ k \in \mathcal{Z}\}$. Here we consider $N = -\infty$ and $M = \infty$. This is clearly not completely possible, but it may be a reasonable approximation for a system using a very long observation record to estimate a sample of a related process in the middle of the records.
2. The filter is causal ($h_k = 0$ for $k < 0$), a constraint that can be incorporated by assuming that $n \geq M$; that is, that samples occurring after the one we wish to estimate are not observed. When $n > M$ the estimator is sometimes called a predictor since it estimates the value of the desired process at a time later than the last observation. Here we assume that we observe the entire past of the Y process; that is, we take $N = -\infty$ and observe $\{Y_k;\ k < M\}$. If, for example, the X process and the Y process are the same and $M = n$, then this case is called the one-step predictor (based on the semi-infinite past).
3. The filter is causal, and we have a finite record of T seconds; that is, we observe $\{Y_k;\ M - T \leq K < M\}$.

As one might suspect, the fewer the constraints, the easier the solution but the less practical the resulting filter. We will develop a general characterization for the optimum filters, but we will provide specific solutions only for certain special cases. We formally state the basic result as a theorem and then prove it.

Theorem 5.2 *Suppose that we are given a set of observations* $\{Y_k;\ k \in \mathcal{K}\}$ *of a zero-mean random process* $\{Y_k\}$ *and that we wish to find a linear estimate* $\widehat{X}_n$ *of a sample* X_n *of a zero-mean random process* $\{X_n\}$ *of the form*

$$\widehat{X}_n = \sum_{k:\, n-k\in\mathcal{K}} h_k Y_{n-k}. \tag{5.109}$$

If the estimation error is defined as

$$\epsilon_n = X_n - \widehat{X}_n,$$

then for a fixed n *no linear filter can yield a smaller expected squared error* $E(\epsilon_n^2)$ *than a filter* h *(if it exists) that satisfies the relation*

$$E(\epsilon_n Y_k) = 0; \quad all\ k \in \mathcal{K}, \tag{5.110}$$

or, equivalently,

$$E(X_n Y_k) = \sum_{i:\, n-i\in\mathcal{K}} h_i E(Y_{n-i} Y_k); \quad all\ k \in \mathcal{K}. \tag{5.111}$$

If $R_Y(k,j) = E(Y_k Y_j)$ *is the autocorrelation function of the* Y *process and* $R_{X,Y}(k,j) = E(X_k Y_j)$ *is the cross-correlation function of the two processes, then (5.111) can be written as*

$$R_{X,Y}(n,k) = \sum_{i:\, n-i\in\mathcal{K}} h_i R_Y(n-i,k); \quad all\ k \in \mathcal{K}. \tag{5.112}$$

If the processes are jointly weakly stationary in the sense that both are individually weakly stationary with a cross-correlation function that depends only on the difference between the arguments, then, with the replacement of k *by* $n-k$*, the condition becomes*

$$R_{X,Y}(k) = \sum_{i:\, n-i\in\mathcal{K}} h_i R_Y(k-i); \quad all\ k : n-k \in \mathcal{K}. \tag{5.113}$$

Comments. Two random variables U and V are said to be orthogonal if $E(UV) = 0$. Therefore (5.110) is known as the *orthogonality principle* because it states that the optimal filter causes the estimation error to be orthogonal to the observations. Note that (5.110) implies not only that the estimation error is orthogonal to the observations, but that it is also orthogonal to all linear combinations of the observations. Relation (5.113) with $\mathcal{K} = (-\infty, n)$ is known as the Wiener–Hopf equation. For this to be useful in practice, we must be able to find a δ-response that solves one of these equations. We shall later find solutions for some simple cases. A more general treatment is beyond the intended scope of this book. Our emphasis here is to demonstrate an example in which determination of an optimal filter for a reasonably general problem requires the solution of an equation given in terms of second-order moments.

Proof Suppose that we have a filter h that satisfies the given conditions. Let g be any other linear filter with the same input observations and let $\tilde{X}_n$ be the resulting estimate. We will show that the given conditions imply that g can yield an expected squared error no better than that of h. Let $\tilde{\epsilon}_n = X_n - \tilde{X}_n$ be the estimation error using g so that

$$E(\tilde{\epsilon}_n^2) = E\left(\left(X_n - \sum_{i:\, n-i \in \mathcal{K}} g_i Y_{n-i}\right)^2\right).$$

Add and subtract the estimate using h satisfying the conditions of the theorem and expand the square to obtain

$$\begin{aligned}
E(\tilde{\epsilon}_n^2) &= E\left(\left(X_n - \sum_{i:\, n-i \in \mathcal{K}} h_i Y_{n-i} + \sum_{i:\, n-i \in \mathcal{K}} h_i Y_{n-i} - \sum_{i:\, n-i \in \mathcal{K}} g_i Y_{n-i}\right)^2\right) \\
&= E\left(\left(X_n - \sum_{i:\, n-i \in \mathcal{K}} h_i Y_{n-i}\right)^2\right) + E\left(\left(\sum_{i:\, n-i \in \mathcal{K}} (h_i - g_i) Y_{n-i}\right)^2\right) \\
&\quad +2E\left(\left(X_n - \sum_{i:\, n-i \in \mathcal{K}} h_i Y_{n-i}\right)\left(\sum_{i:\, n-i \in \mathcal{K}} (h_i - g_i) Y_{n-i}\right)\right).
\end{aligned}$$

The first term on the right is the expected squared error using the filter h, say $E(\epsilon_n^2)$. The last term on the right is the expectation of something squared and is hence nonnegative. Thus we have the lower bound

$$E(\tilde{\epsilon}_n^2) \geq E(\epsilon_n^2) + \sum_{i:\, n-i \in \mathcal{K}} (h_i - g_i) \left\{ E(X_n Y_{n-i}) - \sum_{j:\, n-j \in \mathcal{K}} h_j E(Y_{n-j} Y_{n-i}) \right\},$$

where we have brought one of the sums out, used different dummy variables for the two sums, and interchanged some expectations and sums. From (5.111), however, the bracketed term is zero for each i in the index set being summed over, and hence the entire sum is zero, proving that

$$E(\tilde{\epsilon}_n^2) \geq E(\epsilon_n^2),$$

which completes the proof of the theorem. □

Note that from (5.109) through (5.113) we can write the mean square error for the optimum linear filter as

$$\begin{aligned} E(\epsilon_n^2) &= E\left(\epsilon_n\left(X_n - \sum_{k:\, n-k\in\mathcal{K}} h_k Y_{n-k}\right)\right) = E(\epsilon_n X_n) \\ &= R_X(n,n) - \sum_{k:\, n-k\in\mathcal{K}} h_k R_{X,Y}(n, n-k) \end{aligned}$$

in general and

$$E(\epsilon_n^2) = R_X(0) - \sum_{k:\, n-k\in\mathcal{K}} h_k R_{X,Y}(k)$$

for weakly stationary processes.

Older proofs of the result just given use the calculus of variations, that is, calculus minimization techniques. The method we have used, however, is simple and intuitive and shows that a filter satisfying the given equations actually yields a global minimum to the mean squared error and not only a local minimum as usually obtained by obtained by calculus methods. A popular proof of the basic orthogonality principle is based on Hilbert space methods and the projection theorem, the generalization of the standard geometric result that the shortest line from a point to a plane is the projection of the point on the plane – the line passing through the point which meets the plane at a right angle (is orthogonal to the plane). The projection method also proves that the filter of (5.110) yields a global minimum.

We consider four examples in which the theorem can be applied to construct an estimate. The first two are fairly simple and suffice for a brief reading.

[5.4] Suppose that the processes are jointly weakly stationary, that we are given the entire two-sided realization of the random process $\{Y_n\}$, and that there are no restrictions on the linear filter h. Equation (5.113) then becomes

$$R_{X,Y}(k) = \sum_{i\in\mathcal{Z}} h_i R_Y(k-i); \quad \text{all } k \in \mathcal{Z}.$$

This equation is a simple convolution and can be solved by standard Fourier techniques. Take the Fourier transform of both sides and define the transform of the cross-correlation function $R_{X,Y}$ to be the cross-spectral density $S_{X,Y}(f)$. We obtain $S_{X,Y}(f) = H(f)S_Y(f)$ or

$$H(f) = \frac{S_{X,Y}(f)}{S_Y(f)},$$

which can be inverted to find the optimal δ-response h:

$$h(k) = \int_{-1/2}^{1/2} \frac{S_{X,Y}(f)}{S_Y(f)} e^{i2\pi kf} \, df,$$

which yields an optimum estimate

$$\widehat{X}_n = \sum_{i=-\infty}^{\infty} h_i Y_{n-i}.$$

Thus we have an explicit solution for the optimal linear estimator for this case in terms of the second-order properties of the given processes. Note, however, that the resulting filter is *not* causal in general. Another important observation is that the filter itself does not depend on the sample time n at which we wish to estimate the X process; e.g., if we want to estimate X_{n+1}, we apply the same filter to the shifted observations; that is,

$$\widehat{X}_{n+1} = \sum_{i=-\infty}^{\infty} h_i Y_{n+1-i}.$$

Thus in this example not only have we found a means of estimating X_n for a fixed n, but the same filter also works for *any* n. When one filter works for all estimate sample times by simply shifting the observations, we say that it is a time-invariant or stationary estimator. As one might guess, such time invariance is a consequence of the weak stationarity of the processes.

The most important application of Example [5.4] is to "infinite smoothing," where $Y_n = X_n + V_n$. The noise process $\{V_n\}$ is assumed to be orthogonal to the signal process $\{X_n\}$, i.e., $R_{X,V}(k) = 0$ for all k. Then $R_{X,Y} = R_X$ and hence $R_Y = R_X + R_V$, so that

$$H(f) = \frac{S_X(f)}{S_X(f) + S_V(f)}. \tag{5.114}$$

[5.5] Again assume that the processes are jointly weakly stationary. Assume that we require a causal linear filter h but that we observe the infinite past of the observation process. Assume further that the observation process is white noise; that is, $R_Y(k) = (N_0/2)\delta_k$. Then

$\mathcal{K} = \{n, n-1, n-2, \ldots\}$, and (5.113) becomes the Wiener–Hopf equation

$$R_{X,Y}(k) = \sum_{i:\, n-i \in \mathcal{K}} h_i \frac{N_0}{2} \delta_{k-i} = h_k \frac{N_0}{2}; \;\; k \in \mathcal{Z}_+.$$

This equation easily reduces because of the δ-function to

$$h_k = \frac{2}{N_0} R_{X,Y}(k), \;\; k \in \mathcal{Z}_+.$$

Thus we have for this example the optimal estimator

$$\widehat{X}_n = \sum_{k=0}^{\infty} \frac{2}{N_0} R_{X,Y}(k) Y_{n-k}.$$

As with the previous example, the filter does not depend on n, and hence the estimator is time-invariant.

The case of a white observation process is indeed special, but it suggests a general approach to solving the Wiener–Hopf equation, which we sketch next.

[5.6] Assume joint weak stationarity and a causal filter on a semi-infinite observation sequence as in Example [5.5], but do not assume that the observation process is white. In addition, assume that the observation process is physically realizable so that a spectral factorization of the form of (5.40) exists; that is,

$$S_Y(f) = |G(f)|^2$$

for some causal stable filter with transfer function $G(f)$. As previously discussed, for practical purposes, all random processes have spectral densities of this form. We also assume that the inverse filter, the filter with transfer function $1/G(f)$, is causal and stable. Again, this holds under quite general conditions. Observe in particular that you cannot run into trouble with $G(f)$ being zero on a frequency interval of nonzero length because the condition in the spectral factorization theorem would be violated.

Unlike the earlier examples, this example does not have a trivial solution. We sketch a solution as a modification to the solution of Example [5.5]. The given observation process may not be white, but suppose that we pass it through a linear filter r with transfer function $R(f) = 1/G(f)$ to obtain a new random process, say $\{W_n\}$. Since the inverse filter $1/G(f)$ is assumed stable, then the W process has power spectral density $S_W(f) = S_Y(f)|R(f)|^2 = 1$ for all f;

that is, $\{W_n\}$ is white. One says that the W process is a *whitened* version of the Y process, sometimes called the *innovations process* of $\{Y_n\}$. Intuitively, the W process contains the same information as the Y process from it by passing it through the filter $G(f)$ (at least in principle). Thus we can get an estimate of X_n, from the W process that is just as good as (and no better than) that obtainable from the Y process. Furthermore, if we now filter the W process to estimate X_n, then the overall operation of the whitening filter followed by the estimating linear filter is also a linear filter, producing the estimate from the original observations. Since the inverse filter is causal, a causal estimate based on the W process is also a causal estimate based on the Y process.

Because W is white, the estimate of X_n from $\{W_n\}$ is given immediately by the solution to Example [5.5]; that is, the filter h with the W process as input is given by $h_k = R_{X,W}(k)$ for $k \geq 0$. The cross-correlation of the X and W processes can be calculated using the standard linear filter I/O techniques. It turns out that the required cross-correlation is the inverse Fourier transform of a cross-spectral density given by

$$S_{X,W}(f) = \frac{S_{X,Y}(f)}{G(f^*)} = H(f), \tag{5.115}$$

where the asterisk denotes the complex conjugate. (See Problem 5.22.) Thus the optimal causal linear estimator given the whitened process is

$$h_k = \begin{cases} \int_{-1/2}^{1/2} \frac{S_{X,Y}(f)}{G(f)^*} e^{2\pi jkf} & k \geq 0 \\ 0 & \text{otherwise} \end{cases},$$

and the overall optimal linear estimate has the form shown in Figure 5.3.

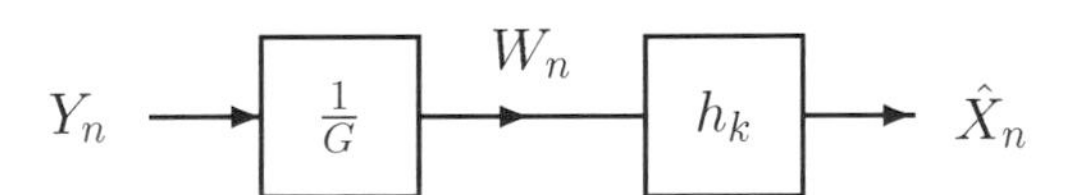

Whitening filter Estimator for X given W

Figure 5.3 Prewhitening method

Although more complicated, we again have a filter that does not depend on n.

This approach to solving the Wiener–Hopf equation is called the "prewhitening" (or "innovations" or "shaping filter") approach and it can be made rigorous under quite general conditions. That is, for all practical purposes, the optimal filter can be written in this cascade form as a whitening filter followed by a LLSE filter given the whitened observations as long as the processes are jointly weakly stationary and we know the observation process's power spectral density.

When the observation interval is finite or when the processes are not jointly weakly stationary, the spectral factorization approach becomes complicated and cumbersome, and alternative methods, usually in the time domain, are required. The final example considers such an estimator.

[5.7] Suppose that the random process we wish to estimate satisfies a difference equation of the form

$$X_{n+1} = \Phi_n X_n + U_n \ , \ \ n \geq 0, \tag{5.116}$$

where the process $\{U_n\}$ is a zero-mean process that is uncorrelated with a possibly time-varying second moment $E(U_n^2) = \Gamma_n$ and X_0 is an initial random variable, independent of the $\{U_n\}$. $\{\Phi_n\}$ is a known sequence of constants. In other words, we know that the random process is defined by a time-varying linear system driven by noise that is uncorrelated but not necessarily stationary. Assume that the observation process has the form

$$Y_n = H_n X_n + V_n, \tag{5.117}$$

a scaled version of the X process plus observation noise, where H_n is a known sequence of constants. We also assume that the observation noise has zero mean and is uncorrelated but not necessarily stationary, say $E(V_n^2) = \Psi_n$. We further assume that the U and V processes are uncorrelated: $E(U_n V_k) = 0$ for all n and k. Intuitively, the random processes are such that new values are obtained by scaling old values and adding some perturbations. Additional noise influences our observations. Suppose that we observe $Y_0, Y_1, \ldots, Y_{n-1}$, what is the best linear estimate of X_n?

In a sense this problem is more restrictive than the Wiener–Hopf formulation of Example [5.6] because we have assumed a particular structure for the process to be estimated and for the observations. On the other hand, it is not a special case of the previous model because the time-varying parameters make it nonstationary and because we restrict the observations to a finite time window (not including the current observation), often a better approximation to reality. Because of these differences the spectral techniques of the standard Wiener–Hopf solution of Example [5.6] do not apply without significant generalization and modification. Hence we consider another approach, called recursive estimation . The technique has its origins in the work of Gauss on plotting the trajectory of heavenly bodies [25]. The method was developed in 1958 by Swerling[65] and subsequently refined and publicized by Kalman and Bucy. The technique is now generally known as *Kalman–Bucy filtering,* . The basic idea is the following: instead of considering how to operate on a complete observation record in order to estimate something at one time, suppose that we already have a good estimate $\widehat{X}_n$ for X_n and that we make a single new observation Y_n. How can we use this new information to update our old estimate in a linear fashion to form a new estimate $\widehat{X}_{n+1}$ of X_{n+1}? For example, can we find sequences of numbers a_n and b_n so that

$$\widehat{X}_{n+1} = a_n \widehat{X}_n + b_n Y_n$$

is a good estimate? One way to view this is that instead of constructing a filter h described by a convolution that operates on the past to produce an estimate for each time, we wish a possibly time-varying filter with feedback that observes its own past outputs or estimates and operates on this and a new observation to produce a new estimate. This is the basic idea of recursive filtering, which is applicable to more general models than that considered here. In particular, the standard developments in the literature consider vector generalizations of the above difference equations. We sketch a derivation for the simpler scalar case.

We begin by trying to apply directly the orthogonality principle of (5.110) through (5.112). If we fix a time n and try to estimate X_n by a linear filter as

$$\widehat{X}_n = \sum_{i=1}^{n} h_i Y_{n-i}, \tag{5.118}$$

then the LLSE filter is described by the time-dependent δ-response, say $h^{(n-1)}$, which, from (5.112), solves the equations

$$R_{X,Y}(n,l) = \sum_{i=1}^{n} h_i^{(n-1)} R_Y(n-i,l); \;\; l = 0, 1, \ldots, n-1, \tag{5.119}$$

where the superscript reflects the fact that the estimate is based on observations through time $n-1$ and the fact that for this very nonstationary problem, the filter will likely depend very much on n. To demonstrate this, consider the estimate for X_{n+1}. In this case we will have a filter of the form

$$\widehat{X}_{n+1} = \sum_{i=1}^{n+1} h_i^{(n)} Y_{n+1-i}, \tag{5.120}$$

where the LLSE filter satisfies

$$R_{X,Y}(n+1,l) = \sum_{i=1}^{n+1} h_i^{(n)} R_Y(n+1-i,l); \;\; l = 0,1,\ldots,n. \tag{5.121}$$

Note that (5.121) is different from (5.119), and hence the δ-responses satisfying the respective equations will also differ. In principle these equations can be solved to obtain the desired filters. Since they will in general depend on n, however, we are faced with the alarming possibility of having to apply for each time n a completely different filter $h^{(n)}$ to the entire record of observations $Y_0, \ldots, Y_n$ up to the current time, clearly an impractical system design. We shall see, however, that a more efficient means of recursively computing the estimate can be found. It will still be based on linear operations, but now they will be time-varying.

We begin by comparing (5.119) and (5.121) more carefully to find a relation between the two filters $h^{(n)}$ and $h(n-1)$. If we consider $l < n$, then (5.116) implies that

$$\begin{aligned} R_{X,Y}(n+1,l) &= E(X_{n+1}Y_l) = E((\Phi_n X_n + U_n)Y_l) \\ &= \Phi_n E(X_n Y_l) + E(U_n Y_l) = \Phi_n E(X_n Y_l) \\ &= \Phi_n R_{X,Y}(n,l). \end{aligned}$$

Reindexing the sum of (5.121) using this relation and restricting ourselves to $l < n$ then yields

$$\begin{aligned} \Phi_n R_{X,Y}(n,l) &= \sum_{i=0}^{n} h_{i+1}^{(n)} R_Y(n-i,l) \\ &= h_1^{(n)} R_Y(n,l) + \sum_{i=1}^{n} h_{i+1}^{(n)} R_Y(n-i,l), \; l = 0,1,\ldots,n-1. \end{aligned}$$

But for $l < n$ we also have that

$$\begin{aligned} R_{X,Y}(n,l) &= E(X_n, Y_l) = E\left(\frac{Y_n - V_n}{H_n} Y_l\right) \\ &= \frac{1}{H_n} E(Y_n Y_l) = \frac{1}{H_n} R_Y(n,l) \end{aligned}$$

or

$$R_Y(n,l) = H_n R_{X,Y}(n,l).$$

Substituting this result, we have with some algebra that

$$R_{X,Y}(n,l) = \sum_{i=1}^{n} \frac{h_{i+1}^{n}}{\Phi_n - h_1^{(n)} H_n} R_Y(n-i,l); \; l = 0,1,\ldots,n-1,$$

which is the same as (5.119) if one identifies

$$h_i^{(n-1)} = \frac{h_{i+1}^{n}}{\Phi_n - h_1^{(n)} H_n}; \; l = 1,\ldots,n.$$

From (5.120) the estimate for X_n is

$$\widehat{X}_n = \sum_{i=1}^{n} h_i^{(n-1)} Y_{n-i} = \frac{1}{\Phi_n - h_1^{(n)} H_n} \sum_{i=2}^{n+1} h_i^{(n)} Y_{n+1-i}.$$

Comparing this with (5.120) yields

$$\widehat{X}_{n+1} = h_1^{(n)} Y_n + (\Phi_n - h_1^{(n)} H_n)\widehat{X}_n,$$

which has the desired form. It remains, however, to find a means of computing the numbers $h_1^{(n)}$. Since this really depends on only one argument n, we now change notation for brevity and henceforth denote this term by x_n; that is,

$$x_n = h_1^{(n)}.$$

To describe the estimator completely we need to find a means of computing x_n and an initial estimate. The initial estimate does not depend on any observations. The LLSE estimate of a random variable without observations is the mean of the random variable (see, e.g., Problem 4.26). Since by assumption the processes all have zero mean, $\widehat{X}_0 = 0$ is the initial estimate.

Before computing x_n, we make several remarks on the estimator and its properties. First, we can rewrite the estimator as

$$\widehat{X}_{n+1} = x_n(Y_n - H_n\widehat{X}_n) + \Phi_n\widehat{X}_n.$$

It is easily seen from the orthogonality principle that if $\widehat{X}_n$ is a LLSE estimate of X_n given $Y_0, Y_1, \ldots, Y_{n-1}$, then $\widehat{Y}_n - H_n\widehat{X}_n = Y_n - \widehat{Y}_n \stackrel{\Delta}{=} \nu_n$ can be interpreted as the "new" information in the observation Y_n in the sense that our best prediction of Y_n based on previously known samples has been removed from Y_n. We can now write

$$\widehat{X}_{n+1} = \kappa_n \nu_n + \Phi_n \widehat{X}_n.$$

This can be interpreted as saying that the new estimate is formed from the old estimate by using the same transformation Φ_n used on the actual samples and then by adding a term depending only on the new information.

It also follows from the orthogonality principle that the sequence ν_n is uncorrelated: since $H_n \widehat{X}_n$ is the LLSE estimate for Y_n based on the previous n samples $Y_0, \ldots, Y_{n-1}$, the error ν_n must be orthogonal to past Y_l from the orthogonality principle. Hence ν_n must also be orthogonal to linear combinations of past Y_l and hence also to past ν_l. It is straightforward to show that $E\nu_n = 0$ for all n and hence orthogonality of the sequence implies that it is also uncorrelated (Problem 5.25). Because of the various properties, the sequence $\{\nu_n\}$ is called the innovations sequence of the observations process. Note the analog with Example [5.6], where the observations were first whitened to form innovations and then the estimate was formed based on the whitened version.

Observe next that the innovations and the estimation error are simply related by the formula

$$\nu_n = Y_n - H_n \widehat{X}_n = H_n X_n + V_n - H_n \widehat{X}_n = H_n (X_n - \widehat{X}_n) + V_n$$

or

$$\nu_n = H_n \epsilon_n + V_n,$$

a useful formula for deriving some of the properties of the filter. For example, we can use this formula to find a recursion for the estimate error:

$$\epsilon_0 = X_0$$

$$\begin{aligned} \epsilon_{n+1} &= X_{n+1} - \widehat{X}_{n+1} \\ &= \Phi_n X_n + U_n - \kappa_n \nu_n - \Phi_n \widehat{X}_n \\ &= (\Phi_n - \kappa_n H_n) \epsilon_n + U_n - \kappa_n V_n; \;\; n = 0, 1. \end{aligned} \tag{5.122}$$

This formula implies that

$$E(\epsilon_n) = 0; \;\; n = 0, 1, \ldots,$$

and hence the estimate is unbiased (i.e., an estimate having an error which has zero mean is defined to be unbiased). It also provides a recursion for finding the expected squared estimation error:

$$E(\epsilon_{n+1}^2) = (\Phi_n - \kappa_n H_n)^2 E(\epsilon_n^2) + E(U_n^2) + \kappa_n^2 E(V_n^2),$$

where we have made use of the assumptions of the problem statement, namely, the uncorrelation of U_n and V_n sequences with each other, with $Y_0, \ldots, Y_{n-1}$, $X_0, \ldots, X_n$, and hence also with ϵ_n. Rearranging terms for later use, we have

$$E(\epsilon_{n+1}^2) = \Phi_n^2 E(\epsilon_n^2) - 2\kappa_n H_n \Phi_n E(\epsilon_n^2) - +\kappa_n^2 (H_n^2 E(\epsilon_n^2) + \Psi_n) + \Gamma_n, \tag{5.123}$$

where $\Gamma_n \stackrel{\Delta}{=} E(U_n^2)$ and $\Psi_n \stackrel{\Delta}{=} E(V_n^2)$.

Since we know that $E(\epsilon_0^2) = E(X_0^2)$, if we knew the κ_n we could recursively evaluate the expected squared errors from the formula and the given problem parameters. We now complete the system design by developing a formula for the κ_n. This is most easily done by using the orthogonality relation and (5.122):

$$0 = E(\epsilon_{n+1} Y_n) = (\Phi_n - \kappa_n H_n) E(\epsilon_n Y_n) + E(U_n Y_n) - \kappa_n E(V_n Y_n).$$

Consider the terms on the right. Proceeding from left to right, the first term involves

$$E(\epsilon_n Y_n) = E(\epsilon_n (H_n X_n + V_n)) = H_n E(\epsilon_n (\epsilon_n + \widehat{X}_n)) = H_n E(\epsilon_n^2),$$

where we have used the fact that ϵ_n is orthogonal to V_n, to $Y_0, \ldots, Y_{n-1}$ and hence to $\widehat{X}_n$. The second term is zero by the assumptions of the problem. The third term requires the evaluation

$$E(V_n Y_n) = E(V_n (H_n X_n + V_n)) = E(V_n^2).$$

Thus we have that

$$0 = (\Phi_n - \kappa_n H_n) H_n E(\epsilon_n^2) - \kappa_n E(V_n^2)$$

or

$$\kappa_n = \frac{\Phi_n H_n E(\epsilon_n^2)}{E(V_n^2) + H_n^2 E(\epsilon_n^2)}.$$

Thus for each n we can solve the recursion for $E(\epsilon_n^2)$ and for the required κ_n to form the next estimate.

We can now combine all of the foregoing mess to produce the final answer. A recursive estimator for the given model is

$$\widehat{X}_0 = 0; \quad E(\epsilon_0^2) = E(X_0^2), \tag{5.124}$$

and for $n = 0, 1, 2, \ldots,$

$$\widehat{X}_{n+1} = \kappa_n (Y_n - H_n \widehat{X}_n) + \Phi_n \widehat{X}_n; \quad n = 0, 1, \ldots, \tag{5.125}$$

where

$$\kappa_n = \frac{\Phi_n H_n E(\epsilon_n^2)}{\Psi_n + H_n^2 E(\epsilon_n^2)} \tag{5.126}$$

and where, from (5.123) and (5.126),

$$E(\epsilon_{n+1}^2) = \Phi_n E(\epsilon_n^2) + \Gamma_n - \frac{(\Phi_n H_n E(\epsilon_n^2))^2}{\Psi_n + H_n^2 E(\epsilon_n^2)}. \tag{5.127}$$

Although these equations seem messy, they can be implemented numerically or in hardware in a straightforward manner. Variations of their matrix generalizations are also well suited to fast implementation. Such algorithms in greater generality are a prime focus of the areas of estimation, detection, signal identification, and signal processing.

5.9.2 Continuous time

Suppose that a real-valued weakly stationary continuous time random process $\{Y(t);\, t \in \Re\}$ is the input to a linear system. This can be considered as an observed process which might, for example, be a desired signal plus noise,

$$Y(t) = X(t) + N(t),$$

where $X(t)$ is an information-bearing term we wish to recover and $N(t)$ is unwanted noise, which is generally assumed to be independent of the signal. Ideally we would like the output of our linear system to be a process $Z(t)$, perhaps the uncorrupted signal $X(t)$ or a delayed version thereof. Usually it is not possible to achieve the ideal signal since there is no general way to remove the added noise without damaging the signal. The practical compromise is to consider the output of the linear system as an approximation $\hat{Z}(t)$ to the ideal $Z(t)$:

$$\hat{Z}(t) = \int_{\infty}^{\infty} h(\tau) Y(t-\tau)\, d\tau, \tag{5.128}$$

where in this discussion we limit consideration to time-invariant filters. The design goal is to make the estimate $\hat{Z}(t)$ look as much like $Z(t)$ as possible. A typical criterion of goodness is to find h to minimize the mean squared error

$$E[|Z(t) - \hat{Z}(t)|^2].$$

A filter h is considered *optimal* within some class of filters if it yields the smallest mean squared error over all filters of the class. The most common classes of filters of interest are as follows

- Causal filters where

$$\hat{Z}(t) = \int_0^\infty h(\tau)Y(t-\tau)\,d\tau,$$

  Here only past data can be considered in the estimate.
- Filters that can operate on the past and a finite portion of the future, i.e., they have a physical delay T before producing an output. Here

$$\hat{Z}(t) = \int_{-T}^\infty h(\tau)Y(t-\tau)\,d\tau.$$

- Filters that are allowed to operate only on a finite amount of past data,

$$\hat{Z}(t) = \int_0^T h(\tau)Y(t-\tau)\,d\tau.$$

- Noncausal filters that can operate on the entire input signal.

In all cases we can write

$$\hat{Z}(t) = \int_{\mathcal{T}} h(\tau)Y(t-\tau)\,d\tau.$$

where $\mathcal{T}$ is a finite, semi-infinite, or infinite interval.

We will find the basic integral equation which must be solved in terms of a general $\mathcal{T}$. Unfortunately the solutions range from simple to extremely complicated depending on which of the above classes of filters is considered. We here consider only a few of the simplest cases.

In general we wish to minimize the mean squared error using a filter h:

$$\begin{aligned}\mathcal{E}_h(t) &\triangleq E[|Z(t) - \hat{Z}(t)|^2] \\ &= E[|Z(t|^2] - 2E[Z(t)\hat{Z}] + E[|\hat{Z}(t)|^2] \\ &= R_Z(t,t) - 2E\left[Z(t)\int_{\mathcal{T}} h(\tau)Y(t-\tau)\,d\tau\right] \\ &\quad +E[\int_{\mathcal{T}} h(\tau)Y(t-\tau)\,d\tau \int_{\mathcal{T}} h(s)Y(t-s)\,ds] \\ &= R_Z(t,t) - 2\int_{\mathcal{T}} h(\tau)E[Z(t)Y(t-\tau)]\,d\tau \\ &\quad + \int_{\mathcal{T}}\int_{\mathcal{T}} h(\tau)E[Y(t-\tau)Y(t-s)]h(s)\,d\tau\,ds \\ &= R_Z(t,t) - 2\int_{\mathcal{T}} h(\tau)R_{ZY}(t,t-\tau)\,d\tau \\ &\quad + \int_{\mathcal{T}}\int_{\mathcal{T}} h(\tau)R_Y(t-\tau,t-s)h(s)\,d\tau\,ds.\end{aligned}$$

If we assume that the random processes $\{Y(t)\}$ and $Z(t)$ are jointly weakly stationary, then this becomes

$$\mathcal{E}_h(t) = \\ R_Z(0) - 2\int_{\mathcal{T}} h(\tau)R_{ZY}(\tau)\,d\tau + \int_{\mathcal{T}}\int_{\mathcal{T}} h(\tau)R_Y(s-\tau)h(s)\,d\tau\,ds, \quad (5.129)$$

an integral equation to be minimized for the optimal h.

Analogous to the discrete time case, we shall see that the optimal filter h is that which satisfies the orgonality condition that the observations are orthogonal to the error:

$$E[(Z(t) - \hat{Z}(t))Y(t-\tau)] = 0 \text{ for all } \tau \in \mathcal{T}, \quad (5.130)$$

which expands into

$$\begin{aligned}&E\left[(Z(t) - \int_{\mathcal{T}} h(s)Y(t-s)\,ds)Y(t-\tau)\right] \\ &= E[Z(t)Y(t-\tau)] - \int_{\mathcal{T}} h(s)E[Y(t-s)\,ds)Y(t-\tau)] \\ &= R_{ZY}(\tau) - \int_{\mathcal{T}} h(s)R_Y(\tau-s)\,ds\end{aligned}$$

so that (5.130) becomes

$$R_{ZY}(\tau) - \int_{\mathcal{T}} h(s)R_Y(\tau-s)\,ds = 0 \text{ for all } \tau \in \mathcal{T}, \quad (5.131)$$

the continuous time version of the Wiener–Hopf equation. To prove that an h satisfying the Wiener–Hopf equation minimizes $\mathcal{E}_h(t)$, suppose that g is any other filter with the resulting estimate

$$\tilde{Z}(t) = -\int_{\mathcal{T}} g(\tau)Y(t-\tau)\,d\tau$$

and consider the resulting mean squared error

$$\begin{aligned}
\mathcal{E}_g(t) &= E[(Z(t)-\tilde{Z}(t))^2] \\
&= E[(Z(t)-\hat{Z}(t)+\hat{Z}(t)-\tilde{Z}(t))^2] \\
&= E[(Z(t)-\hat{Z}(t))^2] + 2E[(Z(t)-\hat{Z}(t))(\hat{Z}(t)-\tilde{Z}(t))] \\
&\quad + E[(\hat{Z}(t)-\tilde{Z}(t))^2].
\end{aligned}$$

The cross-product is

$$\begin{aligned}
&E[(Z(t)-\hat{Z}(t))(\hat{Z}(t)-\tilde{Z}(t))] \\
&= E\left[((Z(t)-\hat{Z}(t))\left(\int_{\mathcal{T}} h(\tau)Y(t-\tau)\,d\tau - \int_{\mathcal{T}} g(\tau)Y(t-\tau)\,d\tau\right)\right] \\
&= E\left[(Z(t)-\hat{Z}(t))\int_{\mathcal{T}} (h(\tau)-g(\tau))Y(t-\tau)\,d\tau\right] \\
&= \int_{\mathcal{T}} (h(\tau)-g(\tau))E[(Z(t)-\hat{Z}(t))Y(t-\tau)]\,d\tau = 0
\end{aligned}$$

so that

$$\mathcal{E}_g = \mathcal{E}_h(t) + E[(\hat{Z}(t)-\tilde{Z}(t))^2] \geq \mathcal{E}_h, \tag{5.132}$$

with equality if $g = h$.

As a simple example, if $\mathcal{T} = \Re$, noncausal filters are allowed, in which case the Wiener–Hopf equation becomes

$$R_{ZY}(\tau) - \int_{-\infty}^{\infty} h(s)R_Y(\tau-s)\,ds = 0 \text{ for all } \tau \in \Re$$

which as in the discrete time case is easily solved by taking a Fourier transform to conclude

$$H(f) = \frac{S_{ZY}(f)}{S_Y(f)}, \tag{5.133}$$

where

$$H(f) = \int_{-\infty}^{\infty} h(t)e^{-j2\pi ft}\, dt \tag{5.134}$$

$$S_{ZY}(f) = \int_{-\infty}^{\infty} R_{ZY}(t)e^{-j2\pi ft}\, dt \tag{5.135}$$

$$S_Y(f) = \int_{-\infty}^{\infty} R_Y(t)e^{-j2\pi ft}\, dt. \tag{5.136}$$

The optimal $h(t)$ is then the inverse Fourier transform of H, which will in general be noncausal.

In general solutions are much more difficult, the most important approach being the use of spectral factorization or rational spectra. An excellent reference remains the classical treatment of the subject in Chapter 11 of Davenport and Root [14].

5.10 Problems

1. Suppose that X_n is an iid Gaussian process with mean m and variance σ^2. Let h be the δ-response $h_0 = 1$, $h_1 = r$, and $h_k = 0$ for all other k. Let $\{W_n\}$ be the output process when the X process is put into the filter described by h; that is,

$$W_n = X_n + rX_{n-1}.$$

Assuming that the processes are two-sided – that is, that they are defined for $n \in \mathcal{Z}$ – find EW_n and $R_W(k, j)$. Is $\{W_n\}$ strictly stationary? Next assume that the processes are one-sided; that is, defined for $n \in \mathcal{Z}_+$. Find EW_n and $R_W(k, j)$. For the one-sided case, evaluate the limits of EW_n and $R_W(n, n + k)$ as $n \to \infty$.
2. We define the following two-sided random processes. Let $\{X_n\}$ be an iid random process with marginal pdf $f_X(x) = e^{-x}$, $x \geq 0$. Let $\{Y_n\}$ be another iid random process, independent of the X process, having marginal pdf $f_Y(y) = 2e^{-2y}$, $y \geq 0$. Define a random process $\{U_n\}$ by the difference equation

$$U_n = X_n + X_{n-1} + Y_n.$$

The process U_n can be thought of as the result of passing X_n through a first-order moving-average filter and then adding noise. Find EU_0 and $R_U(k)$.
3. Let $\{X(t)\}$ be a stationary continuous time random process with zero mean and autocorrelation function $R_X(\tau)$. The process $X(t)$ is put into a linear time-invariant stable filter with impulse response $h(t)$ to form a random process $Y(t)$. A random process $U(t)$ is then defined as $U(t) = Y(t)X(t - T)$, where T is a fixed delay. Find $EU(t)$ in terms of R_X, h, and T. Simplify your answer for the case where $S_X(f) = N_0/2$, all f.
4. Find the output power spectral densities in Problems 5.1 and 5.2.

5. A discrete time random process $\{X_n;\ n \in \mathcal{Z}\}$ is iid and Gaussian, with mean 0 and variance 1. It is the input process for a linear time-invariant (LTI) causal filter with δ-responses h defined by

$$h_k = \begin{cases} 1/K & k = 0, 1, \ldots, K-1 \\ 0 & \text{otherwise} \end{cases},$$

so that the output process $\{Y_n\}$ is defined by

$$Y_n = \sum_{k=0}^{K-1} \frac{1}{K} X_{n-k}.$$

This filter (an FIR filter) is often referred to as a *comb filter*.
A third process $\{W_n\}$ is defined by

$$W_n = Y_n - Y_{n-1}.$$

(a) What are the mean and the power spectral density of the process $\{Y_n\}$?
(b) Find the characteristic function $M_{Y_n}(ju)$ and the marginal pdf $f_{Y_n}(y)$.
(c) Find the Kronecker delta response g of an LTI filter for which

$$W_n = \sum_k g_k X_{n-k}.$$

(d) Find the covariance function of $\{W_n\}$.
(e) Do $n^{-1}\sum_{k=0}^{n-1} Y_k$ and $n^{-1}\sum_{k=0}^{n-1} W_k$ converge in probability as $n \to \infty$? If so, to what?

6. Let $\{X_n\}$ be a random process, where X_i is independent of X_j for $i \neq j$. Each random variable X_n in the process is uniform on the region $[-1/2n, +1/2n]$. That is, $f_{X_n}(x_n) = n$, when $x_n \in [-1/2n, +1/2n]$, and 0 otherwise.
Define $\{Y_n\}$ by

$$Y_n = \begin{cases} 0 & n = 0, 1 \\ nX_n - Y_{n-2} & n = 2, 3, 4, \ldots \end{cases}$$

(a) What is the expected value of X_n?
(b) What is the variance of X_n?
(c) What is the covariance function, $K_X(i, j)$?
(d) Let $S_n = n^{-1}\sum_{j=1}^{n} X_j$. What is the expected value of S_n?
(e) Does $\{X_n\}$ have a weak law of large numbers, i.e., does the sample mean $n^{-1}\sum_{k=0}^{n-1} X_n$ converge in probability to the mean $E[X_n]$? If so, to what value does the sample mean converge? If it has no WLLN, explain why not. Make sure to justify your answer based on the definitions of WLLN and convergence in probability.
(f) Find EY_n.
(g) Find $R_Y(i, j)$.

(h) Find the cdf of $Y_4 + Y_6$.

7. Let $\{X(t);\ t \in \Re\}$ be a stationary continuous time Gaussian random process with zero mean and power spectral density function

$$S_X(f) = \begin{cases} \gamma/2 & 0 \leq |f| \leq W \\ 0 & W < |f| < \infty \end{cases}.$$

Let $\{Z(t);\ t \in \Re\}$ be a stationary continuous time Gaussian random process with zero mean and power spectral density function

$$S_Z(f) = \begin{cases} N_0/2 & 0 \leq |f| \leq B \\ 0 & B < |f| < \infty \end{cases}$$

where we assume that $B >> W$, $\gamma > N_0$, and that the two processes are mutually independent. We consider $X(t)$ to be the "signal" and $Z(t)$ to be the "noise." The receiver observes the process $\{Y(t)\}$, where

$$Y(t) = X(t) + Z(t).$$

(a) Find and sketch the power spectral density of $\{Y(t)\}$.

(b) Find the conditional pdf $f_{Y(t)|X(t)}(y|x)$, the marginal pdf $f_{Y(t)}(y)$, and the conditional pdf $f_{X(t)|Y(t)}(x|y)$.

(c) Find the minimum mean squared estimate $\hat{X}(t)$ given the single observation $Y(t)$ and compute the resulting mean squared error

$$\epsilon^2 = E[(\hat{X}(t) - X(t))^2].$$

(d) Suppose that you are allowed to use the entire observed signal $\{Y(t)\}$ to estimate $X(t)$ at a specific time and you can do this by linearly filtering the observed process. Suppose in particular that you pass the observed process $\{Y(t)\}$ through a linear filter with a transfer function

$$H(f) = \begin{cases} 1 & 0 \leq |f| \leq W \\ 0 & W < |f| < \infty \end{cases}$$

with output $\tilde{X}(t)$, an estimate of $X(t)$. (This filter is not causal, but all the results we derived for second-order input/output relations hold for noncausal filters as well and can be used here.)

Find the resulting mean squared error $E[(\tilde{X}(t) - X(t))^2]$.

Which scheme yields smaller average mean squared error?

Hint: convince yourself that linearity implies that $\tilde{X}(t)$ can be expressed as $X(t)$ plus the output of the filter H when the input is $Z(t)$.

8. Let $\{X_n\}$ be an iid Gaussian random process with zero mean and variance $R_X(0) = \sigma^2$. Let $\{U_n\}$ be an iid binary random process, independent of the X process, with $\Pr(U_n = 1) = \Pr(U_n = -1) = 1/2$. (All processes are assumed to be two-sided in this problem.) Define the random process $Z_n = X_n U_n$,

$Y_n = U_n + X_n$, and $W_n = U_0 + X_n$, all n. Find the mean, covariance, and power spectral density of each of these processes. Find the cross-covariance functions between the processes.

9. Let $\{U_n\}, \{X_n\}$, and $\{Y_n\}$ be the same as in Problem 5.8. The process $\{Y_n\}$ can be viewed as a binary signal corrupted by additive Gaussian noise. One possible method of trying to remove the noise at a receiver is to quantize the received Y_n to form an estimate $\widehat{U}_n = q(Y_n)$ of the original binary sample, where

$$q(r) = \begin{cases} +1 & \text{if } r \geq 0 \\ -1 & \text{if } r < 0 \end{cases}.$$

Write an integral expression for the error probability $P_e = \Pr(\widehat{U}_n \neq U_n)$. Find the mean, covariance, and power spectral density of the $\widehat{U}_n$ process. Are the processes $\{U_n\}$ and $\{\widehat{U}_n\}$ equivalent – that is, do they have the same process distributions? Define an error process ϵ_n by $\epsilon_n = 0$ if $\widehat{U}_n = U_n$ and $\epsilon_n = 1$ if $\widehat{U}_n \neq U_n$. Find the marginal pmf, mean, covariance, and power spectral density of the error process.

10. *Cascade filters.* Let $\{g_k\}$ and $\{a_k\}$ be the δ-responses of two discrete time causal linear filters ($g_k = a_k = 0$ for $k < 0$) and let $G(f)$ and $A(f)$ be the corresponding transfer functions, e.g.,

$$G(f) = \sum_{k=0}^{\infty} g_k e^{-j2\pi kf}.$$

Assume that $g_0 = a_0 = 1$. Let $\{Z_n\}$ be a weakly stationary uncorrelated random process with variance σ^2 and zero mean. Consider the cascade of two filters formed by first putting an input Z_n into the filter g to form the process X_n, which is in turn put into the filter a to form the output process Y_n.

(a) Let $\{d_k\}$ denote the δ-response of the overall cascade filter, that is,

$$Y_n = \sum_{k=0}^{\infty} d_k Z_{n-1}.$$

Find an expression for d_k in terms of $\{g_k\}$ and $\{a_k\}$. As a check on your answer you should have $d_0 = g_0 a_0 = 1$.

(b) Let $D(f)$ be the transfer function of the cascade filter. Find $D(f)$ in terms of $G(f)$ and $A(f)$.

(c) Find the power spectral density $S_Y(f)$ in terms of σ^2, G, and A.

(d) Prove that

$$E\left(Y_n^2\right) = \int_{-1/2}^{1/2} S_Y(f) df \geq \sigma^2.$$

Hint: show that if $d_0 = 1$ (from part (a)), then

$$\int_{-1/2}^{1/2} |D(f)|^2 \, df = 1 + \int_{-1/2}^{1/2} |1 - D(f)|^2 \, df \geq 1.$$

11. *One-step prediction.* This problem develops a basic result of estimation theory. No prior knowledge of estimation theory is required. Results from Problem 5.10 may be quoted without proof (even if you did not complete it). Let $\{X_n\}$ be as in problem 5.10; that is, it is a discrete time zero-mean random process with power spectral density $\sigma^2 |G(f)|^2$, where $G(f)$ is the transfer function of a causal filter with δ-response $\{g_k\}$ with $g_0 = 1$. Form the process $\{\widehat{X}_n\}$ by putting X_n into a causal linear time-invariant filter with δ-response h_k

$$\widehat{X}_n = \sum_{k=1}^{\infty} h_k X_{n-k}.$$

Suppose that the linear filter tries to estimate the value of X_n based on the values of X_i for all $i < n$ by choosing the δ-response $\{h_k\}$ optimally. That is, the filter estimates the next sample based on the present value and the entire past. Such a filter is called a *one-step predictor.* Define the error process $\{\epsilon_n\}$ by

$$\epsilon_n = X_n - \widehat{X}_n.$$

(a) Find expressions for the power spectral density $S_\epsilon(f)$ in terms of $S_X(f)$ and $H(f)$. Use this result to evaluate $E\epsilon_n^2$.

(b) Evaluate $S_\epsilon(f)$ and $E(\epsilon_n^2)$ for the case where $1 - H(f) = 1/G(f)$.

(c) Use part (d) of Problem 5.10 to show that the prediction filter $H(f)$ of (b) in this problem yields the smallest possible value of $E(\epsilon_n^2)$ for any prediction filter. You have just developed the optimal one-step prediction filter for the case of a process that can be modeled as a weakly stationary uncorrelated sequence passed through a linear filter. As discussed in the text, most discrete time random processes can be modeled in such a fashion, at least through second-order properties.

(d) *Spectral factorization.* Suppose that $\{X_n\}$ has a power spectral density $S_X(f)$ that satisfies

$$\int_{-1/2}^{1/2} \ln S_X(f) \, df < \infty.$$

Expand $\ln S_X(f)$ in a Fourier series and write the expression for $\exp(\ln S_X(f))$ in terms of the series to find $G(f)$. Find the δ-response of the optimum prediction filter in terms of your result. Find the mean square error. (*Hint:* you will need to know what evenness of $S_X(f)$ implies for the coefficients in the requested series and what the Taylor series of an exponential is.)

12. *Binary filters.* All of the linear filters considered so far were linear in the sense

of real arithmetic. It is sometimes useful to consider filters that are linear in other algebraic systems, e.g., in binary or modulo 2 arithmetic as defined in (3.64)–(3.65). Such systems are more appropriate, for example, when considering communications systems involving only binary arithmetic, such as binary codes for noise immunity on digital communication links. A binary first-order autoregressive filter with input process $\{X_n\}$ and output process $\{Y_n\}$ is defined by the difference equation

$$Y_n = Y_{n-1} \oplus X_n, \text{ all } n.$$

Assume that the $\{X_n\}$ is a Bernoulli process with parameter p. In this case the process $\{Y_n\}$ is called a binary first-order autoregressive source.

(a) Show that for nonnegative integers k, the autocorrelation function of the process $\{Y_n\}$ satisfies

$$R_Y(k) = E(Y_j Y_{j+k}) = \frac{1}{2} \Pr\left(\sum_{i=1}^{k} X_i = \text{an even number}\right).$$

(b) Use the result of (a) to evaluate R_Y and K_Y. *Hint:* this is most easily done using a trick. Define the random variable

$$W_k = \sum_{i=1}^{k} X_i.$$

This is a binomial random variable. Use this fact and the binomial theorem to show that

$$\Pr(W_k \text{ is odd}) - \Pr(W_k \text{ is even}) = -(1-2p)^k.$$

Alternatively, find a linear recursion relation for $p_k = Pr(W_k \text{ is odd})$ using conditional probability (i.e., find a formula giving p_k in terms of p_{k-1}) and then solve for p_k.

(c) Find the power spectral density of the process $\{Y_n\}$.

13. Let $\{X_n\}$ be a Bernoulli random process with parameter p and let $\oplus$ denote mod 2 addition as considered in Problem 5.12. Define the first-order binary moving-average process $\{W_n\}$ by the difference equation

$$W_n = X_n \oplus X_{n-1}.$$

This is a mod 2 convolution and an example of what is called a *convolutional code* in communication and information theory. Find $p_{W_n}(w)$ and $R_W(k, j)$. Find the power spectral density of the process $\{W_n\}$.

14. Let $\{X(t)\}$ be a continuous time zero-mean Gaussian random process with spectral density $S_X(f) = N_0/2$, all f. Let $H(f)$ and $G(f)$ be the transfer functions of two linear time-invariant filters with impulse responses $h(t)$ and $g(t)$, respectively. The process $\{X(t)\}$ is passed through the filter $h(t)$ to obtain a process

$\{Y(t)\}$ and is also passed through the filter $g(t)$ to obtain a process $\{V(t)\}$; that is,

$$Y(t) = \int_0^\infty h(\tau)X(t-\tau)d\tau,$$

$$V(t) = \int_0^\infty g(\tau)X(t-\tau)d\tau.$$

(a) Find the cross-correlation function $R_{Y.V}(t,s) = E(Y_t V_s)$.

(b) Under what assumptions on H and G are Y_t and V_t independent random variables?

15. Let $\{X(t)\}$ and $\{Y(t)\}$ be two continuous time zero-mean stationary Gaussian processes with a common autocorrelation function $R(\tau)$ and common power spectral densities $S(f)$. Assume that $X(t)$ and $Y(t)$ are independent for all t, s. Assume also that $E[X(t)Y(s)] = 0$ all t, s and that $\sigma^2 = R(0)$. For a fixed frequency f_0, define the random process

$$W(t) = X(t)\cos(2\pi f_0 t) + Y(t)\sin(2\pi f_0 t).$$

Find the mean $E(W(t))$ and autocorrelation $R_W(t,s)$. Is $\{W(t)\}$ weakly stationary?

16. Say that we are given an iid binary random process $\{X_n\}$ with alphabet ± 1, each having probability one/2. We form a continuous time random process $\{X(t)\}$ by assigning

$$X(t) = X_n; \;\; t \in [(n-1), nT),$$

for a fixed time T. This process can also be described as follows: let $p(t)$ be a pulse that is 1 for $t \in [0,T)$ and 0 elsewhere. Define

$$X(t) = \sum_k X_k p(t-kT).$$

This is an example of pulse amplitude modulation (PAM). If the process $X(t)$ is then used to phase-modulate a carrier, the resulting process is called a phase-shift-keyed (PSK) modulation of the carrier by the process $\{X(t)\}$. PSK is a popular technique for digital communications. Define the PSK process

$$U(t) = a_0 \cos(2\pi f_0 t + \delta X(t)).$$

Observe that neither of these processes is stationary, but we can force them to be at least weakly stationary by the trick of inserting uniform random variables in appropriate places. Let Z be a random variable, uniformly distributed $[0,T]$ and independent of the original iid process. Define the random process

$$Y(t) = X(t+Z).$$

Let Θ be a random variable uniformly distributed on $[0, 1/f_0]$ and independent of Z and of the original iid random process. Define the process

$$V(t) = U(t + \Theta).$$

Find the mean and autocorrelation functions of the processes $\{Y(t)\}$ and $\{V(t)\}$.

17. Let $\{X(t)\}$ be a Gaussian random process with zero mean and autocorrelation function

$$R_X(\tau) = \frac{N_0}{2} e^{-|\tau|}.$$

Find the power spectral density of the process. Let $Y(t)$ be the process formed by DSB-SC modulation of $X(t)$. Letting Θ be uniformly distributed in (5.37), sketch the lower spectral density of the modulated process.

18. A continuous time two-sided weakly stationary Gaussian random process $\{S(t)\}$ with zero mean and power spectral density $S_S(f)$ is put into a noisy communication channel. First, white Gaussian noise $\{W(t)\}$ with power spectral density $N_0/2$ is added, where the two random processes are assumed to be independent of one another, and then the sum $S(t) + W(t)$ is passed through a linear filter with impulse response $h(t)$ and transfer function $H(f)$ to form a received process $\{Y(t)\}$. Find an expression for the power spectral density $S_Y(f)$. Find an expression for the expected square error $E[(S(t) - Y(t))^2]$ and the so-called signal-to-noise ratio (SNR)

$$\frac{E(S(t)^2)}{E[(S(t) - Y(t))^2]}.$$

Suppose that you know that $S_Y(f)$ can be factored into the form $|G(f)|^2$, where $G(f)$ is a stable causal filter with a stable causal inverse. What is the best choice of $H(f)$ in the sense of maximizing the signal-to-noise ratio? What is the best causal $H(f)$?

19. Show that (5.2) converges in mean square if the filter is stable and the input process has finitely bounded mean and variance. Show that convergence with probability one is achieved if the convergence of (A.43) is fast enough for the δ-response.

20. Show that the sum of equation (5.7) converges for the two-sided weakly stationary case if the filter is stable and the input process has finitely bounded variance.

21. Equation (5.79) shows how the power spectral density is affected by differentiation of a process. Provide a formal argument for the integration counterpart of this result; that is, if $\{X(t)\}$ is a stationary two-sided continuous time random process and $Y(t) = \int_{-\infty}^{t} X(s)\, dx$, then, subject to suitable technical conditions, $S_Y(f) = S_X(f)/f^2$.

22. Prove that (5.115) holds under the conditions given.

23. Suppose that $\{Y_N\}$ is as in Example [5.1] and that $W_n = Y_n + U_n$, where U_n is a zero-mean white noise process with second moment $E(U^2) = N_0/2$. Solve the

Wiener–Hopf equation to obtain a LLSE of Y_{n+m} given $\{W_i;\ i \leq n\}$ for $m > n$. Evaluate the resulting mean squared error.

24. Prove the claim that if $\{X_n\}$ and $\{Y_n\}$ are described by (5.116) and (5.117) and if $\widehat{X}_n$ is a LLSE estimate of X_n given the previous n samples $Y_0, Y_1, \ldots, Y_{n-1}$, then $\widehat{Y}_n = H_n \widehat{X}_n$ is a LLSE estimate of Y_n given the same observations.
25. Prove the claim that the innovations sequence $\{\nu_n\}$ of Example [5.7] is uncorrelated and has zero mean. (Fill in the details of the arguments used in the text.)
26. Let $\{Y_N\}$ be as in Example [5.2]. Find the LLSE for Y_{n+m} given $\{Y_0, Y_1, \ldots, Y_n\}$ for an arbitrary positive integer m. Evaluate the mean square error. Repeat for the process of Example [5.3] (the same process with $r = 1$).
27. Specialize the recursive estimator formulas of (5.124) through (5.127) to the case where $\{X_n\}$ is the $\{Y_n\}$ process of Example [5.2], where H_n is a constant, say a, and where $\Psi_n - N_0/2$, all n. Describe the behavior of the estimator as $n \to \infty$.
28. Find an expression for the mean square error in Example [5.4]. Specialize to infinite smoothing.
29. In the section on linear estimation we assumed that all processes had zero-mean functions. In this problem we remove this assumption. Let $\{X_n\}$ and $\{Y_n\}$ be random processes with mean functions $\{m_X(n)\}$ and $\{m_Y(n)\}$, respectively. We estimate X_n by adding a constant to (5.109); i.e.,

$$\widehat{X}_n = a_n + \sum_{k:n-k \in \mathcal{K}} h_k Y_{n-k}.$$

 (a) Show that the minimum mean square estimate of X_n is $\widehat{X}_n = m_X(n)$ if no observations are used.
 (b) Modify and prove Theorem 5.2 to allow for the nonzero means.
30. Suppose that $\{X_n\}$ and $\{Z_n\}$ are zero-mean, mutually independent, iid, two-sided Gaussian random processes with correlations

$$R_X(k) = \sigma_x^2 \delta_k; \quad R_Z(k) = \sigma_z^2 \delta_k;$$

These processes are used to construct new processes as follows:

$$\begin{aligned} Y_n &= Z_n + rY_{n-1} \\ U_n &= X_n + Z_n \\ W_n &= U_n + rU_{n-1}. \end{aligned}$$

Find the covariance and power spectral densities of $\{U_n\}$ and $\{W_n\}$. Find $E[(X_n - W_n)^2]$.
31. Suppose that $\{Z_n\}$ and $\{W_n\}$ are two mutually independent two-sided zero-mean iid Gaussian processes with variances σ_Z^2 and σ_W^2, respectively. Z_n is put into a linear time-invariant filter to form an output process $\{X_n\}$ defined by

$$X_n = Z_n - rZ_{n-1},$$

where $0 < r < 1$. (Such a filter is sometimes called a *pre-emphasis filter* in speech processing.) This process is then used to form a new process

$$Y_n = X_n + W_n,$$

which can be viewed as a noisy version of the pre-emphasized Z_n process. Lastly, the Y_n process is put through a "de-emphasis filter" to form an output process U_n defined by

$$U_n = rU_{n-1} + Y_n.$$

(a) Find the autocorrelation R_Z and the power spectral density S_Z. Recall that for a weakly stationary discrete time process with zero mean $R_Z(k) = E(Z_n Z_{n+k})$ and

$$S_Z(f) = \sum_{k=-\infty}^{\infty} R_Z(k) e^{-j2\pi fk},$$

the discrete time Fourier transform of R_Z.

(b) Find the autocorrelation R_X and the power spectral density S_X.

(c) Find the autocorrelation R_Y and the power spectral density S_Y.

(d) Find the overall mean squared error $E[(U_n - Z_n)^2]$.

32. Suppose that $\{X_n;\ n \in \mathcal{Z}\}$ is a discrete time iid Gaussian random processes with zero mean and variance $\sigma_X^2 = E[X_0^2]$. We consider this an input signal to a signal processing system. Suppose also that $\{W_n;\ n \in \mathcal{Z}\}$ is a discrete time iid Gaussian random processes with zero mean and variance σ_W^2 and that the two processes are mutually independent. We consider W_n to be noise. Suppose that X_n is put into a linear filter with unit δ-response h, where

$$h_k = \begin{cases} 1 & k = 0 \\ -1 & k = -1 \\ 0 & \text{otherwise} \end{cases}$$

to form an output $U = X * h$, the convolution of the input signal and the unit δ-response. The final output signal is then formed by adding the noise to the filtered input signal, $Y_n = U_n + W_n$.

(a) Find the mean, power spectral density, and marginal pdf for U_n.

(b) Find the mean, covariance, and power spectral density for $\{Y_n\}$.

(c) Find $E[Y_n X_n]$.

(d) Does the mean ergodic theorem hold for $\{Y_n\}$?

33. Suppose that $\{X(t);\ t \in \mathcal{R}\}$ is a weakly stationary continuous time Gaussian random process with zero mean and autocorrelation function

$$R_X(\tau) = E[X(t)X(t+\tau)] = \sigma_X^2 e^{-|\tau|}.$$

(a) Define the random process $\{Y(t);\ t \in \mathcal{R}\}$ by

$$Y(t) = \int_{t-T}^{t} X(\alpha)\, d\alpha,$$

where $T > 0$ is a fixed parameter. (This is a short-term integrator.) Find the mean and power spectral density of $\{Y(t)\}$.

(b) For fixed $t > s$, find the characteristic function and the pdf for the random variable $X(t) - X(s)$.

(c) Consider the following *nonlinear* modulation scheme: define

$$W(t) = e^{j(2\pi f_0 t + cX(t) + \Theta)},$$

where f_0 is a fixed frequency, Θ is a uniform random variable on $[0, 2\pi]$, Θ is independent of all of the $X(t)$, and c is a modulation constant. (This is a mathematical model for phase modulation.)
Define the expectation of a complex-valued random variable in the natural way, that is, if $Z = \Re(Z) + j\Im(Z)$, then $E(Z) = E[\Re(Z)] + jE[\Im(Z)]$.)
Define the autocorrelation of a complex-valued random process $W(t)$ by

$$R_W(t, s) = E(W(t)W(s)^*),$$

where $W(s)^*$ denotes the complex conjugate of $W(s)$.
Find the mean $E(W(t))$ and the autocorrelation function $R_W(t, s) = E[W(t)W(s)^*]$.
Hint: the autocorrelation is admittedly a trick question (but a very useful trick). Keep part (b) in mind and think about characteristic functions.

34. A random variable X is described by a pmf

$$p_X(k) = \begin{cases} ca^k & k = 0, 1, \dots \\ 0 & \text{else} \end{cases} \tag{5.137}$$

where $0 < a < 1$. A random variable Z is described by a pmf

$$p_Z(k) = \frac{1}{2},\ k = \pm 1. \tag{5.138}$$

(a) Find the mean, variance and characteristic function of Z.
(b) Evaluate c and find the mean, variance, and characteristic function of X.
(c) Now suppose that $\{X_n\}$ and $\{Z_n\}$ are two mutually independent iid random processes with marginal pmf's p_X of (5.137) and p_Z of (5.138), respectively. Form a new random process Y_n defined by

$$Y_n = X_n Z_n \quad \text{all } n. \tag{5.139}$$

Find the mean and covariance function for $\{Y_n\}$. Is $\{Y_n\}$ weakly stationary? If so, find its power spectral density.
(d) Find the marginal pmf p_{Y_n}.
(e) Find the probability $\Pr(X_n \geq 2X_{n-1})$.

(f) Find the conditional expectations $E[Y_n|Z_n]$ and $E[X_n|Y_n]$.
(g) Find the probability $\Pr(Y_n \geq 2Y_{n-1})$.

35. Suppose that $\{X_n\}$ is an iid random process with marginal pdf

$$f_X(x) = \begin{cases} \lambda e^{-\lambda x} & x \geq 0 \\ 0 & \text{otherwise} \end{cases}.$$

Let N be a fixed positive integer.
(a) Find the probability that at least one of the samples $X_0, \ldots, X_{N-1}$ exceeds a fixed positive value γ.
(b) What is the probability that all of the samples $X_0, \ldots, X_{N-1}$ exceed a fixed positive value γ?
(c) Define a new process $U_n = ZX_n$, where Z is a binary random variable with the marginal pmf of (5.138) and and the Z is independent of all the X_n. Find the mean $\overline{U_n}$ and covariance K_U of $\{U_n\}$.
Is U_n weakly stationary? Is it iid?
(d) Does the sample average

$$S_n = \frac{1}{n} \sum_{k=0}^{n-1} U_k$$

converge in probability? If yes, to what?
(e) Find a *simple* nontrivial numerical upper bound to the probability

$$\Pr(|U_n - \overline{U}| > 10\sigma_U),$$

where σ_U^2 is the variance of U_0.

36. Suppose that $\{X_n\}$ is a weakly stationary random process with zero mean and autocorrelation $R_X(k) = \sigma^2 \alpha^{|k|}$ for all integer k, where $|\alpha| < 1$. A new random process $\{Y_n\}$ is defined by the relation $Y_n = X_n + \beta X_{n-1}$.
(a) Find the autocorrelation function $R_Y(k)$ and the average power $E[Y_k^2]$.
(b) For what value of β is $\{Y_n\}$ a white noise process? In other words, what is the value of β for which $S_Y(f)$ is a constant? This is an example of a *whitening filter.*
(c) Suppose that β is chosen as in the previous part so that Y_n is white. (You do not need the actual value of β for this; you can leave things in terms of β if you did not do the previous part.) Assume also that $\{X_n\}$ is a Gaussian random process. Find the variance and the pdf of the random variable

$$S_N = \frac{1}{N} \sum_{i=0}^{N-1} Y_i,$$

where N is a fixed positive integer.

37. Suppose that $\{X_n;\ n \in \mathcal{Z}\}$ is a Bernoulli random process with parameter p, i.e., it is an iid binary process with $p_X(1) = 1 - p_X(0) = p$. Suppose that Z is

a binary random variable with the pmf of (5.138) and that Z and the X_n are independent of each other. Define for integers $n > k \geq 0$ the random variables

$$W_{k.n} = \sum_{i=k+1}^{n} X_i.$$

Define a one-sided random process $\{Y_n;\ n = 0, 1, \ldots\}$ as follows:

$$Y_n = \begin{cases} Z & n = 0 \\ Y_{n-1}(-1)^{X_n} & n = 1, 2, \ldots \end{cases}.$$

Note that for any $n > k \geq 0$,

$$Y_n = Y_k(-1)^{W_{k.n}}. \tag{5.140}$$

(a) Find the mean $m_Y = E[Y_n]$. Show that $p_{Y_n}(1)$ can be expressed as a very simple function of m_Y and use this fact to evaluate $p_{Y_n}(y)$ for any nonnegative integer n.
(b) Find the mean, variance, and characteristic function of $W_{k.n}$.
(c) If you fix a positive integer k, do the random variables

$$\frac{W_{k.n}}{n-k}$$

converge in mean square as $n \to \infty$? If so, to what?
(d) Write an expression for the conditional pmf $p_{Y_n|Y_k}(l|m)$ for $n > k \geq 0$ in terms of the random variable $W_{k.n}$. Evaluate this probability.
Hint: the key issue of this problem is to get the general expression, i.e., a sum with correct limits and summand, correct. The actual evaluation is a bit tricky, so do not waste time on it if you do not see the trick.
(e) Find the covariance function $K_Y(k, n)$ of $\{Y_n\}$.
Hint: one way (not the only way) to do this part is to consider the case $n > k \geq 0$, use (5.140) and and the fact

$$-1 = e^{j\pi}, \tag{5.141}$$

and try to make your formula look like the characteristic function for $W_{k.n}$.

Problems 38–41 courtesy of the EE Department of the Technion.

38. Consider a process $\{Y_t;\ t \in \Re\}$ that can take on only the values $\{-1, +1\}$ and suppose that

$$p_{Y_t}(+1) = p_{Y_t}(-1) = 0.5$$

for all t. Suppose also that for $\tau > 0$

$$p_{Y_{t+\tau}|Y_t}(1|-1) = p_{Y_{t+\tau}|Y_t}(-1|+1) = \begin{cases} \tau/(2T) & \tau \leq T \\ 1/2 & \tau > T \end{cases}.$$

(a) Find the autocorrelation function R_Y of the process $\{Y_t;\ t \in \Re\}$.
(b) Find the power spectral density $S_Y(f)$.

39. A known deterministic signal $\{s(t);\ t \in \Re\}$ is transmitted over a noisy channel and the received signal is $\{X(t);\ t \in \Re\}$, where $X(t) = As(t) + W(t)$, where $\{W(t);\ t \in \Re\}$ is a Gaussian white noise process with power spectral density $S_W(f) = N_0/2;\ f \in \Re$ and A is a random variable independent of $W(t)$ for all t. The receiver, which is assumed to know the transmitted signal, computes the statistic $Y_T = \int_0^T X(t)\, dt$.
(a) Find the conditional pdf $f_{Y_T|A}(y|a)$.
(b) Assuming that A is $\mathcal{N}(0, \sigma_A^2)$, find the MMSE estimate of A given y_T.
(c) Find the MMSE resulting in the previous part.

40. Suppose that $\{Y(t);\ t \in \Re\}$ is a weakly stationary random process with zero mean and autocorrelation function $R_Y(\tau)$ and that A is a random variable that is independent of $Y(t)$ for all t. Define the random process $\{X(t);\ t \in \Re\}$ by $X(t) = A + Y(t)$. Consider the estimator for A defined by

$$\hat{A} = \frac{1}{T} \int_0^T X(t)\, dt. \tag{5.142}$$

(a) Show that $E(\hat{A}) = E(A)$.
(b) Show that the mean squared error is given by

$$\begin{aligned} E[(\hat{A} - A)^2] &= (\frac{1}{T})^2 \int_0^T 2 \int_0^T R_Y(t-s)\, dt\, ds \\ &= \frac{2}{T} \int_0^T (1 - \frac{\tau}{T}) R_Y(\tau)\, d\tau. \end{aligned}$$

41. Let $X(t) = S(t) + N(t)$ where $S(t)$ is a deterministic signal that is 0 outside the interval $[-T, 0]$ and $N(t)$ is white noise with zero mean and power spectral density $N_0/2$. The random process $X(t)$ is passed through a linear filter with impulse response $h(t) = S(-t)$, a time-reversed version of the signal. Let $Y(t)$ denote the filter output process.
(a) Find $E[Y(t)]$.
(b) Find the covariance $K_Y(t, t+\tau)$.
(c) Express the covariance function in terms of the mean function.

6

A menagerie of processes

The basic tools for describing and analyzing random processes have all been developed in the preceding chapters along with a variety of examples of random processes with and without memory. The goal of this chapter is to use these tools to describe a menagerie of useful random processes, usually by taking a simple random process and applying some form of signal processing such as linear filtering in order to produce a more complicated random process. In Chapter 5 the effect of linear filtering on second-order moments was considered. In this chapter we look in more detail at the resulting output process and consider other forms of signal processing as well. In the course of the development a few new tools and several variations on old tools for deriving distributions are introduced. Much of this chapter can be considered as practice of the methods developed in the previous chapters, with names often being given to the specific examples developed. In fact several processes with memory have been encountered previously: the binomial counting process and the discrete time Wiener process, in particular. The goal now is to extend the techniques used in these special cases to more general situations and to introduce a wider variety of processes.

The development of examples begins with a continuation of the study of the output processes of linear systems with random process inputs. The goal is to develop the detailed structure of such processes and of other processes with similar behavior that cannot be described by a linear system model. In Chapter 5, we confined interest to second-order properties of the output random process, properties that can be found under quite general assumptions about the input process and filter. In order to get more detailed probabilistic descriptions of the output process, in this chapter we restrict the input process for the discrete time case to be an iid random process and study the resulting output process and the continuous time analog to such a process. By restricting the structure of the output process in this manner, we shall see that in some cases we can find complete descriptions of the process and not just the first and second moments. The random processes obtained in this way provide many important and useful models that are frequently encountered in the signal processing literature, including moving-average, autoregressive, autoregressive moving-average (ARMA), independent increment, counting, random walk, Markov, Wiener, Poisson, and Gaussian processes. Similar techniques are used for the development of a variety of random processes with markedly different behavior, the key tools being characteristic functions and conditional probability distributions. This chapter contains extensive practice in derived distributions and in specifying random processes.

6.1 Discrete time linear models

Many complicated random processes are well modeled as a linear operation on a simple process. For example, a complicated process with memory might be constructed by passing a simple iid process through a linear filter. In this section we define some general linear models that will be explored in some detail in the rest of the chapter.

Recall that if we have a random process $\{X_n;\, n \in \mathcal{T}\}$ as input to a linear system described by a convolution, then as in (5.2) there is a δ-response h_k such that the output process $\{Y_n\}$ is given by

$$Y_n = \sum_{k:\, n-k \in \mathcal{T}} X_{n-k} h_k. \tag{6.1}$$

A linear filter with such a description – that is, one that can be defined as a convolution – is sometimes called a *moving-average filter* since the output is a weighted running average of the inputs. If only a finite number of the h_k are not zero, then the filter is called a finite-order moving-average filter (or an FIR filter, for "finite impulse response," in contrast to an IIR filter or "infinite impulse response.") The order of the filter is equal to the maximum minus the minimum value of k for which the h_k are nonzero. For example, if $Y_n = X_n + X_{n-1}$, we have a first-order moving-average filter. Although some authors reserve the term *moving-average filter* for a finite-order filter, we will use the broader definition. A block diagram for such a filter is given in Figure 6.1.

Several other names are used to describe finite-order moving-average filters. Since the output is determined by the inputs without any feedback from past or future outputs, the filter is sometimes called a feedforward or tapped delay line or transversal filter. If the filter has a well-defined transfer

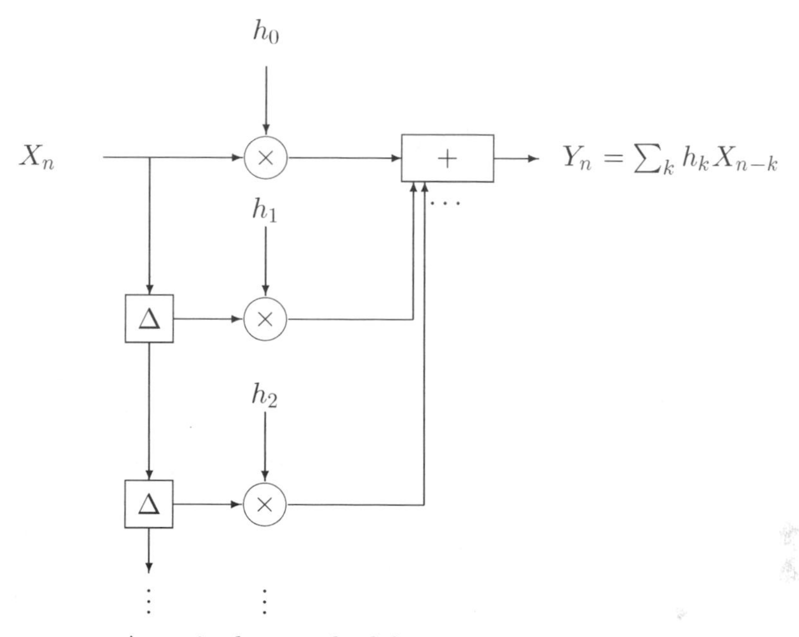

Figure 6.1 Moving average filter

function $H(f)$ (e.g., it is stable) and if the transfer function is analytically continued to the complex plane by making the substitution $z = e^{j2\pi f}$, then the resulting complex function contains only zeros and no poles on the unit circle in the complex plane. For this reason such a filter is sometimes called an "all-zeros" filter. This nomenclature really only applies to the Fourier transform or to the z-transform confined to the unit circle. If one considers arbitrary z, then the filter can have zeros at $z = 0$.

In Chapter 5 we considered only linear systems involving moving-average filters, that is, systems that could be represented as a convolution. This was because the convolution representation is well suited to second-order I/O relations. In this chapter, however, we will find that other representations are often more useful. Recall that a convolution is simply one example of a difference equation. Another form of difference equation describing a linear system is obtained by convolving the outputs to get the inputs instead of vice versa. For example, the output process may satisfy a difference equation of the form

$$X_n = \sum_k a_k Y_{n-k}. \tag{6.2}$$

For convenience it is usually assumed that $a_0 = 1$ and $a_k = 0$ for negative k and hence that the equation can be expressed as

$$Y_n = X_n - \sum_{k=1}^{\infty} a_k Y_{n-k}. \tag{6.3}$$

As in the moving-average case, the limits of the sum depend on the index set; e.g., the sum could be from $k = -\infty$ to ∞ in the two-sided case with $\mathcal{T} = \mathcal{Z}$ or from $k = -\infty$ to n in the one-sided case with $\mathcal{T} = \mathcal{Z}_+$.

The numbers $\{a_k\}$ are called *regression coefficients*, and the corresponding filter is called an *autoregressive filter*. If $a_k \neq 0$ for only a finite number of k, the filter is said to be finite-order autoregressive. The order is equal to the maximum minus the minimum value of k for which a_k is nonzero. For example, if $X_n = Y_n + Y_{n-1}$, we have a first-order regressive filter. As with the moving-average filters, for some authors the "finite" is implicit, but we will use the more general definition. A block diagram for such a filter is given in Figure 6.2.

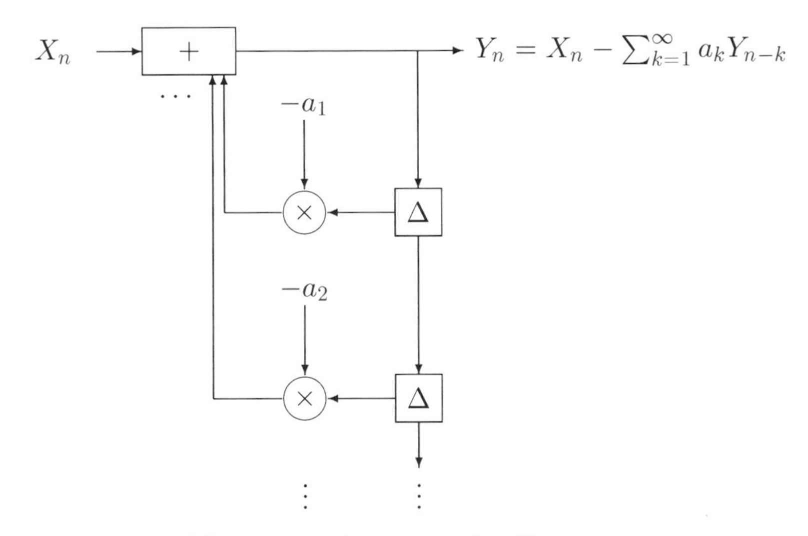

Figure 6.2 Autoregressive filter

Note that, in contrast with a finite-order moving-average filter, a finite-order autoregressive filter contains only feedback terms and no feedforward terms – the new output can be found solely from the current input and past of future outputs. Hence it is sometimes called a feedback filter. If we consider a deterministic input and transform both sides of (6.2), then we find that the transfer function of an autoregressive filter has the form

$$H(f) = \frac{1}{\sum_k a_k e^{-j2\pi kf}},$$

where we continue to assume that $a_0 = 1$. Note that the analytic continuation of the transfer function into the complex plane with the substitution $z = e^{j2\pi f}$ for a finite-order autoregressive filter has poles but no zeros on the unit circle in the complex plane. Hence a finite-order autoregressive filter is sometimes called an all-poles filter. An autoregressive filter may or may not be stable, depending on the location of the poles.

More generally, one can describe a linear system by a general difference equation combining the two forms – moving-average and autoregressive – as in (A.47):

$$\sum_k a_k y_{n-k} = \sum_i b_i x_{n-i}.$$

Filters with this description are called ARMA(for "autoregressive moving-average filters. ARMA filters are said to be finite-order if only a finite number of the a_k's and b_k's are not zero. A finite-order ARMA filter is depicted in Figure 6.3.

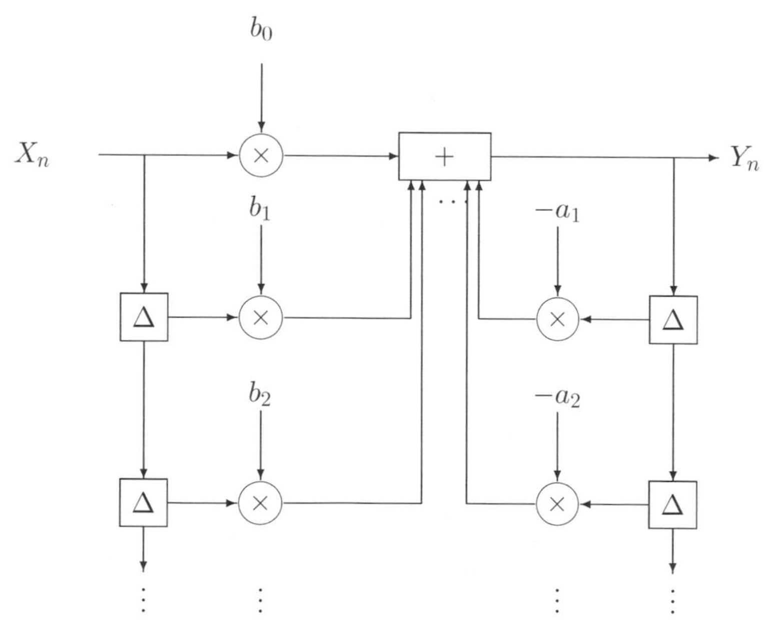

Figure 6.3 Moving-average filter

Once again, it should be noted that some authors use finite-order implicitly, a convention that we will not adopt. Applying a deterministic input and using (A.45), we find that the transfer function of an ARMA filter has the form

$$H(f) = \frac{\sum_i b_i e^{-j2\pi i f}}{\sum_k a_k e^{-j2\pi k f}}, \tag{6.4}$$

where we continue to assume that $a_0 = 1$.

As we shall see by example, one can often describe a linear system by any of these filters, and hence one often chooses the simplest model for the desired application. For example, an ARMA filter representation with only three nonzero a_k and two nonzero b_k would be simpler than either a pure autoregressive or pure moving-average representation, which would in general require an infinite number of parameters. The general development of representations of one type of filter or process from another is an area of complex analysis that is outside the scope of this book. We shall, however, see some simple examples where different representations are easily found.

We are now ready to introduce three classes of random processes that are collectively called *linear models* since they are formed by putting an iid process into a linear system.

A discrete time random process $\{Y_n\}$ is called an *autoregressive random process* if it is formed by putting an iid random process into an autoregressive filter. Similarly, the process is said to be a *moving-average random process* or *ARMA random process* if it is formed by putting an iid process into a moving-average or ARMA filter, respectively. If a finite-order filter is used, the *order* of the process is the same as the order of the filter.

Since iid processes are uncorrelated, the techniques of Chapter 5 immediately yield the power spectral densities of these processes in the two-sided weakly stationary case and yield in general the second-order moments of moving-average processes. In fact, some books and papers which deal only with second-order moment properties define an autoregressive (moving-average, ARMA) process more generally as the output of an autoregressive (moving-average, ARMA) filter with a weakly stationary uncorrelated input. We use the stricter definition in order to derive actual distributions in addition to second-order moments. We shall see that we can easily find marginal probability distributions for moving-average processes. Perhaps surprisingly, however, the autoregressive models will prove to be much more useful for finding more complete specifications, that is, joint probability distributions for the output process. The basic ideas are most easily demonstrated in the simple, and familiar, Example [5.3], summing successive outputs of an iid process.

6.2 Sums of iid random variables

We begin by recalling a simple but important example from Chapters 3 and 5 considered in examples [3.35], [3.37], and [5.3]. These examples can be used to exemplify both autoregressive and moving-average filters. Let $\{X_n;\ n = 1, 2, \ldots\}$ be an iid process with mean m and variance σ^2 (with discrete or continuous alphabet). Consider a linear filter with Kronecker δ-response

$$h_k = \begin{cases} 1 & k = 0, 1, 2 \ldots \\ 0 & \text{otherwise} \end{cases}. \tag{6.5}$$

This is the discrete time integrator and it is not stable. The output process is then given as the sum of iid random variables:

$$Y_n = \begin{cases} 0 & n = 0 \\ \sum_{i=1}^{n} X_i & n = 1, 2, \ldots \end{cases}. \tag{6.6}$$

The two best-known members of this class are the binomial counting process and the Wiener process (or discrete time diffusion process), which were encountered in Chapter 3

We have changed notation slightly from Example [5.3] since here we force $Y_0 = 0$. Observe that if we further let $X_0 = 0$, then by definition $\{Y_n;\ n \in \mathcal{Z}_+\}$ is a moving-average random process by construction with the moving-average filter $h_k = 1$ for all nonnegative k.

Since an iid input process is also uncorrelated, we can apply example [5.3] (with a slight change due to the different indexing) and evaluate the first and second moments of the Y process as

$$EY_n = mn;\ \ n = 1, 2, \ldots$$

and

$$K_Y(k, j) = \sigma^2 \min(k, j);\ \ k, j = 1, 2, \ldots$$

For later use we state these results in a slightly different notation: since $EY_1 = m$, since $K_Y(1, 1) = \sigma_{Y_1}^2 = \sigma^2$, and since the formulas also hold for $n = 0$ and for $k = j = 0$, we have that

$$EY_t = tEY_1;\ \ t \geq 0 \tag{6.7}$$

and

$$K_Y(t,s) = \sigma_{Y_1}^2 \min(t,s); \;\; t,s \geq 0. \tag{6.8}$$

We explicitly consider only those values of t and s that are in the appropriate index set, here the nonnegative integers. An alternative representation to the linear system representation defined by (3.136) is obtained by rewriting the sum as a linear difference equation with initial conditions:

$$Y_n = \begin{cases} 0 & n = 0 \\ Y_{n-1} + X_n & n = 1, 2, 3, \ldots \end{cases}. \tag{6.9}$$

Observe that in this guise, $\{Y_n\}$ is a first-order autoregressive process (see (6.2)) since it is obtained by passing the iid X process through a first-order autoregressive filter with $a_0 = 1$ and $a_1 = -1$. Observe again that this filter is not stable, but it does have a transfer function, $H(f) = 1/(1 - e^{-j2\pi f})$ (which, with the substitution $z = e^{j2\pi f}$, has a pole on the unit circle).

We have seen from Section 3.12 how to find the marginal distributions for such sums of iid processes and have seen from Section 3.7 how to find the conditional distributions and hence a complete specification. The natural question at this point is how general the methods and results referred to are. Toward this end we consider generalizations in several directions. First we consider a direct generalization to continuous time processes, the class of processes with independent and stationary increments. We next consider partial generalizations to discrete time moving-average and autoregressive processes.

6.3 Independent stationary increment processes

We now generalize the class of processes formed by summing iid random variables in a way that works for both continuous and discrete time. The generalization is accomplished by focusing on the *changes* in a process rather than on the values of the process. The general class, that of processes with independent and stationary increments, reduces in the discrete time case to the class considered in the previous sections: processes formed by summing outputs of an iid process.

The change in value of a random process in moving forward in any given time interval is called a *jump* or *increment* of the process. The specific class of processes that we now consider consists of random processes whose jumps or increments in nonoverlapping time intervals are independent random variables whose probability distributions depend only on the time differences over which the jumps occur. In the discrete time case, the nth output of such processes can be regarded as the sum of the first n random variables produced by an iid random process. Because the jumps in nonoverlapping time intervals then consist of sums of different iid random variables, the jumps are obviously independent. This general class of processes is of interest for three reasons. First, the class contains two of the most important examples of random processes: the Wiener process and the Poisson counting process. Second, members of the class form building blocks for many other random process models. For example, in Chapter 5 we presented an intuitive derivation of the properties of continuous time Gaussian white noise. A rigorous development would be based on the Wiener process, which we can treat rigorously with elementary tools. Third, these processes provide a useful vehicle for practice with several important and useful tools of probability theory: characteristic functions, conditional pmf's, conditional pdf's, and nonelementary conditional probability. In addition, independent increment processes provide specific examples of several general classes of processes: Markov processes, counting processes, and random walks.

Independent and stationary increment processes are generally not themselves weakly stationary since, as has already been seen in the discrete time case, their probabilistic description changes with time. They possess, however, some stationarity properties. In particular the distributions of the jumps or increments taken over fixed-length time intervals are stationary even though the distributions of the process are not.

The increments or jumps or differences of a random process are obtained by picking a collection of ordered sample times and forming the pairwise differences of the samples of the process taken at these times. For example, given a discrete time or continuous time random process $\{Y_t;\, t \in \mathcal{T}\}$, one can choose a collection of sample times $t_0, t_1, \ldots, t_k$ ($t_i \in \mathcal{T}$, all i) where we assume that the sample times are *ordered* in the sense that

$$t_0 < t_1 < t_2 < \ldots < t_k.$$

Given this collection of sample times, the corresponding *increments* of the process $\{Y_t\}$ are the differences

$$Y_{t_i} - Y_{t_{i-1}};\ \ i = 1, 2, \ldots, k.$$

Note that the increments very much depend on the choice of the sample times; one would expect quite different behavior when the samples are widely separated than when they are nearby. We can now define the general class of processes with independent increments for both the discrete and continuous time cases.

A random process $\{Y_t\}$; $t \in \mathcal{T}$ is said to have *independent increments* or to be an *independent increment random process* if for all choices of k and sample times $\{t_i;\ i = 1, \ldots, k\}$, the increments $Y_{t_i} - Y_{t_{i-1}}$; $i = 1, 2, \ldots, k$ are independent random variables. An independent increment process is said to have *stationary increments* if the distribution of the increment $Y_{t+\delta} - Y_{s+\delta}$ does not depend on δ for all allowed values of $t > s$ and δ. (Observe that this is really only a first-order stationarity requirement on the increments, not by definition a strict stationarity requirement, but the language is standard. In any case, if the increments are independent and stationary in this sense, then they are also strictly stationary.)

We shall call a random process an *independent stationary increment* or isi process if it has independent and stationary increments.

We shall always make the additional assumption that 0 is the smallest possible time index; that is, that $t \geq 0$ for all $t \in \mathcal{T}$, and that $Y_0 = 0$ as in the discrete time case. We shall see that such processes are not stationary and that they must "start" somewhere or, equivalently, be one-sided random processes. We simply define the starting time as 0 for convenience and fix the starting value of the random process as 0, again for convenience. If these initial conditions are changed, the development changes only in notational details.

A discrete time random process is an isi process if and only if it can be represented as a sum of iid random variables, i.e., if it has the form considered in the preceding sections. To see this, observe that if $\{Y_n\}$ has independent and stationary increments, then by choosing sample times $t_i = i$ and defining $X_n = Y_n - Y_{n-1}$ for $n = 1, 2, \ldots$, the X_n must be independent from the independent increment assumption, and they must be identically distributed from the stationary increment assumption. Thus we have that

$$Y_n = \sum_{k=1}^{n} (Y_k - Y_{k-1}) = \sum_{k=1}^{n} X_k,$$

and hence Y_n has the form of (6.6). Conversely, if Y_n is the sum of iid random variables, then increments will always have the form

$$Y_t - Y_s = \sum_{i=s+1}^{t} X_i;\ \ t > s,$$

that is, the form of sums of disjoint collections of iid random variables, and hence will be independent. Furthermore, the increments will clearly be stationary since they are sums of iid random variables; in particular, the distribution of the increment will depend only on the number of samples added and not on the starting time. Thus all of the development for sums of iid processes could have been entitled "discrete time processes with independent and stationary increments."

Unfortunately, there is no such nice construction of continuous time independent increment processes. The natural continuous time analog would be to integrate a memoryless process, but as with white noise, such memoryless processes are not well defined. One can do formal derivations analogous to the discrete time case and sometimes (but not always) arrive at correct answers. We will use alternative and more rigorous tools when dealing with the continuous time processes. We do note, however, that while we cannot express a continuous time process with independent increments as the output of a linear system driven by a continuous time memoryless process, for any collection of sample times $t_0 = 0, t_1, t_2, \ldots, t_k$ we can write

$$Y_{t_n} = \sum_{i=1}^{n} (Y_{t_i} - Y_{t_{i-1}}). \tag{6.10}$$

and that the increments in the parentheses are independent – that is, we can write Y_{t_n} as a sum of independent increments (in many ways, in fact) – and the increments are identically distributed if the time interval widths are identical for all increments.

Since discrete time isi processes can always be expressed as the sum of iid random variables, their first and second moments always have the form of (6.7) and (6.8). In Section 6.4 it is shown that (6.7) and (6.8) also hold for continuous time processes with stationary and independent increments!

We again emphasize that an independent increment process may have stationary increments, but we already know from the moment calculations of (6.7) and (6.8) that the process itself cannot be weakly stationary. Since the mean and covariance grow with time, independent increment processes clearly only make sense as one-sided processes.

6.4 ⋆Second-order moments of isi processes

In this section we show that several important properties of the discrete time independent increment processes hold for the continuous time case. In the next section we generalize the specification techniques and give two examples of such processes – the continuous time Wiener process and the Poisson counting process. This section is devoted to the proof that (6.7) and

(6.8) hold for continuous time processes with independent and stationary increments. The proof is primarily algebraic and can easily be skipped.

Consider a continuous time random process $\{Y_t;\, t \in \mathcal{T}\}$, where $\mathcal{T} = [0, \infty)$, having independent stationary increments and initial condition $Y_0 = 0$. The techniques used in this section can also be used for an alternative derivation of the discrete time results.

First observe that given any time t and any positive delay or lag $\tau > 0$, we have that

$$Y_{t+\tau} = (Y_{t+\tau} - Y_t) + Y_t, \tag{6.11}$$

and hence, by the linearity of expectation,

$$EY_{t+\tau} = E[Y_{t+\tau} - Y_t] + EY_t.$$

Since the increments are stationary, however, the increment $Y_{t+\tau} - Y_t$ has the same distribution and hence the same expectation as the increment $Y_\tau - Y_0 = Y_\tau$, and therefore

$$EY_{t+\tau} = EY_\tau + EY_t.$$

This equation has the general form

$$g(t + \tau) = g(\tau) + g(t). \tag{6.12}$$

An equation of this form is called a *linear functional equation* and has a unique solution of the form $g(t) = ct$, where c is a constant that is determined by some boundary condition. Thus, in particular, the solution to (6.12) is

$$g(t) = g(1)t. \tag{6.13}$$

Thus the mean of a continuous time independent increment process with stationary increments is given by

$$EY_t = tm,\ t \in \mathcal{T}, \tag{6.14}$$

where the constant m is determined by the boundary condition

$$m = EY_1.$$

Thus (6.7) extends to the continuous time case.

Since $Y_0 = 0$, we can rewrite (6.11) as

$$Y_{t+\tau} = (Y_{t+\tau} - Y_t) + (Y_t - Y_0), \tag{6.15}$$

that is, we can express $Y_{t+\tau}$ as the sum of two independent increments. The variance of the sum of two independent random variables, however, is just the sum of the two variances. In addition, the variance of the increment $Y_{t+\tau} - Y_t$ is the same as the variance of $Y_\tau - Y_0 = Y_\tau$ since the increments are stationary. Thus (6.15) implies that

$$\sigma^2_{Y_{t+\tau}} = \sigma^2_{Y_\tau} + \sigma^2_{Y_t},$$

which is again a linear functional equation and hence has the solution

$$\sigma^2_{Y_t} = t\sigma^2 \tag{6.16}$$

where the appropriate boundary condition is

$$\sigma^2 = \sigma^2_{Y_1}.$$

Knowing the variance immediately yields the second moment:

$$E(Y_t^2) = \sigma^2_{Y_t} + (EY_t)^2 = t\sigma^2 + (tm)^2. \tag{6.17}$$

Consider next the autocorrelation function $R_Y(t,s)$. Choose $t > s$ and write Y_t as the sum of two increments as

$$Y_t = (Y_t - Y_s) + Y_s,$$

and hence

$$R_Y(t,s) = E[Y_t Y_s] = E[(Y_t - Y_s)Y_s] + E[Y_s^2]$$

using the linearity of expectation. The term $E[(Y_t - Y_s)Y_s]$ is, however, the expectation of the product of two independent random variables since the increments $Y_t - Y_s$ and $Y_s - Y_0$ are independent. Thus from Theorem 4.3 the expectation of the product is the product of the expectations. Furthermore, the expectation of the increment $Y_t - Y_s$ is the same as the expectation of the increment $Y_{t-s} - Y_0 = Y_{t-s}$ since the increments are stationary. Thus we have from this, (6.14), and (6.17) that

$$R_Y(t,s) = (t-s)msm + s\sigma^2 + (sm)^2 = s\sigma^2 + (tm)(sm).$$

Repeating the development for the case $t \le s$ then yields

$$R_Y(t,s) = \sigma^2 \min(t,s) + (tm)(sm), \tag{6.18}$$

which yields the covariance

$$K_Y(t,s) = \sigma^2 \min(t,s); \;\; t,s \in \mathcal{T}, \tag{6.19}$$

which extends (6.8) to the continuous time case.

6.5 Specification of continuous time isi processes

The specification of processes with independent and stationary increments is almost the same in continuous time as it is in discrete time. The only real difference is that in continuous time we must consider more general collections of sample times. In discrete time the specification was constructed using the marginal probability function of the underlying iid process, which implies the pmf of the increments. In continuous time we have no underlying iid process so we instead assume that we are given a formula for the cdf (pdf or pmf) of the increments; that is, for any $t > s$ we have a cdf

$$F_{Y_t - Y_s}(y) = F_{Y_{|t-s|} - Y_0}(y) = F_{Y_{|t-s|}}(y) \tag{6.20}$$

or, equivalently, we have the corresponding pmf $p_{Y_t - Y_s}(y)$ for a continuous amplitude process.

To specify a continuous time process we need a formula for the joint probability functions for all n and all ordered sample times $t_1, t_2, \ldots, t_n$ (that is, $t_i < t_j$ if $i < j$). As in the discrete time case, we consider conditional probability functions. To allow both discrete and continuous alphabet, we first focus on conditional cdf's and find the conditional cdf $P(Y_{t_n} \leq y_n | Y_{t_{n-1}} = y_{n-1}, Y_{t_{n-2}} = y_{n-2}, \ldots)$. Then, using (6.10) we can apply the techniques used in discrete time by simply replacing the sample times i by t_i for $i = 0, 1, \ldots, n$. That is, we define the random variables $\{X_n\}$ by

$$X_n = Y_{t_n} - Y_{t_{n-1}}. \tag{6.21}$$

Then the $\{X_n\}$ are independent (but not identically distributed unless the times between adjacent samples are all equal), and

$$Y_{t_n} = \sum_{1=1}^{n} X_i \tag{6.22}$$

and

$$\begin{aligned} &P(Y_{t_n} \leq y_n | Y_{t_{n-1}} = y_{n-1}, Y_{t_{n-2}} = y_{n-2}, \ldots) \\ &= F_{X_n}(y_n - y_{n-1}) = F_{Y_{t_n} - Y_{t_{n-1}}}(y_n - y_{n-1}). \end{aligned} \tag{6.23}$$

This conditional cdf can then be used to evaluate the conditional pmf or pdf as

$$\begin{aligned} p_{Y_{t_n}|Y_{t_{n-1}}, \ldots, Y_{t_1}}(y_n | y_{n-1}, \ldots, y_1) &= p_{X_n}(y_n - y_{n-1}) \\ &= p_{Y_{t_n} - Y_{t_{n-1}}}(y_n - y_{n-1}) \end{aligned} \tag{6.24}$$

or

$$\begin{aligned} f_{Y_{t_n}|Y_{t_{n-1}},\ldots,Y_{t_1}}(y_n|y_{n-1},\ldots,y_1) &= f_{X_n}(y_n - y_{n-1}) \\ &= f_{Y_{t_n}-Y_{t_{n-1}}}(y_n - y_{n-1}), \end{aligned} \tag{6.25}$$

respectively. These can then be used to find the joint pmf's or pdf's as before as

$$f_{Y_{t_1},\ldots,Y_{t_n}}(y_1,\ldots,y_n) = \prod_{i=1}^{n} f_{Y_{t_i}-Y_{t_{i-1}}}(y_i - y_{i-1})$$

or

$$p_{Y_{t_1},\ldots,Y_{t_n}} = \prod_{i=1}^{n} p_{Y_{t_i}-Y_{t_{i-1}}}(y_i - y_{i-1}),$$

respectively. Since we can write the joint probability functions for any finite collection of sample times in terms of the given probability function for the increments, the process is completely specified.

The most important point of these relations is that if a process has independent and stationary increments and we are given a cdf or pmf or pdf for $Y_t = Y_t - Y_0$, then the process is completely defined via the specification just given. Knowing the probabilistic description of the jumps and that the jumps are independent and stationary completely describes the process.

As in discrete time, a continuous time random process $\{Y_t\}$ is called a Markov process if and only if for all n and all ordered sample times $t_1 < t_2 < \ldots < t_n$ we have for all $y_n, y_{n-1}, \ldots$ that

$$\begin{aligned} P(Y_{t_n} \le y_n | Y_{t_{n-1}} = y_{n-1}, Y_{t_{n-2}} = y_{n-2}, \ldots) \\ = P(Y_{t_n} \le y_n | Y_{t_{n-1}} = y_{n-1}) \end{aligned} \tag{6.26}$$

or equivalently,

$$f_{Y_{t_n}|Y_{t_{n-1}},\ldots,Y_{t_1}}(y_n|y_{n-1},\ldots,y_1) = f_{Y_{t_n}|Y_{t_{n-1}}}(y_n|y_{n-1})$$

for continuous alphabet processes and

$$p_{Y_{t_n}|Y_{t_{n-1}},\ldots,Y_{t_1}}(y_n|y_{n-1},\ldots,y_1) = p_{Y_{t_n}|Y_{t_{n-1}}}(y_n|y_{n-1})$$

for discrete alphabet processes. Analogous to the discrete time case, continuous time independent increment processes are Markov processes.

We close this section with the two most famous examples of continuous time independent increment processes.

[6.1] *The continuous time Wiener process*

The Wiener process is a continuous time independent increment process with stationary increments such that the increment densities are Gaussian with zero mean; that is, for $t > 0$,

$$f_{Y_t}(y) = \frac{e^{-y^2/(2t\sigma^2)}}{\sqrt{2\pi t\sigma^2}}; \quad y \in \mathcal{R}.$$

The form of the variance follows necessarily from the previously derived form for all independent increment processes with stationary increments. The specification for this process and the Gaussian form of the increment pdf's imply that the Wiener process is a Gaussian process.

[6.2] *The Poisson counting process*
The Poisson counting process is a continuous time, discrete alphabet, independent increment process with stationary increments such that the increments have a Poisson distribution; that is, for $t > 0$,

$$p_{Y_t}(k) = \frac{(\lambda t)^k}{k!} e^{-\lambda t}; \quad k = 0, 1, 2, \ldots$$

6.6 Moving-average and autoregressive processes

We have seen in the preceding sections that for discrete time random processes the moving-average representation can be used to yield the second-order moments and can be used to find the marginal probability function of independent increment processes. The general specification for independent increment processes, however, was found using the autoregressive representation. In this section we consider results for more general processes using virtually the same methods.

First assume that we have a moving-average process representation described by (6.1). We can use characteristic function techniques to find a simple form for the marginal characteristic function of the output process. In particular, assuming convergence conditions are satisfied where needed and observing that Y_n is a weighted sum of independent random variables, the characteristic function of marginal distribution of the output random process is calculated as the product of the transforms

$$M_{Y_n}(ju) = \prod_k M_{h_k X_{n-k}}(ju).$$

The individual transforms are easily shown to be

$$M_{h_k X_{n-k}}(ju) = E\left[e^{juh_k X_{n-k}}\right] = M_{X_{n-k}}(juh_k) = M_X(juh_k).$$

Thus

$$M_{Y_n}(ju) = \prod_k M_X(juh_k), \tag{6.27}$$

where the product is, as usual, dependent on the index sets on which $\{X_n\}$ and $\{h_k\}$ are defined.

Equation (6.27) can be inverted in some cases to yield the output cdf and pdf or pmf. Unfortunately, however, in general this is about as far as one can go in this direction, even for an iid input process. Attempts to find joint or conditional distributions of the output process by distributions or other techniques will generally be frustrated by the complexity of the calculations required except for a few special cases, including the important Gaussian case.

Part of the difficulty in finding conditional distributions lies in the moving-average representation. The techniques used successfully for the independent increment processes relied on an autoregressive representation of the output process. We will now show that the methods used work for more general autoregressive process representations. We will consider specifically causal autoregressive processes represented as in (6.3) so that

$$Y_n = X_n - \sum_{k>0} a_k Y_{n-k}.$$

By the independence and causality conditions, the $\{Y_{n-k}\}$ in the sum are independent of X_n. Hence we have a representation for Y_n as the sum of two independent random variables, X_n and the weighted sum of the Ys. The latter quantity is treated as if it were a constant in calculating conditional probabilities for Y_n. Thus the *conditional* probability of an event for Y_n can be specified in terms of the *marginal* probability of an easily determined event for X_n. Specifically, the conditional cdf for Y_n is

$$\begin{aligned}\Pr[Y_n \le y_n | y_{n-1}, y_{n-2}, \ldots] &= \Pr\left[X_n \le \sum_{k\ge 0} a_k y_{n-k}\right] \\ &= F_X\left(\sum_{k\ge 0} a_k y_{n-k}\right),\end{aligned} \tag{6.28}$$

where $a_0 = 1$. The conditional pmf or pdf can now be found. For example, if the input random process is continuous alphabet, the conditional output pdf is found by differentiation to be

$$f_{Y_n|Y_{n-1},Y_{n-2},\ldots}(y_n|y_{n-1}, yn-2,\ldots) = f_X\left(\sum_{k\geq 0} a_k y_{n-k}\right). \tag{6.29}$$

Finally, the complete specification can be obtained by a product of pmf's or pdf's by the chain rule as in (3.142) or (3.152). The discrete time independent increment result is obviously a special case of this equation. For more general processes, we need only require that the sum converge in (6.29) and that the corresponding conditional pdf's be appropriately defined (using the general conditional probability approach). We next consider an important example of the ideas of this section.

6.7 The discrete time Gauss–Markov process

As an example of the development of the preceding section, consider the filter given in Example [5.1]. Let $\{X_n\}$ be an iid Gaussian process with mean m and variance σ^2. The moving-average representation is

$$Y_n = \sum_{k=0}^{\infty} X_{n-k} r^k, \tag{6.30}$$

from which (6.27) can be applied to find that

$$\begin{aligned} M_{Y_n}(ju) &= \prod_k \exp\left(j(ur^k)m - \frac{1}{2}(ur^k)^2\sigma_X^2\right) \\ &= \exp\left(jum\sum_k r^k - \frac{1}{2}u^2\sigma_X^2\sum_k r^{2k}\right). \end{aligned}$$

that is, a Gaussian random variable with mean $m_Y = m\sum_k r^k = m/(1-r)$ and variance $\sigma_Y^2 = \sigma_X^2 \sum_k r^{2k} = \sigma_X^2/(1-r^2)$, agreeing with the moments found by the second-order theory in Example [5.1].

To find a complete specification for this process, we now turn to an autoregressive model. From (6.30) it follows that Y_n must satisfy the difference equation

$$Y_n = X_n + rY_{n-1}. \tag{6.31}$$

Hence $\{Y_n\}$ is a first-order autoregressive source with $a_0 = 1$ and $a_1 = -r$. Note that as with the Wiener process, this process can be represented as a first-order autoregressive process or as an infinite-order, moving-average process. In fact, the Wiener process is the one-sided version of this process with $r = 1$.

Application of (6.29) yields

$$\begin{aligned} f_{Y^n}(y^n) &= f_{Y_n}(y_n|y_{n-1}, y_{n-2}, \ldots) f_{Y_{n-1}}(y_{n-1}|y_{n-2}, y_{n-3}, \ldots) \cdots f_{Y_1}(y_1) \\ &= f_{Y_1}(y_1) \prod_{i=2}^{n} f_X(y_i - ry_{i-1}) \\ &= \frac{e^{-y_1^2/(2\sigma_Y^2)}}{\sqrt{2\pi\sigma_Y^2}} \prod_{i=2}^{n} \frac{e^{-(y_i - ry_{i-1})^2/2\sigma^2}}{\sqrt{2\pi\sigma^2}} \\ &= \frac{e^{-y_1^2/(2\sigma_Y^2)}}{\sqrt{2\pi\sigma_Y^2}} \frac{e^{-\sum_{i=2}^{n}(y_i - ry_{i-1})^2/2\sigma^2}}{(2\pi\sigma^2)^{(n-1)/2}}. \end{aligned} \tag{6.32}$$

6.8 Gaussian random processes

We have seen how to calculate the mean, covariance function, or spectral density of the output process of a linear filter driven by an input random process whose mean, covariance function, or spectral density is known. In general, however, it is not possible to derive a complete specification of the output process. We have seen one exception: the output random process of an autoregressive filter driven by an iid input random process can be specified through the conditional pmf's or pdf's, as in (6.29). In this section we develop another important exception by showing that the output process of a linear filter driven by a Gaussian random process – not necessarily iid – is also Gaussian. Thus simply knowing the output mean and autocorrelation or covariance functions – the only parameters of a Gaussian distribution – provides a complete specification. The underlying idea is that of Theorem 4.4: a linear operation on a Gaussian vector yields another Gaussian random vector. The output vector mean and matrix covariance of the theorem are in fact just the vector and matrix versions of the linear system second-moment I/O relations (5.3) and (5.7). The output of a discrete time FIR linear filter can be expressed as a linear operation on the input as in (4.26), that is, a finite-dimensional matrix times an input vector plus a constant. Therefore we can immediately extend Theorem 4.4 to FIR filtering and argue that all finite-dimensional distributions of the output process are Gaussian and hence the process itself must be Gaussian. It is also possible to extend Theorem 4.4 to include more general impulse responses and to continuous time by using appropriate limiting arguments. We will not prove such extensions. Instead we will merely state the result as a corollary:

Corollary 6.1 *If a Gaussian random process $\{X_t\}$ is passed through a linear filter, then the output is also a Gaussian random process with mean and covariance given by (5.3) and (5.7).*

6.9 The Poisson counting process

An engineer encounters two types of random processes in practice. The first is the random process whose probability distribution depends largely on design parameters: the type of modulation used, the method of data coding, etc. The second probability distributions that depend on naturally occurring phenomena over which the engineer has little if any control: noise in physical devices, speech waveforms, the number of messages in a telephone system as a function of time, etc. Gaussian processes provide one example of these. This section is devoted to another example: the Poisson process. Here the basic Poisson counting process is derived from physical assumptions and a variety of properties are developed. Gaussian and Poisson processes provide classes of random processes that characterize (at least approximately) the majority of naturally occurring random processes. The development of Poisson processes provides further examples of many of the techniques developed so far.

Our intent here is to remove some of the mystery of the functional forms of two important distributions by showing how these apparently complicated distributional forms arise from nature. Therefore, the development presented is somewhat brief, without consideration of all the mathematical details.

The Poisson counting process was introduced as an example of specification of an independent and stationary increment process. In this section the same process is derived from a more physical argument.

Consider modeling a continuous time counting process $\{N_t;\, t \geq 0\}$ with the following properties:

1. $N_0 = 0$ (the initial condition).
2. The process has independent and stationary increments. Hence the changes, called jumps, during nonoverlapping time intervals are independent random variables. The jumps in a given time interval are memoryless, and their amplitude does not depend on what happened before that interval.
3. In the limit of *very small* time intervals, the probability of an increment of 1, that is, of increasing the total count by 1, is proportional to the length of the time interval. The probability of an increment greater than 1 is negligible in comparison, e.g., is proportional to powers greater than 1 of the length of the time interval.

These properties well describe many physical phenomena such as the emission of electrons and other subatomic particles from irradiated objects (remember vacuum tubes?), the arrival of customers at a store, phone calls at an exchange, or data packets at a switch in a network, and other phenomena where events such as arrivals or discharges occur randomly in time. The properties naturally capture the intuition that such events do not depend on the past and that for a very tiny interval, the probability of such an event is proportional to the length of the interval. For example, if you are waiting for a phone call, the probability of its happening during a short period of τ seconds is proportional to τ. The probability of two or more phone calls in a very small period τ is, however, negligible in comparison.

The third property can be quantified as follows: let λ be the proportionality constant. Then for a small enough time interval Δt,

$$\Pr(N_{t+\Delta t} - N_t = 1) \cong \lambda \Delta t$$

$$\Pr(N_{t+\Delta t} - N_t = 0) \cong 1 - \lambda \Delta t$$

$$\Pr(N_{t+\Delta t} - N_t > 1) \cong 0. \tag{6.33}$$

The relations of (6.33) can be stated rigorously by limit statements, but we shall use them in the more intuitive form given.

We use the properties 1 to 3 to derive the probability mass function $p_{N_t - N_0}(k) = p_{N_t}(k)$ for an increment $N_t - N_0$, from the starting time at time 0 up to time $t > 0$ with $N_0 = 0$. For convenience we temporarily change notation and define

$$p(k,t) = p_{N_t - N_0}(k); \quad t > 0.$$

Let Δt be a differentially small interval as in (6.33), and we have that

$$p(k, t + \Delta t) = \sum_{n=0}^{k} Pr(N_t = n) \Pr(N_{t+\Delta t} - N_t = k - n | N_t = n).$$

Since the increments are independent, the conditioning can be dropped so that, using (6.33),

$$\begin{aligned} p(k, t + \Delta t) &= \sum_{n=0}^{k} Pr(N_t = n) \Pr(N_{t+\Delta t} - N_t = k - n) \\ &\cong p(k,t)(1 - \lambda \Delta t) + p(k-1, t) \lambda \Delta t, \end{aligned}$$

which with some algebra yields

$$\frac{p(k,t+\Delta t)-p(k,t)}{\Delta t}=p(k-1,t)\lambda-p(k,t)\lambda.$$

In the limit as $\Delta t \to 0$ this becomes the differential equation

$$\frac{d}{dt}p(k,t)+\lambda p(k,t)=\lambda p(k-1,t),\ \ t>0.$$

The initial condition for this differential equation follows from the initial condition for the process, $N_0 = 0$; i.e.,

$$p(k,0)=\begin{cases}0 & k\neq 0\\ 1 & k=0\end{cases}.$$

since this corresponds to $\Pr(N_0 = 0) = 1$. The solution to the differential equation with the given initial condition is

$$p_{N_t}(k)=p(k,t)=\frac{(\lambda t)^k e^{-\lambda t}}{k!};\ \ k=0,1,2,\ldots;\ t\geq 0. \tag{6.34}$$

(This is easily verified by direct substitution.)

The pmf of (6.34) is the Poisson pmf, and hence the given properties produce the Poisson counting process. Note that (6.34) can be generalized using the stationarity of the increments to yield the pmf for k jumps in an arbitrary interval $(s,t), t \geq s$ as

$$p_{N_t-N_s}(k)=\frac{(\lambda(t-s))^k e^{-\lambda(t-s)}}{k!};\ \ k=0,1,\ldots;\ t\geq s. \tag{6.35}$$

As developed in this chapter, these pmf's and the given properties yield a complete specification of the Poisson counting process.

Note that sums of Poisson random variables are Poisson. This follows from the development of this section. That is, for any $t > s > r$, all three of the indicated quantities in $(N_t - N_s) + (N_s - N_r) = N_t - N_r$ are Poisson. Thus the Poisson distribution is infinitely divisible in the sense of Section 4.13. Of course the infinite divisibility of Poisson random variables can also be verified by characteristic functions as in (4.98). Poisson random variables satisfy the requirements of the central limit theorem and hence it can be concluded that with appropriate normalization, the Poisson cdf approaches the Gaussian cdf asymptotically.

6.10 Compound processes

So far the various processes with memory have been constructed by passing iid processes through linear filters. In this section a more complicated construction of a new process is presented which is not a simple linear operation. A compound process is a random process constructed from two other random processes rather than from a single input process. It is formed by summing consecutive outputs of an iid discrete time random process, but the number of terms included in the sum is determined by a counting random process, which can be discrete or continuous time. As an example where such processes arise, suppose that on a particular telephone line the number of calls arriving in t minutes is a random variable N_t. The resulting N_t calls have duration $X_1, X_2, \ldots, X_{N_t}$. What is the total amount of time occupied by the calls? It is the random variable

$$Y_t = \sum_{k=1}^{N_t} X_k.$$

Since this random variable is defined for all positive t, $\{Y_t\}$ is a random process, depending on two separate processes: a counting process $\{N_t\}$ and an iid process $\{X_n\}$. In this section we explore the properties of such processes. The main tool used to investigate such processes is conditional expectation.

Suppose that $\{N_t;\ t \geq 0\}$ is a discrete or continuous time counting process. Thus t is assumed to take on either nonnegative real values or nonnegative integer values. Suppose that $\{X_k\}$ is an iid process and that the X_n are mutually independent of the N_t. Define the compound process $\{Y_t;\ t \geq 0\}$ by

$$Y_t = \begin{cases} 0 & t = 0 \\ \sum_{k=1}^{N_t} X_k & t > 0 \end{cases}. \tag{6.36}$$

What can be said about the process Y_t? From iterated expectation we have that the mean of the compound process is given by

$$\begin{aligned} EY_t = E[E(Y_t|N_t)] &= E\left[E(\sum_{k=1}^{N_t} X_k|N_t)\right] \\ &= E[N_t E(X)] = E(N_t)E(X). \end{aligned} \tag{6.37}$$

Thus, for example, if N_t is a binomial counting process with parameter p, and $\{X_n\}$ is a Bernoulli process with parameter ϵ, then $E(Y_k|N_k) = N_k\epsilon$ and hence $EY_k = \epsilon E(N_k) = \epsilon kp$. If N_t is a Poisson counting process with parameter λ, then $E(Y_t) = E(N_t)E(X) = \lambda t E(X)$.

Other moments follow in a similar fashion. For example, the characteristic function of Y_t can be evaluated using iterated expectation as

$$M_{Y_t}(ju) = E(e^{juY_t}) = E[E(e^{juY_t}|N_t)] = E[M_X(ju)^{N_t}], \tag{6.38}$$

where we have used the fact that conditioned on N_t, Y_t is the sum of N_t iid random variables with characteristic function M_X. To further evaluate this, we assume a distribution for N_t. Suppose first that N_t is a binomial counting process. Then

$$\begin{aligned} E[M_X(ju)^{N_k}] &= \sum_{n=0}^{k} p_{N_k}(n) M_X(ju)^n \\ &= \sum_{n=0}^{k} \binom{n}{k} p^k (1-p)^{n-k} M_X(ju)^n \\ &= (pM_X(ju) + (1-p))^k. \end{aligned} \tag{6.39}$$

Suppose instead that N_t is a continuous time Poisson counting process. Then

$$\begin{aligned} E[M_X(ju)^{N_t}] &= \sum_{n=0}^{\infty} p_{N_t}(n) M_X(ju)^n = \sum_{n=0}^{\infty} \frac{(\lambda t)^n e^{-\lambda t}}{n!} M_X(ju)^n \\ &= e^{-\lambda t} \sum_{n=0}^{\infty} \frac{(\lambda t M_X(ju))^n}{n!} = e^{-\lambda t(1-M_X(ju))}, \end{aligned} \tag{6.40}$$

where we have invoked the Taylor series expansion for an exponential.

Both of these computations involve very complicated processes, yet they result in closed form solutions of modest complication. Since the characteristic functions are known, the marginal distributions of such processes follow. Further properties of compound processes are explored in the problems.

6.11 Composite random processes

Often in practice one encounters random processes that behave locally like a specific random process, but whose parameters change with time. For example, speech waveforms are often modeled locally as a Gaussian autoregressive process, but in different intervals of time the covariance and mean may differ. Different sounds such as phonemes have different random process models, each of which occupies a certain amount of time in the overall output. Because of this variation, such processes are often described in the literature as being nonstationary, but in fact they can be modeled by stationary processes which have distinct local and global properties. This is accomplished by modeling the process as a *composite* or *switched* random process.

A composite random process consists of a family of random processes $\{\{X_t^{(\theta)},\, t \in \mathcal{T}\};\, \theta \in \Re\}$, which are usually assumed jointly stationary and ergodic, together with a "switch" random process $\{S_t,\, t \in \mathcal{T}\}$, which is usually assumed to be jointly stationary with and independent of all of the $X_t^{(\theta)}$. The composite source $\{Y_t,\, t \in \mathcal{T}\}$ is defined by

$$Y_t = X_t^{(S_t)}, \tag{6.41}$$

that is, the switch S_t selects which of the processes $X_t^{(\theta)}$ to produce as the output of the composite process. If the switch process is a randomly selected constant, then the composite process simplifies to become a *mixture process*, which can be viewed as Nature picking a stationary ergodic random process at random and then sending it for ever. For example, she might randomly choose the bias of a coin and flip it forever. A mixture model is sometimes called a *doubly stochastic* random process because of the random selection of a parameter followed by the random generation of a sequence using that parameter.

Generally composite models incorporate an assumption that the switch is slowly varying and stays connected to each random process for a long – but not infinite – period of time. This means that as with speech models, short-term sample averages will approximate the conditional expectations given the random process currently connected to the switch, but very long sample averages will average over models. It is common to allow the switch to change only at specific time intervals, corresponding to an analysis window in speech processing where a different autoregressive process model is fitted to each 100 ms or so of sampled speech, or to a tile or block in image processing consisting of a square array of pixels. Within the window one assumes the switch is fixed and one is observing a single stationary and ergodic random process.

When the collection of X processes and the switch are jointly stationary, it can be shown that the resulting composite process is also stationary, pointing out the fact that it is possible to have stationary models with distinct local and global behavior.

6.12 ⋆Exponential modulation

Lest the reader erroneously assume that all random process derived distribution techniques apply only to linear operations on processes, we next consider an example of a class of processes generated by a nonlinear operation on another process. While linear techniques rarely work exactly for nonlinear systems, the systems that we shall consider form an important exception where one can find second-order moments and sometimes even complete specifications. The primary examples of processes generated in this way are phase modulation and frequency modulation by Gaussian random processes and the Poisson random telegraph wave.

Let $\{X(t)\}$ be a random process and define a new random process

$$Y(t) = a_0 e^{j(a_1 t + a_2 X(t) + \Theta)}, \tag{6.42}$$

where a_0, a_1, and a_2 are fixed real constants and where Θ is a uniformly distributed random phase angle on $[0, 2\pi]$. The process $\{Y(t)\}$ is called an exponential modulation of $\{X(t)\}$. Observe that it is a nonlinear function of the input process. Note further that, as defined, the process is a complex-valued random process. Hence we must modify some of our techniques. In some, but not all, of the interesting examples of exponentially modulated random processes we will wish to focus on the real part of the modulated process, which we will call

$$\begin{aligned} U(t) = \mathrm{Re}(Y(t)) &= \frac{1}{2}Y(t) + \frac{1}{2}Y(t)^* \\ &= a_0 \cos(a_1 t + a_2 X(t) + \Theta). \end{aligned} \tag{6.43}$$

In this form, exponential modulation is called *phase modulation* (PM) of a carrier of angular frequency a_1 by the input process $\{X(t)\}$. If the input process is formed by integrating another random process, say $\{W(t)\}$, then the U process is called *frequency modulation* (FM) of the carrier by the W process. Phase and frequency modulation are extremely important examples of complex exponential modulation.

A classic example of a random process arising in communications that can be put in the same form is obtained by setting $\Theta = 0$ (with probability one), choosing $a_1 = 0$, $a_2 = \pi$, and letting the input process be the Poisson counting process $\{N(t)\}$, that is, by considering the random process

$$V(t) = a_0(-1)^{N(t)}. \tag{6.44}$$

This is a real-valued random process that changes value with every jump in the Poisson counting process. Because of the properties of the Poisson counting process, this process is such that jumps in nonoverlapping time windows are independent, the probability of a change of value in a differentially small interval is proportional to the length of the interval, and the probability of more than one change is negligible in comparison. It is usually convenient to consider a slight change in this process, which makes it somewhat better behaved. Let Z be a binary random variable, independent of $N(t)$ and taking values of $+1$ or -1 with equal probability. Then the random process $Y(t) = ZV(t)$ is called the *random telegraph wave* and has long served as a fundamental example in the teaching of second-order random process theory. The purpose of the random variable Z is to remove an obvious nonstationarity at the origin and make the resulting process equally likely to have either of its two values at time zero. This has the obvious effect of making the process zero-mean. In the form given, it can be treated as simply a special case of exponential modulation.

We develop the second-order moments of exponentially modulated random processes and then apply the results to the preceding examples. We modify our definitions slightly to apply to the complex-valued random variables defined by the vector consisting of the expectations of the real and imaginary parts; that is, if $X = \mathrm{Re}(X) + j\mathrm{Im}(X)$, with $\mathrm{Re}(X)$ and $\mathrm{Im}(X)$ the real and imaginary parts of X, respectively, then

$$EX = (E\mathrm{Re}(X), E\mathrm{Im}(X)).$$

In other words, the expectation of a complex vector is defined to be the vector of ordinary scalar expectations of the components. The autocorrelation function of a complex random process is defined somewhat differently as

$$R_Y(t, s) = E[Y(t)Y(s)^*],$$

which reduces to the usual definition if the process is real valued. The autocorrelation in this more general situation is not in general symmetric, but it is Hermitian in the sense that

$$R_Y(s, t) = R_Y(t, s)^*.$$

Being Hermitian is, in fact, the appropriate generalization of symmetry for developing a useful transform theory, and it is for this reason that the autocorrelation function includes the complex conjugate of the second term.

It is an easy exercise to show that for the general exponentially modulated random process of (6.42) we have that

$$EY(t) = 0.$$

This can be accomplished by separately considering the real and imaginary parts and using (3.124), exactly as was done in the AM case of Chapter 5. The use of the auxiliary random variable $\mathcal{Z}$ in the random telegraph wave definition means that both examples have zero mean. Note that it is *not* true that $Ee^{j(a_1t+a_2X(t)+\Theta)}$ equals $e^{j(a_1t+a_2EX(t)+E\Theta)}$; that is, expectation does not in general commute with nonlinear operations.

To find the autocorrelation of the exponentially modulated process, we begin with some simple algebra,

$$\begin{aligned} E[Y(t)Y(s)^*] &= a_0^2 E[e^{j(a_1(t-s)+a_2(X(t)-X(s)))}] \\ &= a_0^2 e^{ja_1(t-s)} E[e^{ja_2(X(t)-X(s))}], \end{aligned}$$

which implies that

$$R_Y(t,s) = a_0^2 e^{ja_1(t-s)} M_{X(t)-X(s)}(ja_2). \tag{6.45}$$

Thus the autocorrelation of the nonlinearly modulated process is given simply in terms of the characteristic function of the increment between the two sample times. This is often a computable quantity, and when it is, we can find the second-order properties of such processes without approximation or linearization. This is a simple result of the fact that the autocorrelation of an exponentially modulated process is given by an expectation of the exponential of the difference of two samples and hence by the characteristic function of the difference.

There are two examples in which the computation of the characteristic function of the difference of two samples of a random process is particularly easy: a Gaussian input process and an independent increment input process.

If the input process $\{X(t)\}$ is Gaussian with zero mean (for convenience) and autocorrelation function $R_X(t,s)$, then the random variable $X(t) - X(s)$ is also Gaussian (being a linear combination of Gaussian random variables) with mean zero and variance

$$\sigma^2_{X(t)-X(s)} = E[(X(t) - X(s))^2] = R_X(t,t) + R_X(s,s) - 2R_X(t,s).$$

Thus we have shown that if $\{X(t)\}$ is a zero-mean Gaussian random process with autocorrelation function R_X and if $\{Y(t)\}$ is obtained by exponentially modulating $\{X(t)\}$ as in (6.42), then

$$\begin{aligned} R_Y(t,s) &= a_0^2 e^{ja_1(t-s)} M_{X(t)-X(s)}(ja_2) \\ &= a_0^2 e^{ja_1(t-s)} \exp\left(-\frac{1}{2}a_2^2(R_X(t,t) + R_X(s,s) - 2R_X(t,s))\right). \end{aligned}$$

Observe that this autocorrelation is not symmetric, but it is Hermitian.

Thus, for example, if the input process is stationary, then so is the modulated process, and

$$R_Y(\tau) = a_0^2 e^{ja_1\tau} e^{-a_2^2(R_X(0)-R_X(\tau))}. \tag{6.46}$$

We emphatically note that the modulated process is not Gaussian.

We can use this result to obtain the second-order properties for phase modulation as follows:

$$\begin{aligned}
&R_U(t,s)\\
&= E[U(t)U(s)^*]\\
&= E\left[\frac{Y(t)+Y(t)^*}{2}\left(\frac{Y(s)+Y(s)^*}{2}\right)^*\right]\\
&= \frac{1}{4}\left(E[Y(t)Y(s)^*] + E[Y(t)Y(s)] + E[Y(t)^*Y(s)^*] + E[Y(t)^*Y(s)]\right).
\end{aligned}$$

Note that both of the middle terms on the right have the form

$$a_0^2 e^{\pm ja_1(t+s)} E[e^{\pm j(a_2(X(t)+X(s))+2\Theta)}],$$

which evaluates to 0 because of the uniform phase angle. The remaining terms are $R_Y(t,s)$ and $R_Y(t,s)^*$ from the previous development, and hence

$$\begin{aligned}
R_U(t,s) &= \frac{1}{2}a_0^2\cos(a_1(t-s))\\
&\quad\times\exp\left(\frac{1}{2}a_2^2(R_X(t,t)+R_X(s,s)-2R_X(t,s))\right), \qquad (6.47)
\end{aligned}$$

so that in the stationary case,

$$R_U(\tau) = \frac{1}{2}a_0^2\cos(a_1\tau)e^{-a_2(R_X(0)-R_X(\tau))}. \qquad (6.48)$$

As expected, this autocorrelation is symmetric.

Returning to the exponential modulation case, we consider the second example of exponential modulation of independent increment processes. Observe that this overlaps the preceding example in the case of the Wiener process. We also note that phase modulation by independent increment processes is of additional interest because in some examples independent increment processes can be modeled as the integral of another process. For example, the Poisson counting process is the integral of a random telegraph wave with alphabet 0 and 1 instead of -1 and $+1$. (This is accomplished by forming the process $(X(t)+1)/2$ with $X(t)$ the ± 1 random telegraph wave.) In this case the real part of the output process is the FM modulation of the process being integrated.

If $\{X(t)\}$ is a random process with independent and stationary increments, then the characteristic function of $X(t)-X(s)$ with $t>s$ is equal to that of $X(t-s)$. Thus we have from (6.45) that for $t>s$ and $\tau=t-s$,

$$R_Y(\tau) = a_0^2 e^{ja_1\tau} M_{X(\tau)}(ja_2).$$

We can repeat this development for the case of negative lag to obtain

$$R_Y(\tau) = a_0^2 e^{ja_1\tau} M_{X(|\tau|)}(ja_2). \tag{6.49}$$

Observe that this autocorrelation is not symmetric; that is, it is not true that $R_Y(-\tau) = R_Y(\tau)$ (unless $a_1 = 0$). It is, however, Hermitian.

Equation (6.49) provides an interesting oddity: even though the original input process is not weakly stationary (since it is an independent increment process), the exponentially modulated output *is* weakly stationary. For example, if $\{X(t)\}$ is a Poisson counting process with parameter λ, then the characteristic function is

$$\begin{aligned} M_{X(\tau)}(ju) &= \sum_{k=0}^{\infty} \frac{(\lambda\tau)^k e^{-\lambda\tau}}{k!} e^{juk} = e^{-\lambda\tau} \sum_{k=0}^{\infty} \frac{(\lambda\tau e^{ju})^k}{k!} \\ &= e^{\lambda\tau(e^{ju}-1)}; \quad \tau \geq 0. \end{aligned}$$

Thus if we choose $a_1 = 0$ and $a_2 = \pi$, then the modulated output process is the random telegraph wave with alphabet $\pm a_0$ and hence is a real process. Equation (6.49) becomes

$$R_Y(\tau) a_0^2 e^{-2\lambda|\tau|}. \tag{6.50}$$

Note that the autocorrelation (and hence also the covariance) decays exponentially with the delay.

A complete specification of the random telegraph wave is possible and is left as an exercise.

6.13 ⋆Thermal noise

Thermal noise is one of the most important sources of noise in communications systems. It is the "front-end" noise in receivers that is caused by the random motion of electrons in a resistance. The amplifiers then magnify the noise along with the possibly tiny signals. Thus the dominant noise is in the receiver itself and not in the atmosphere – although some noise can be contributed by the atmosphere, for example by lighting. The receiver noise can be comparable in amplitude to the desired signal. In this section we sketch the development of a model of thermal noise. The development provides an interesting example of a process with both Poisson and Gaussian characteristics.

Say we have a uniform conducting cylindrical rod at temperature T. Across this rod we connect an ammeter. The random motion of electrons in the rod will cause a current $I(t)$ to flow through the meter. We wish to develop a random process model for the current based on the underlying physics. The following are the relevant physical parameters:

$$
\begin{aligned}
A &= \text{cross-sectional area of the rod} \\
L &= \text{length of the rod} \\
q &= \text{electron charge} \\
n &= \text{number of electrons per cubic centimeter} \\
\alpha &= \text{average number of electron collisions with} \\
&\quad\ \text{heavier particles per second (about } 10^3) \\
m &= \text{mass of an electron} \\
\rho &= \text{resistivity of the rod } = \frac{m\alpha}{nq^2} \\
R &= \text{resistance of the rod } = \frac{\rho L}{A} \\
\kappa &= \text{Boltzmann's constant}
\end{aligned}
$$

The current measured will be due to electrons moving in the longitudinal direction of the rod, which we denote x. Let $V_{x.k}(t)$ denote the component of velocity in the x direction of the kth electron at time t. The total current $I(t)$ is then given by the sums of the individual electron currents as

$$I(t) = \sum_{k=1}^{nAL} i_k(t) = \sum_{k=1}^{nAL} \frac{q}{L/V_{x.k}(t)} = \sum_{k=1}^{nAL} \frac{q}{L} V_{x.k}(t).$$

We assume that (1) the average velocity $EV_{x.k}(t) = 0$, all k, t; (2) $V_{x.k}(t)$ and $V_{x.j}(s)$ are independent random variables for all $k \neq j$; and (3) the $V_{x.k}(t)$ have the same distribution for all k.

The autocorrelation function of $I(t)$ is found as

$$
\begin{aligned}
R_I(\tau) = E[I(t)I(t+\tau)] &= \sum_{k=1}^{nAL} \frac{q^2}{L^2} E[V_{x.k}(t)V_{x.k}(t+\tau)] \\
&= \frac{nAq^2}{L} E[V_x(t)V_x(t+\tau)], \qquad (6.51)
\end{aligned}
$$

where we have dropped the subscript k since by assumption the distribution, and hence the autocorrelation function of the velocity, does not depend on it.

Next assume that, since collisions are almost always with heavier particles, the electron velocities before and after collisions are independent – the velocity after impact depends only on the momentum of the heavy particle that the electron hits. We further assume that the numbers of collisions in disjoint time intervals are independent and satisfy (6.33) with a change of parameter:

$$\Pr(\text{no collisions in } \Delta t) \cong (1 - \alpha\Delta t)$$
$$\Pr(\text{one collision in } \Delta t) \cong \alpha\Delta t.$$

This implies that the number of collisions is Poisson and that from (6.35)

$$\Pr(\text{a particle has } k \text{ collisions in } [t, t+\tau)) = e^{-\alpha\tau}\frac{(\alpha\tau)^k}{k!}\ ; k = 0, 1, 2, \ldots$$

Thus if $\tau \geq 0$ and $N_{t,\tau}$ is the number of collisions in $[t, t+\tau)$, then, using iterated expectation and the independence with mean zero of electron velocities when one or more collisions have occurred,

$$\begin{aligned} E[V_x(t)V_x(t+\tau)] &= E(E[V_x(t)V_x(t+\tau)|N_{t,\tau}]) \\ &= E(V_x(t)^2)\Pr(N_{t,\tau} = 0) + (EV_x(t))^2 \Pr(N_{t,\tau} \neq 0) \\ &= E(V_x(t)^2)e^{-\alpha\tau}. \end{aligned} \tag{6.52}$$

It follows from the equipartition theorem for electrons in thermal equilibrium at temperature T that the electron velocity variance is

$$E(V_x(t)^2) = \frac{\kappa T}{m}. \tag{6.53}$$

Therefore, after some algebra, we have from (6.51) through (6.53) that

$$R_I(\tau) = \frac{\kappa T}{R}\alpha e^{-\alpha|\tau|}.$$

Thevinin's theorem can the be applied to model the conductor as a voltage source with voltage $E(t) = RI(t)$. The autocorrelation function of $E(t)$ is

$$R_E(\tau) = \kappa T R \alpha e^{-\alpha|\tau|},$$

an autocorrelation function that decreases exponentially with the delay τ. Observe that as $\alpha \to \infty$, $R_E(\tau)$ becomes a taller and narrower pulse with constant area $2\kappa TR$; that is, it looks more and more like a Dirac delta function with area $2\kappa TR$. Since the mean is zero, this implies that the process $E(t)$ is such that samples separated by very small amounts are approximately uncorrelated. Thus thermal noise is approximately white noise. The central limit theorem can be used to show that the finite-dimensional distributions of the process are approximately Gaussian. Thus we can conclude that an approximate model for thermal noise is a Gaussian white noise process!

6.14 Ergodicity

To consider general random processes with memory and state a general form of the strong law of large numbers, we require an additional idea – ergodicity. The notion of ergodicity is often described incorrectly in engineering-oriented texts on random processes. There is, however, some justification for doing so; the definition is extremely abstract and not very intuitive. The intuition comes with the consequences of assuming both ergodicity and stationarity, and it is these consequences that are often used as a definition. For completeness we provide the rigorous definition. We later consider briefly examples of processes that violate the definition. Before possibly obscuring the key issues with abstraction, it is worth pointing out a few basic facts.

- The concept of ergodicity does not require stationarity; a nonstationary process can be ergodic.
- Many perfectly good models of physical processes are not ergodic, yet they have a form of law of large numbers. In other words, nonergodic processes can be perfectly good and useful models.
- The definition is in terms of the process distribution of the random process. There is no finite-dimensional equivalent definition of ergodicity as there is for stationarity. This fact makes it more difficult to describe and interpret ergodicity.
- Processes that are iid are ergodic, so that ergodicity can be thought of as a generalization of iid.

Ergodicity is defined in terms of a property of events: an event F is said to be *τ-invariant* if $\{x_t;\ t \in \mathcal{T}\} \in F$ implies that also $\{x_{t+\tau};\ t \in \mathcal{T}\} \in F$, i.e., if a sequence or waveform is in F, then so is the sequence or waveform formed by shifting by τ. As an example, consider the discrete time random process event F consisting of all binary sequences having a limiting relative frequency of 1s of exactly p. Then this event is τ-invariant for all integer τ since changing the starting time of the sequence by a finite amount does not affect limiting relative frequencies.

A random process $\{X_t;\ t \in \mathcal{T}\}$ is *ergodic* if for any τ all τ-invariant events F have probability one or zero. In the discrete time case it suffices to consider only $\tau = 1$.

In the authors' view, the concept of ergodicity is the most abstract idea of this book, but its importance in practice makes it imperative that the idea at least be introduced and discussed. The reader interested in delving more deeply into the concept is referred to Billingsley's classic book *Ergodic Theory and Information* [4] for a deep look at ergodicity and its implications for discrete time discrete alphabet random processes. Rather than try to provide further insight into the abstract definition, we instead turn to its implications, and then interpret from the implications what it means for a process to be ergodic or not.

The importance of stationarity and ergodicity is largely due to the following classic result of Birkhoff and Khinchine.

Theorem 6.1 *The strong law of large numbers (The pointwise ergodic theorem)*
Given a discrete time stationary random process $\{X_n;\ n \in \mathcal{Z}\}$ with finite expectation $E(X_n) = m_X$, then there is a random variable $\hat{X}$ with the property that

$$\lim_{n\to\infty} \frac{1}{n} \sum_{n=0}^{\infty} X_n = \hat{X} \text{ with probability one,} \tag{6.54}$$

that is, the limit exists. If the process is also ergodic, then $\hat{X} = m_X$ and hence

$$\lim_{n\to\infty} \frac{1}{n} \sum_{n=0}^{\infty} X_n = m_X \text{ with probability one.} \tag{6.55}$$

The conditions also imply convergence in mean square (an L_2 or mean ergodic theorem); that is,

$$\underset{n\to\infty}{\text{l.i.m.}} \frac{1}{n} \sum_{n=0}^{\infty} X_n = \hat{X}, \tag{6.56}$$

but we shall focus on the form for convergence with probability one. There are also continuous time versions of the theorem to the effect that under suitable conditions

$$\lim_{T\to\infty} \frac{1}{T} \int_0^{\infty} X(t)\, dt = \hat{X} \text{ with probability one,} \tag{6.57}$$

but these are much more complicated to describe because special conditions are needed to ensure the existence of the time-average integrals.

The strong law of large numbers shows that for stationary and ergodic processes, time averages converge with probability one to the corresponding expectation. Suppose that a process is stationary but not ergodic. Then the theorem is that time averages still converge, but possibly not to the expectation. Consider the following example of a random process which exhibits this behavior. Suppose that Nature at the beginning of time flips a fair coin. If the coin ends up heads, she sends thereafter a Bernoulli process with parameter p_1, that is, an iid sequence of coin flips with a probability p_1 of getting a head. If the original coin comes up tails, however, Nature sends thereafter a

Bernoulli process with parameter $p_0 \neq p_1$. In other words, you the observer are looking at the output of one of two iid processes, but you do not know which one. This is an example of a *mixture* random process as introduced earlier. Another way to view the process is as follows: let $\{U_n\}$ denote the Bernoulli process with parameter p_1 and $\{W_n\}$ denote the Bernoulli process with parameter p_0. Then the mixture process $\{X_n\}$ is formed by connecting a switch at the beginning of time to either the $\{U_n\}$ process or the $\{W_n\}$ process, and soldering the switch shut. The point is you either see $\{U_n\}$ forever with probability 1/2, or you see $\{W_n\}$ forever. A little elementary conditional probability shows that for any dimension k,

$$p_{X_0,\ldots,X_{k-1}}(\mathbf{x}) = \frac{p_{U_0,\ldots,U_{k-1}}(\mathbf{x}) + p_{W_0,\ldots,W_{k-1}}(\mathbf{x})}{2}. \tag{6.58}$$

Thus, for example, the probability of getting a head in the mixture process is $p_{X_0}(1) = (p_0 + p_1)/2$. Similarly, the probability of getting two heads in a row is $p_{X_0,X_1}(1,1) = (p_0^2 + p_1^2)/2$. Since the joint pmf's for the two Bernoulli processes are not changed by shifting, neither is the joint pmf for the mixture process. Hence the mixture process is stationary and from the strong law of large numbers its relative frequencies will converge to something. Is the mixture process ergodic? It is certainly not iid. For example, the probability of getting two heads in a row was found to be $p_{X_0,X_1}(1,1) = (p_0^2 + p_1^2)/2$, which is not the same as $p_{X_0}(1)p_{X_1}(1) = [(p_0 + p_1)/2]^2$ (unless $p_0 = p_1$), so that X_0 and X_1 are not independent. It could conceivably be ergodic, but is it? Suppose that $\{X_n\}$ were indeed ergodic, than the strong law would say that the relative frequency of heads would have to converge to the probability of a head, i.e., to $(p_0 + p_1)/2$. But this is clearly not true since if you observe the outputs $\{X_n\}$ you are observing a Bernoulli process of bias either p_0 or p_1 and hence you should expect to compute a limiting relative frequency of heads that is either p_0 or p_1, depending on which of the Bernoulli processes you are looking at. In other words, your limiting relative frequency is a random variable, which depends on Nature's original choice of which process to let you observe. As a more direct invocation of the definition of ergodicity, consider the set $F = \{x : \text{relative frequency of heads is } p_0\}$. This set is invariant. The probability of F is the probability that the coin with bias p_0 was chosen, or 1/2. Since the invariant set has a probability that is neither 0 or 1, the process cannot be ergodic.

This explains one possible behavior leading to the general strong law: you observe a mixture of stationary and ergodic processes, that is, you observe a randomly selected stationary and ergodic process, but you do not know a priori which process it is. Since, conditioned on this selection, the strong law holds, relative frequencies will converge, but they do not converge to an overall expectation. They converge to a random variable, which is in fact just *the conditional expectation given knowledge of which stationary and ergodic random process is actually being observed.* Thus the strong law of large numbers can be useful in such a stationary but nonergodic case since one can estimate which stationary ergodic process is actually being observed by measuring the relative frequencies.

A perhaps surprising fundamental result of random processes is that this special example is in a sense typical of all stationary nonergodic processes. The result is called the *ergodic decomposition theorem* and it states that under quite general assumptions, any nonergodic stationary process is in fact a mixture of stationary and ergodic processes and hence you are always observing a stationary and ergodic process, you just do not know in advance which one. In our coin example, you know you are observing one of two Bernoulli processes, but we could equally consider an infinite mixture by selecting p from a uniform distribution on $(0, 1)$. You do not know p in advance, but you can estimate it from relative frequencies. The interested reader can find a development of the ergodic decomposition theorem and its history in chapter 7 of [32].

The previous discussion implies that ergodicity is not required for the strong law of large numbers to be useful. The next question is whether stationarity is required. Again the answer is no. Given that the main concern is the convergence of sample averages and relative frequencies, it should be reasonable to expect that random processes could exhibit transient or short-term behavior that violated the stationarity definition, yet eventually died out so that if one waited long enough the process would look increasingly stationarity. In fact one can make precise the notion of asymptotically stationary (in several possible ways) and the strong law extends to this case. Again the interested reader is referred to chapter 7 of [32]. The point is that the notions of stationarity and ergodicity should not be taken too seriously since ergodicity can easily be dispensed with and stationarity can be significantly weakened and still leave processes for which laws of large numbers hold so that time averages and relative frequencies have well-defined limits.

6.15 Random fields

We close this collection of random processes with an extension of the idea to that of a multidimensional random process, where now the multidimensional aspect refers to the index set (time in the traditional case) and not the alphabet. We focus on the special case of two dimensions where the random objects are called *random fields* [66, 15, 19]. These models have recently become popular for a variety of image processing applications including segmentation, analysis, de-noising, smoothing, and coding. Hidden Markov models, which have played a key role in the development of automatic speech recognition, have been extended to model images as hidden Markov random fields. The reader is referred to [71] for a survey and extensive reference list. We here provide only the basic definition and a few comments.

A random field extends the idea of a random process by allowing the index which corresponds to time in the traditional situation be two-dimensional, e.g., to correspond to two spatial coordinates. A *random field* is a family of random variables $X = \{X_s : s \in \mathcal{T}\}$, where

- $\mathcal{T}$ is a two-dimensional index set such as $\Re^2$ or $\{(i,j)|1 \leq i \leq N_1, 1 \leq j \leq N_2\}$.
- For each point or *pixel* $s \in \mathcal{T}$, X_s is a real-valued random variable.
- The random field X is characterized by a joint probability distribution P_X When the alphabet is discrete, the joint distribution is completely described by a joint probability mass function $p_Y(y) = P_Y(\{y\}) = P(Y = y)$ for which

  $$P_Y(F) = P(Y \in F) = \sum_F p_Y(y).$$

  In the case of a continuous alphabet, the distribution is usually described by a a joint probability density function (pdf) p_Y for which

  $$P_Y(F) = P(Y \in F) = \int_F dy\, p_Y(y)$$

  for all events F.

The idea of a random field is simple enough. For the case of a continuous index set one can think of the random field as providing a model for a random photograph, where every spot or pixel has a random intensity described by a joint pdf for the entire image. For the discrete case the index set typically corresponds to points on a square lattice, each representing an image intensity. The bulk of the theory of random fields is devoted to developing models that are both mathematically interesting and useful in practice. The two most important models are the Gibbs random field and the Markov random field. The precise definitions would take too much additional notation to present here, but roughly speaking a Gibbs random field focuses on joint distributions with a pdf that is constructed from exponentials of functions of the pixels. A Markov random field is constructed from conditional distributions for a pixel intensity that is conditionally independent of values at distant pixels given the values within a close neighborhood of pixels. A fundamental theorem of the theory states that the two models are equivalent and both models are used in a wide variety of applications. The Markov model is considerably more complicated than in the traditional case because there is no natural ordering to two-dimensional indexes as there is with one-dimensional indexes and hence there is no notion of "past" and "future" in general. The ideas of "near" and "neighbors" are used instead, and the resulting formulas are generally more complicated than in the one-dimensional case.

6.16 Problems

1. Let $\{X_n\}$ be an iid process with a Poisson marginal pmf with parameter λ. Let $\{Y_n\}$ denote the induced sum process as in (6.6). Find the pmf for Y_n and find $\sigma^2_{Y_n}$, EY_n, and $K_Y(t,s)$.
2. Let $\{X_n\}$ be an iid process. Define a new process $\{U_n\}$ by

$$U_n = X_n - X_{n-1};\ \ n = 1, 2, \ldots$$

   Find the characteristic function and the pmf for U_n. Find $R_U(t,s)$. Is $\{U_n\}$ an independent increment process?
3. Find the characteristic function $M_{U_n}(ju)$ for the $\{U_n\}$ process of Problem 5.2.
4. Let $\{X_n\}$ be a ternary iid process with $p_{X_n}(+1) = p_{X_n}(-1) = \epsilon/2$ and $p_{X_n}(0) = 1 - \epsilon$. Fix an integer N and define the "sliding average"

$$Y_n = \frac{1}{N}\sum_{i=0}^{N-1} X_{n-i}.$$

   (a) Find $EX_n, \sigma^2_{X_n}, M_{X_n}(ju)$, and $K_X(t,s)$.
   (b) Find $EY_n, \sigma^2_{Y_n}, M_{Y_n}(ju)$.
   (c) Find the *cross-correlation* $R_{X,Y}(t,s) \triangleq E[X_t Y_s]$.
   (d) Given $\delta > 0$ find a simple upper bound to $\Pr(|Y_n| > \delta)$ in terms of N and ϵ.
5. All ships travel at the same speed through a wide canal. Eastbound ships arrive as a Poisson process with an arrival rate of λ_{E} ships per day. Westbound ships arrive as an independent Poisson process with an arrival rate of λ_{W} ships per day. An indicator at a point in the canal is always pointing in the direction of travel of the most recent ship to pass. Each ship takes T days to traverse the canal.
   (a) What is the probability that the next ship passing by the indicator causes it to change its direction?
   (b) What is the probability that an eastbound ship will see no westbound ships during its eastward journey through the canal?
   (c) If we begin observing at an arbitrary time, determine the probability mass

function of the total number of ships we observe up to and including the seventh eastbound ship we see.

(d) If we begin observing at an arbitrary time, determine the pdf of the time until we see the seventh eastbound ship.

(e) Given that the pointer is pointing west:
- What is the probability that the next ship to pass it will be westbound?
- What is the pdf of the remaining time until the pointer changes direction?

(f) An observation post in the middle of the canal counts the number of ships as they pass in either direction. Suppose that from the time the counter was first installed the number of ships that have passed after t days is N_t. What are the mean and variance of N_t? Does N_t/t converge in probability? If so, to what?

6. $\{X_n;\ n = 1, 2, \ldots\}$ is a binary random process with outputs of 0 or 1. We know the following:

$$p_{X_1}(1) = 1$$
$$p_{X_{n+1}|X_n}(1 \mid 1) = \frac{3}{4}$$
$$p_{X_{n+1}|X_n}(0 \mid 0) = \frac{2}{3}.$$

(a) Determine the probability that the first 0 will occur on the kth toss, $k = 2, 3, \ldots$

(b) Find the probability $P(X_{5000} = 1)$.

(c) Find the probability $P(X_{5000} = 1 \text{ and } X_{5002} = 1)$.

(d) Given that $X_{5001} = X_{5002} = X_{5003} = \cdots = X_{5001+m}$, where $m \geq 1$, what is the probability that the common value is 1? Simplify the result as much as possible and discuss the meaning of the result as m gets large.

(e) We are told that the 365th 1 in the sequence just occurred at time $n = 500$. Determine the expected value of the number of subsequent time units until we observe the 379th 1.

(f) Find the numerical values of the following limits:
- $\lim_{n\to\infty} E[X_{n+1} - X_n]$
- $\lim_{n\to\infty} E[(X_{n+1} - X_n)^2]$.

7. A stationary continuous time random process $\{X(t)\}$ switches randomly between the values of 0 and 1. We have that

$$\Pr(X(t) = 1) = \Pr(X(t) = 0) = \frac{1}{2},$$

and if N_t is the number of changes of output during $(0, t]$, then

$$p_{N_t}(k) = \frac{1}{1 + \alpha t}\left(\frac{\alpha t}{1 + \alpha t}\right)^k;\quad k = 0, 1, 2, \ldots,$$

where $\alpha > 0$ is a fixed parameter. (This is called the Bose–Einstein distribution.)

(a) Find $M_{N_t}(ju)$, EN_t, and $\sigma^2_{N_t}$.
(b) Find $EX(t)$ and $R_X(t,s)$.

8. Given two random processes $\{X_t\}$, called the signal process, and $\{N_t\}$, called the noise process, define the process $\{Y_t\}$ by
$$Y_t = X_t + N_t .$$
The $\{Y_t\}$ process can be considered as the output of a *channel with additive noise* where the $\{X_t\}$ process is the input. This is a common model for dealing with noisy linear communication systems; e.g., the noise may be due to atmospheric effects or to front-end noise in a receiver. Assume that the signal and noise processes are independent; that is, any vector of samples of the N process. Find the characteristic function, mean, and variance of Y_t in terms of those for X_t and N_t. Find the covariance of the output process in terms of the covariances of the input and noise process.
9. Find the inverse of the covariance matrix of the discrete time Wiener process, that is, the inverse of the matrix $\{\min(k,j);\ k = 1,2,\ldots,n,\ j = 1,2,\ldots,n\}$.
10. Let $\{X(t)\}$ be a Gaussian random process with zero mean and autocorrelation function
$$R_X(\tau) = \frac{N_0}{2} e^{-|\tau|} .$$
Is the process Markov? Find its power spectral density. Let $Y(t)$ be the process formed by DSB-SC modulation of $X(t)$ as in (5.37) with $a_0 = 0$. If the phase angle Θ is assumed to be 0, is the resulting modulated process Gaussian? Letting Θ be uniformly distributed, sketch the power spectral density of the modulated process. Find $M_{Y(0)}(ju)$.
11. Let $\{X(t)\}$ and $\{Y(t)\}$ be the two continuous time random processes of Exercise 5.15 and let
$$W(t) = X(t)\cos(2\pi f_0 t) + Y(t)\sin(2\pi f_0 t),$$
as in that problem. Find the marginal probability density function $f_{W(t)}$ and the joint pdf $f_{W(t).W(s)}(u,\nu)$. Is $\{W(t)\}$ a Gaussian process? Is it strictly stationary?
12. Let $\{N_k\}$ be the binomial counting process and define the discrete time random process $\{Y_n\}$ by
$$Y_n = (-1)^{N_n} .$$
(This is the discrete time analog to the random telegraph wave.) Find the autocorrelation, mean, and power spectral density of the given process. Is the process Markov?
13. Find the power spectral density of the random telegraph wave. Is this process a Markov process? Sketch the spectrum of an amplitude modulated random telegraph wave.
14. Suppose that (U,W) is a Gaussian random vector with $EU = EW = 0$, $E(U^2) =$

$E(W^2) = \sigma^2$, and $E(UW) = \rho\sigma^2$. (The parameter ρ has magnitude less than or equal to 1 and is called the *correlation coefficient.*) Define the new random variables

$$S = U + W, \quad D = U - W$$

(a) Find the marginal pdf's for S and D.

(b) Find the joint pdf $f_{S.D}(\alpha, \beta)$ or the joint characteristic function $M_{S.D}(ju, j\nu)$. Are S and D independent?

15. Suppose that K is a random variable with a Poisson distribution, that is, for a fixed parameter λ

$$Pr(K = k) = p_k(k) = \frac{\lambda^k e^{-\lambda}}{k!}; \quad k = 0, 1, 1, \ldots$$

(a) Define a new random variable N by $N = K + 1$. Find the characteristic function $M_N(ju)$, the expectation EN, and the pmf $p_N(n)$ for the random variable N.

We define a one-sided discrete time random process $\{Y_n; n = 1, 2, \ldots\}$ as follows: Y_n has a binary alphabet $\{-1, 1\}$. Y_0 is equally likely to be -1 or $+1$. Given Y_0 has some value, it will stay at that value for a total of T_1 time units which has the same distributions N, and then it will change sign. It will stay at the new sign for a total of T_2 time units, where T_2 has the same distribution as N and is independent of T_1, and then change sign again. It will continue in this way, that is, it will change sign for the kth time at time

$$S_k = \sum_{i=1}^{k} T_i,$$

where the T_i form an iid sequence with the marginal distribution found in part (a).

(b) Find the characteristic function $M_{S_k}(ju)$ and the pmf $p_{S_k}(m)$ for the random variable S_k. Is $\{S_k\}$ an independent increment process?

16. Suppose that $\{Z_n\}$ is a two-sided Bernoulli process, that is, an iid sequence of binary $\{0, 1\}$ random variables with $\Pr(Z_n = 1) = \Pr(Z_n = 0)$. Define the new processes

$$X_n = (-1)^{Z_n},$$

$$Y_n = \sum_{i=0}^{n} 2^{-i} X_i; \quad n = 0, 1, 2, \ldots,$$

and

$$V_n = \sum_{i=0}^{\infty} 2^{-i} X_{n-i}; \quad n \in \mathcal{Z}.$$

(a) Find the mean and autocorrelation function of the $\{X_n\}$ process and the $\{V_n\}$ process. If possible, find the power spectral densities.
(b) Find the characteristic functions for both Y_n and V_n.
(c) Is $\{Y_n\}$ an autoregressive process? a moving-average process? Is it weakly stationary? Is V_n an autoregressive process? a moving-average process? Is it weakly stationary? (*Note:* answers to parts (a) and (b) are sufficient to answer the stationarity questions, no further computations are necessary.)
(d) Find the conditional pmf

$$p_{V_n|V_{n-1},V_{n-2},\ldots,V_0}(\nu_n|\nu_{n-1},\ldots,\nu_0).$$

Is $\{V_n\}$ a Markov process?

17. Suppose that $\{Z_n\}$ and $\{W_n\}$ are two mutually independent two-sided zero-mean iid Gaussian processes with variances σ_Z^2 and σ_W^2, respectively. Z_n is put into a linear time-invariant filter to form an output process $\{X_n\}$ defined by

$$X_n = Z_n - rZ_{n-1},$$

where $0 < r < 1$. (Such a filter is sometimes called a *pre-emphasis filter* in speech processing.) This process is then used to form a new process

$$Y_n = X_n + W_n,$$

which can be viewed as a noisy version of the pre-emphasized Z_n process. Lastly, the Y_n process is put through a "de-emphasis filter" to form an output process U_n defined by

$$U_n = rU_{n-1} + Y_n.$$

(a) Find the autocorrelation R_Z and the power spectral density S_Z. Recall that for a weakly stationary discrete time process with zero mean $R_Z(k) = E(Z_n Z_{n+k})$ and

$$S_Z(f) = \sum_{k=-\infty}^{\infty} R_Z(k)e^{-j2\pi fk},$$

the discrete time Fourier transform of R_Z.
(b) Find the autocorrelation R_X and the power spectral density S_X.
(c) Find the autocorrelation R_Y and the power spectral density S_Y.
(d) Find the conditional pdf $f_{Y_n|X_n}(y|x)$.
(e) Find the pdf f_{U_n,W_n} (or the corresponding characteristic function $M_{U_n,W_n}(ju, jv)$).
(f) Find the overall mean squared error $E[(U_n - Z_n)^2]$.

18. Suppose that $\{N_t\}$ is a binomial counting process and that $\{X_n\}$ is an iid process that is mutually independent of $\{N_t\}$. Assume that the X_n have zero mean and

variance σ^2. Let Y_k denote the compound process

$$Y_k = \sum_{i=1}^{N_k} X_i.$$

Use iterated expectation to evaluate the autocorrelation function $R_Y(t,s)$.

19. Suppose that $\{W_n\}$ is a discrete time Wiener process. What is the minimum mean squared estimate of W_n given $W_{n-1}, W_{n-2}, \ldots$? What is the linear least squares estimator?
20. Let $\{X_n\}$ be an iid binary random process with $\Pr(X_n = \pm 1) = 1/2$ and let $\{N_t\}$ be a Poisson counting process. A continuous time random walk $Y(t)$ can be defined by

$$Y_t = \sum_{i=1}^{N_t} X_i;\ t > 0.$$

Find the expectation, covariance, and characteristic function of Y_t.

21. Are compound processes independent increment processes?
22. Suppose that $\{N_t;\ t \geq 0\}$ is a process with independent and stationary increments and that

$$p_{N_t}(k) = \frac{(\lambda t)^k e^{-\lambda t}}{k!};\ k = 0, 1, 2, \ldots$$

Suppose also that $\{L_t;\ t \geq 0\}$ is a process with independent and stationary increments and that

$$p_{L_t}(k) = \frac{(\nu t)^k e^{-\nu t}}{k!};\ k = 0, 1, 2, \ldots$$

Assume that the two processes N_t and L_t are mutually independent of each other. Define for each t the random variable $I_t = N_t + L_t$. The variable I_t might model, for example, the number of requests for CPU cycles arriving from two independent sources, each of which produces requests according to a Poisson process.

(a) What is the characteristic function for I_t? What is the corresponding pmf?
(b) Find the mean and covariance function of $\{I_t\}$.
(c) Is $\{I_t;\ t \geq 0\}$ an independent increment process?
(d) Suppose that Z is a discrete random variable, independent of N_t, with probability mass function

$$p_Z(k) = \frac{a^k}{(1+a)^{k+1}},\ k = 0, 1, \ldots$$

Find the probability $P(Z = N_t)$.

(e) Suppose that $\{Z_n\}$ is an iid process with marginal pmf $p_Z(k)$ as in the previous part. Define the compound process

$$Y_t = \sum_{k=0}^{N_t} Z_k.$$

Find the mean $E(Y_t)$ and variance $\sigma^2_{Y_t}$.

23. Suppose that $\{X_n;\ n \in \mathcal{Z}\}$ is a discrete time iid Gaussian random process with zero mean and variance $\sigma_X^2 = E[X_0^2]$. This is an input signal to a signal processing system. Suppose also that $\{W_n;\ n \in \mathcal{Z}\}$ is a discrete time iid Gaussian random processes with zero mean and variance σ_W^2. The two processes are mutually independent. W_n is considered to be noise. Suppose that X_n is put into a linear filter with unit δ-response h, where

$$h_k = \begin{cases} 1 & k = 0 \\ -1 & k = -1 \\ 0 & \text{otherwise} \end{cases}$$

to form an output $U = X * h$, the convolution of the input signal and the δ-response. The final output signal is formed by adding the noise to the filtered input signal, $Y_n = U_n + W_n$.

(a) Find the mean, power spectral density, and marginal pdf for U_n.

(b) Find the joint pdf $f_{U_1,U_2}(\alpha, \beta)$. You can leave your answer in terms of an inverse matrix $\mathbf{\Lambda}^{-1}$, but you must accurately describe $\mathbf{\Lambda}$.

(c) Find the mean, covariance, and power spectral density for Y_n.

(d) Find $E[Y_n X_n]$.

(e) Does the mean ergodic theorem hold for $\{Y_n\}$?

24. Suppose that $\{X(t);\ t \in \mathcal{R}\}$ is a weakly stationary continuous time Gaussian random process with zero mean and autocorrelation function

$$R_X(\tau) = E[X(t)X(t+\tau)] = \sigma_X^2 e^{-|\tau|}.$$

(a) Define the random process $\{Y(t);\ t \in \mathcal{R}\}$ by

$$Y(t) = \int_{t-T}^{t} X(\alpha)\, d\alpha,$$

where $T > 0$ is a fixed parameter. (This is a short-term integrator.) Find the mean and power spectral density of $\{Y(t)\}$.

(b) For fixed $t > s$, find the characteristic function and the pdf for the random variable $X(t) - X(s)$.

25. Consider the process $\{X_k;\ k = 0, 1, \ldots\}$ defined by $X_0 = 0$ and

$$X_{k+1} = aX_k + W_k,\ k \geq 0 \tag{6.59}$$

where a is a constant, $\{W_k;\ k = 0, 1, \ldots\}$ is a sequence of iid Gaussian random variables with $E(W_k) = 0$ and $E(W_k^2) = \sigma^2$.

(a) Calculate $E(X_k)$ for $k \geq 0$.
(b) Show that X_k and W_k are uncorrelated for $k \geq 0$.
(c) By squaring both sides of (6.59) and taking expectation, obtain a recursive equation for $K_X(k,k)$.
(d) Solve for $K_X(k,k)$ in term of a and σ. *Hint:* distinguish between $a = 1$ and $a \neq 1$.
(e) Is the process $\{X_k;\ k = 1, 2, \ldots\}$ weakly stationary?
(f) Is the process $\{X_k;\ k = 1, 2, \ldots\}$ Gaussian?
(g) For $-1 < a < 1$, show that

$$P(|X_n| > 1) \leq \frac{\sigma^2}{1 - a^2}.$$

26. A distributed system consists of N sensors which view a common random variable corrupted by different observation noises. In particular, suppose that the ith sensor measures a random variable

$$W_i = X + Y_i,\ i = 0, 1, 2, \ldots, N - 1,$$

where the random variables $X, Y_1, \ldots, Y_N$ are all mutually independent Gaussian random variables with zero mean. The variance of X is 1 and the variance of Y_i is r^i for a fixed parameter $|r| < 1$. The observed data are gathered at a central processing unit to form an estimate of the unknown random variable X as

$$\hat{X}_N = \frac{1}{N} \sum_{i=0}^{N-1} W_i.$$

(a) Find the mean, variance, and probability density function of the estimate $\hat{X}_N$.
(b) Find the probability density function $f_{\epsilon_N}(\alpha)$ of the error

$$\epsilon_N = X - \hat{X}_N.$$

(c) Does $\hat{X}_N$ converge in probability to the true value X?

27. Suppose that $\{N_t;\ t \geq 0\}$ is a process with independent and stationary increments and that

$$p_{N_t}(k) = \frac{(\lambda t)^k e^{-\lambda t}}{k!};\ \ k = 0, 1, 2, \ldots$$

(a) What is the characteristic function for N_t?
(b) What is the characteristic function for the increment $N_t - N_s$ for $t > s$?
(c) Suppose that Y is a discrete random variable, independent of N_t, with probability mass function

$$p_Y(k) = (1 - p)p^k,\ k = 0, 1, \ldots$$

Find the probability $P(Y = N_t)$.

(d) Suppose that we form the discrete time process $\{X_n\ n = 1, 2, \ldots\}$ by

$$X_n = N_{2n} - N_{2(n-1)}.$$

What is the covariance of X_n?

(e) Find the conditional probability mass function

$$p_{X_n|N_{2(n-1)}}(k|m).$$

(f) Find the expectation

$$E\left(\frac{1}{N_t + 1}\right).$$

28. Suppose that $\{X(t)\}$ is a continuous time weakly stationary Gaussian random process with zero mean and autocorrelation function $R_X(\tau) = e^{-2\alpha|\tau|}$, where $\alpha > 0$. The signal is passed through an RC filter (a filter with a single resistor and capacitor) with transfer function

$$H(f) = \frac{\beta}{\beta + j2\pi f},$$

where $\beta = 1/RC$, to form an output process $\{Y)t)\}$.

(a) Find the power spectral densities $S_X(f)$ and $S_Y(f)$.
(b) Evaluate the average powers $E[X^2(t)]$ and $E[Y^2(t)]$.
(c) What is the marginal pdf $f_{Y(t)}$?
(d) Now form a discrete time random process $\{W_n\}$ by $W_n = X(nT)$, for all integer n. This is called *sampling* with a *sampling period* of T. Find the mean, autocorrelation function, and, if it exists, the power spectral density of $\{W_n\}$.
(e) Is $\{Y(t)\}$ a Gaussian random process? Is $\{W_n\}$ a Gaussian random process? Are they stationary in the strict sense?
(f) Let $\{N_t\}$ be a Poisson counting process. Let $i(t)$ be the deterministic waveform defined by

$$i(t) = \begin{cases} 1 \text{ if } t \in [0, \delta] \\ 0 \text{ otherwise} \end{cases}$$

– that is, a flat pulse of duration δ. For $k = 1, 2, \ldots$, let t_k denote the time of the kth jump in the counting process (that is, t_k is the smallest value of t for which $N_t = k$). Define the random process $\{Y(t)\}$ by

$$Y(t) = \sum_{k=1}^{N_t} i(t - t_k).$$

This is a special case of a class of processes known as filtered Poisson processes. This particular example is a model for shot noise in vacuum tubes. Draw some sample waveforms of this process. Find $M_{Y(t)}(ju)$ and $p_{Y(t)}(n)$.

Hint: you need not consider any properties of the random variables $\{t_k\}$ to solve this problem.

29. In the physically motivated development of the Poisson counting process, we fixed time values and looked at the random variables giving the counts and the increments of counts at the fixed times. In this problem we explore the reverse description: what if we fix the counts and look at the times at which the process achieves these counts? For example, for each strictly positive integer k, let r_k denote the time that the kth count occurs; that is, $r_k = \alpha$ if and only if

$$N_\alpha = k; \;\; N_t < k; \;\; \text{all } t < \alpha.$$

Define $r_0 = 0$. For each strictly positive integer k, define the *interarrival times* τ_k by

$$\tau_k = r_k - r_{k-1},$$

and hence

$$r_k = \sum_{i=1}^{k} \tau_i .$$

(a) Find the pdf for r_k for $k = 1, 2, \ldots$

Hint: first find the cdf by showing that

$$F_{r_k}(\alpha) = \Pr(k\text{th count occurs before or at time } \alpha) = \Pr(N_\alpha \geq k).$$

Then use the Poisson pmf to write an expression for the sum. Differentiate to find the pdf. You may have to do some algebra to reduce the answer to a simple form not involving any sums. This is most easily done by writing a difference of two sums in which all terms but one cancel. The final answer is called the *Erlang* family of pdf's. You should find that the pdf r_1 is an exponential density.

(b) Use the basic properties of the Poisson counting process to prove that the interarrival times are iid.

Hint: prove that

$$F_{\tau_n|\tau_1,\ldots,\tau_{n-1}}(\alpha|\beta_1, \ldots, \beta_{n-1}) = F_{\tau_n}(\alpha) = 1 - e^{-\lambda\alpha}; \;\; n = 1, 2, \ldots; \;\; \alpha \geq 0.$$

Appendix A
Preliminaries

The theory of random processes is constructed on a large number of abstractions. These abstractions are necessary to achieve generality with precision while keeping the notation used manageably brief. Students will probably find learning facilitated if, with each abstraction, they keep in mind (or on paper) a concrete picture or example of a special case of the abstraction. From this the general situation should rapidly become clear. Concrete examples and exercises are introduced throughout the book to help with this process.

A.1 Set theory

In this section the basic set theoretic ideas that are used throughout the book are introduced. The starting point is an *abstract space*, or simply a *space*, consisting of *elements* or *points*, the smallest quantities with which we shall deal. This space, often denoted by Ω, is sometimes referred to as the universal set. To describe a space we may use braces notation with either a list or a description contained within the braces $\{\ \}$. Examples are:

[A.0] The abstract space consisting of no points at all, that is, an empty (or trivial) space. This possibility is usually excluded by assuming explicitly or implicitly that the abstract space is *nonempty*, that is, it contains at least one point.

[A.1] The abstract space with only the two elements *zero* and *one* to denote, for example, the possible receptions of a radio receiver of binary data at one particular time instant of signaling. Equivalently, we could give different names to the elements and have a space $\{0, 1\}$, the binary numbers, or a space with the elements *heads* and *tails*. Clearly the structure of all of these spaces is the same; only the names have been changed. They are different, however, in that one is *numeric*, and hence we can perform arithmetic operations on the outcomes, while the other is not. Spaces which do not have numeric points (or points labeled by numeric vectors, sequences, or waveforms) are sometimes referred to as *categorical.* Notationally we describe these spaces as $\{zero,\ one\}$, $\{0, 1\}$, and $\{heads,\ tails\}$, respectively.

[A.2] Given a fixed positive integer k, the abstract space consisting of all possible binary k-tuples, that is, all 2^k k-dimensional binary vectors. This space could model the possible sequences of k flips of the same coin or a single flip of k coins. Note the Example [A.1] is a special case of Example [A.2].

[A.3] The abstract space with elements consisting of all infinite sequences of *ones* and *zeros* or 1*s* and 0*s* denoting, for example, the sequence of possible receptions of a radio receiver of binary data over all signaling times. The sequences could be one-sided in the sense of beginning at time zero and continuing forever, or they could be two-sided in the sense of beginning in the infinitely remote past (time $-\infty$) and continuing into the infinitely remote future.

[A.4] The abstract space consisting of all ASCII (American Standard Code for Information Interchange) codes for characters (letters, numerals, and control characters such as line feed, rub out, etc.). These might be in decimal, hexadecimal, or binary form. In general, we can consider this space as just a space $\{a_i, i = 1, \ldots, N\}$ containing a finite number of elements (which here might well be called symbols, letters, or characters).

[A.5] Given a fixed positive integer k, the space of all k-dimensional vectors with coordinates in the space of Example [A.4]. This could model all possible contents of an ASCII buffer used to drive a serial printer.

[A.6] The abstract space of all infinite (single-sided or double-sided) sequences of ASCII character codes.

[A.7] The abstract space with elements consisting of all possible voltages measured at the output of a radio receiver at one instant of time. Since all physical equipment has limits to the values of voltage (called "dynamic range") that it can support, one model for this space is a subset of the real line such as the closed interval $[-V, V] = \{r : -V \leq r \leq V\}$, i.e., the set of all real numbers r such that $-V \leq r \leq +V$. If, however, the dynamic range is not precisely known or if we wish to use a single space as a model for several measurements with different dynamic ranges, then we might wish to use the entire real line $\Re = (-\infty, \infty) = \{r : -\infty < r < \infty\}$. The fact that the space includes "impossible" as well as "possible" values is acceptable in a model.

[A.8] Given a positive integer k, the abstract space of all k-dimensional vectors with coordinates in the space of Example [A.7]. If the real line is chosen as the coordinate space, then this is k-dimensional Euclidean space.

[A.9] The abstract space with elements being all infinite sequences of members of the space of Example [A.7], e.g., all single-sided real-valued sequences of the form $\{x_n, n = 0, 1, 2, \ldots\}$, where $x_n \in \Re$ for all $n = 1, 2, \ldots$

[A.10] Instead of constructing a new space as sequences of elements from another space, we might wish to consider a new space consisting of all waveforms whose instantaneous values are elements in another space, e.g., the space of all waveforms $\{x(t); t \in (-\infty, \infty)\}$, where $x(t) \in \Re$, all t. This would model, for instance, the space of all possible voltage–time waveforms at the output of a radio receiver. Examples of members of this space are $x(t) = \cos \omega t$, $z(t) = e^{st}$, $x(t) = 1$, $x(t) = t$, and so on. As with sequences, the waveforms may begin in the remote past or they might be defined for t running from 0 to ∞.

The preceding examples focus on three related themes that are considered throughout the book: Examples [A.1], [A.4], and [A.7] present models for the possible values of a single measurement. The mathematical model for such a measurement with an unknown outcome is called a *random variable.* Such simple spaces describe the possible values that a random variable can assume. Examples [A.2], [A.5], and [A.8] treat vectors (finite collections or finite sequences) of individual measurements. The mathematical model for such a vector-valued measurement is called a *random vector.* Since a vector is made up of a finite collection of scalars, we can also view this random object as a collection (or family) of random variables. These two viewpoints — a single random vector-valued measurement and a collection of random scalar-valued measurements — will both prove useful. Examples [A.3], [A.6], and [A.9] consider infinite sequences of values drawn from a common alphabet and hence the possible values of an infinite sequence of individual measurements. The mathematical model for this is called a *random process* (or a *random sequence* or a *random time series*). Example [A.10] considers a waveform taking values in a given coordinate space. The mathematical model for this is also called a *random process.* When it is desired to distinguish between random sequences and random waveforms, the first is called a *discrete time random process* and the second is called a *continuous time random process.*

In Chapter 3 we define precisely what is meant by a random variable, a random vector, and a random process. For now, random variables, random vectors, and random processes can be viewed simply as abstract spaces such as in the preceding examples for scalars, vectors, and sequences or waveforms together with a probabilistic description of the possible outcomes, that is, a means of quantifying how likely certain outcomes are. It is a crucial observation at this point that the three notions are intimately connected: random vectors and processes can be viewed as collections or families of random variables. Conversely, we can obtain the scalar random variables by observing the coordinates of a random vector or random process. That is, if we "sample" a random process once, we get a random variable. Thus we shall often be interested in several different, but related, abstract spaces. For example, the individual scalar outputs may be drawn from one space, say A, which could be any of the spaces in Examples [A.1], [A.4], or [A.7]. We then may also wish to look at all possible k-dimensional vectors with coordinates in A, a space that is often denoted by A^k, or at spaces of infinite sequences of waveforms of A. These latter spaces are called *product spaces* and play an important role in modeling random phenomena.

Usually one will have the option of choosing any of a number of spaces as a model for the outputs of a given random variable. For example, in flipping a coin one could use the binary space $\{head,\ tail\}$, the binary space $\{0, 1\}$ (obtained by assigning 0 to *head* and 1 to *tail*), or the entire real line $\Re$. Obviously the last space is much larger than needed, but it still captures all of the possible outcomes (along with many "impossible" ones). Which view and which abstract space is the "best" will depend on the problem at hand, and the choice will usually be made for reasons of convenience.

Given an abstract space, we consider groupings or collections of the elements that may be (but are not necessarily) smaller than the whole space and larger than single points. Such groupings are called *sets*. If every point in one set is also a point in a second set, then the first set is said to be a *subset* of the second. Examples (corresponding respectively to the previous abstract space examples) are:

[A.11] The empty set $\emptyset$ consisting of no points at all. Thus we could rewrite Example [A.0] as $\Omega = \emptyset$. By convention, the empty set is considered to be a subset of all other sets.

[A.12] The set consisting of the single element *one*. This is an example of a *one-point set* or *singleton set*.

[A.13] The set of all k-dimensional binary vectors with exactly one coordinate with value zero.

[A.14] The set of all infinite sequences of *ones* and *zeros* with exactly 50% of the symbols being *one* (as defined by an appropriate mathematical limit).

[A.15] The set of all ASCII characters for capital letters.

[A.16] The set of all four-letter English words.

[A.17] The set of all infinite sequences of ASCII characters excluding those representing control characters.

[A.18] Intervals such as the set of all voltages lying between 1 volt and 20 volts. These are useful subsets of the real line. They come in several forms, depending on whether or not the end points are included. Given $b > a$, define the "open" interval $(a, b) = \{r : a < r < b\}$, and given $b \geq a$, define the "closed" interval $[a, b] = \{r : a \leq r \leq b\}$. That is, we use a bracket if the end point is included and a parenthesis if it is not. We will also consider "half-open" or "half-closed" intervals of the form $(a, b] = \{r : a < r \leq b\}$ and $[a, b) = \{r : a \leq r < b\}$. (We use quotation marks around terms like open and closed because we are not rigorously defining them, we are implicitly defining them by their most important examples, intervals of the real line.)

[A.19] The set of all vectors of k voltages such that the largest value is less than 1 volt.

[A.20] The set of all sequences of voltages that are all nonnegative.

[A.21] The set of all voltage–time waveforms that lie between 1 and 20 volts for all time.

Given a set F of points in an abstract space Ω, we write $\omega \in F$ for "the point ω is contained in the set F" and $\omega \notin F$ for "the point ω is not contained in the set F." The symbol $\in$ is referred to as the *element inclusion symbol*. We often describe a set using this notation in the form $F = \{\omega : \omega \text{ has some property}\}$. Thus $F = \{\omega : \omega \in F\}$. For example, a set in the abstract space $\Omega = \{\omega : -\infty < \omega < \infty\}$ (the real line $\Re$) is $\{\omega : -2 \leq \omega < 4.6\}$. The abstract space itself is a grouping of elements and hence is also called a set. Thus $\Omega = \{\omega : \omega \in \Omega\}$.

If a set F is a subset of another set G, that is, if $\omega \in F$ implies that also $\omega \in G$, then we write $F \subset G$. The symbol $\subset$ is called the *set inclusion symbol*. Since a set is included within itself, every set is a subset of itself.

An individual element or point ω_0 in F can be considered both as a point or element in the space and as a one-point set or singleton set $\{\omega_0\} = \{\omega :$

$\omega = \omega_0\}$. Note, however, that the braces notation is more precise when we are considering the one-point set and that $\omega_0 \in \Omega$ whereas $\{\omega_0\} \subset \Omega$.

The three basic operations on sets are *complementation*, *intersection*, and *union*. The definitions are given next. Refer also to Figure A.1 as an aid in visualizing the definitions. In Figure A.1 Ω is pictured as the outside box and the sets F and G are pictured as arbitrary blobs within the box. Such diagrams are called *Venn diagrams*.

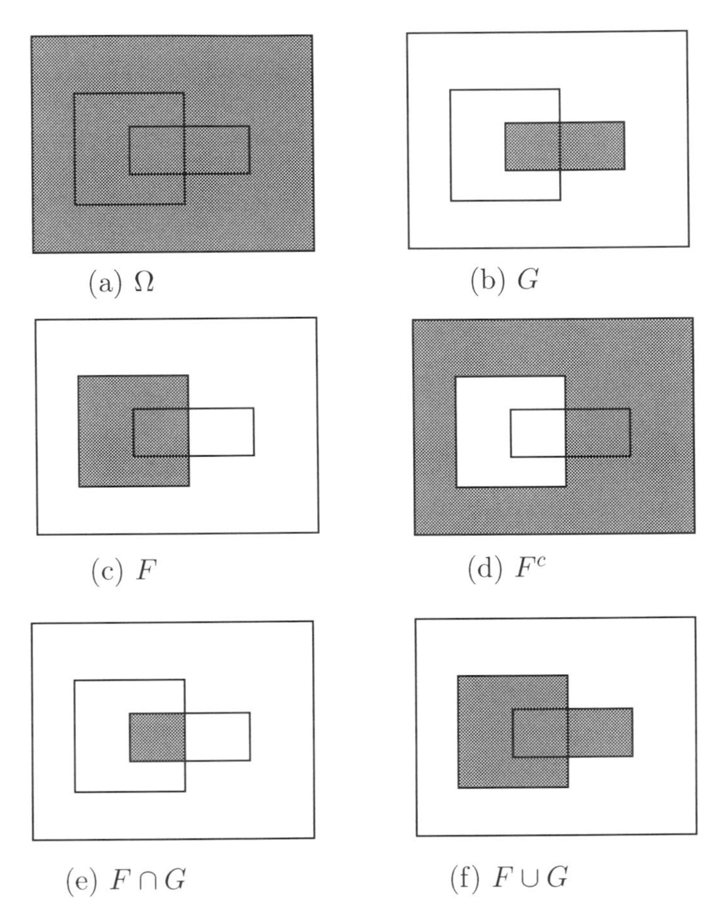

Figure A.1 Basic set operations

Given a set F, the complement of F is denoted by F^c, which is defined by

$$F^c = \{\omega : \omega \notin F\},$$

that is, the complement of F contains all of the points of Ω that are not in F.

Given two sets F and G, the *intersection* of F and G is denoted by $F \cap G$, which is defined by

$$F \cap G = \{\omega : \omega \in F \text{ and } \omega \in G\},$$

that is, the intersection of two sets F and G contains the points which are in both sets.

If F and G have no points in common, then $F \cap G = \emptyset$, the null set, and F and G are said to be *disjoint* or *mutually exclusive.*

Given two sets F and G, the *union* of F and G is denoted by $F \cup G$, which is defined by

$$F \cup G = \{\omega : \omega \in F \text{ or } \omega \in G\},$$

that is, the union of two sets F and G contains the points that are either in one set or the other, or both.

Observe that the intersection of two sets is *always* a subset of each of them, e.g., $F \cap G \subset F$. The union of two sets, however, is not a subset of either of them (unless one set is a subset of the other). Both of the original sets are subsets of their union, e.g., $F \subset F \cup G$.

In addition to the three basic set operations, there are two others that come in handy. Both can be defined in terms of the three basic operations. Refer to Figure A.2 as a visual aid in understanding the definitions.

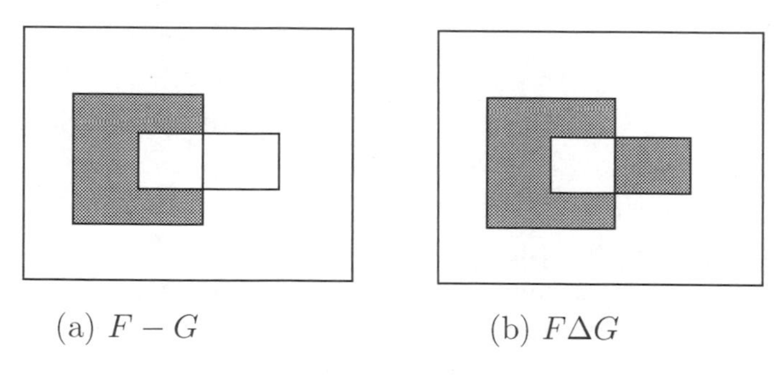

(a) $F - G$ (b) $F \Delta G$

Figure A.2 Set difference operations

Given two sets F and G, the *set difference* of F and G is denoted by $F - G$, which is defined as

$$F - G = \{\omega : \omega \in F \text{ and } \omega \notin G\} = F \cap G^c;$$

that is, the difference of F and G contains all of the points in F that are not also in G. Note that this operation is not completely analogous to the "minus" of ordinary arithmetic because there is no such thing as a "negative set."

Given two sets F and G, their *symmetric difference* is denoted by $F\Delta G$, which is defined as

$$\begin{aligned} F\Delta G &= \{\omega : \omega \in F \text{ or } \omega \in G \text{ but not both}\} \\ &= (F - G) \cup (G - F) = (F \cap G^c) \cup (F^c \cap G) \\ &= (F \cup G) - (F \cap G); \end{aligned}$$

that is, the symmetric difference between two sets is the set of points that are in one of the two sets but are not common to both sets. If both sets are the same, the symmetric difference consists of no points, that is, it is the empty set. If $F \subset G$, then obviously $F\Delta G = G - F$.

Observe that two sets F and G will be equal if and only if $F \subset G$ and $G \subset F$. This observation is often useful as a means of proving that two sets are identical: first prove that each point in one set is in the other and hence the first set is subset of the second. Then prove the opposite inclusion. Surprisingly, this technique is frequently much easier than a direct proof that two sets are identical by a pointwise argument of commonality.

We will often wish to combine sets in a series of operations and to reduce the expression for the resulting set to its simplest and most compact form. Although the most compact form frequently can be seen quickly with the aid of a Venn diagram, as in Figures A.1 and A.2, to be completely rigorous, the use of set theory or set algebra to manipulate the basic operations is required. Table A.1 collects the most important such identities. The first seven relations can be taken as axioms in an algebra of sets and used to derive all other relations, including the remaining relations in the table. Some examples of such derivations follow the table. Readers who are familiar with Boolean algebra will find a one-to-one analogy between the algebra of sets and Boolean algebra. DeMorgan's "laws" (A.6) and (A.10) are useful when complementing unions of intersections. Relation (A.16) is useful for writing the union of overlapping sets as a union of disjoint sets. A set and its complement are always disjoint by relation (A.5).

A.2 Examples of proofs

Relation (A.8). From the definition of intersection and Figure A.1 we verify the truth of (A.8). Algebraically, we show the same thing from the basic

$$F \cup G = G \cup F \text{ commutative law} \tag{A.1}$$

$$F \cup (G \cup H) = (F \cup G) \cup H \text{ associative law} \tag{A.2}$$

$$F \cap (G \cup H) = (F \cap G) \cup (F \cap H) \text{ distributive law} \tag{A.3}$$

$$(F^c)^c = F \tag{A.4}$$

$$F \cap F^c = \emptyset \tag{A.5}$$

$$(F \cap G)^c = F^c \cup G^c \text{ DeMorgan's ``law''} \tag{A.6}$$

$$F \cap \Omega = F \tag{A.7}$$

$$F \cap G = G \cap F \text{ commutative law} \tag{A.8}$$

$$F \cap (G \cap H) = (F \cap G) \cap H \text{ associative law} \tag{A.9}$$

$$(F \cup G)^c = F^c \cap G^c \text{ DeMorgan's other ``law''} \tag{A.10}$$

$$F \cup F^c = \Omega \tag{A.11}$$

$$F \cup \emptyset = F \tag{A.12}$$

$$F \cup (F \cap G) = F = F \cap (F \cup G) \tag{A.13}$$

$$F \cup \Omega = \Omega \tag{A.14}$$

$$F \cap \emptyset = \emptyset \tag{A.15}$$

$$F \cup G = F \cup (F^c \cap G) = F \cup (G - F) \tag{A.16}$$

$$F \cup (G \cap H) = (F \cup G) \cap (F \cup H) \text{ distributive law} \tag{A.17}$$

$$\Omega^c = \emptyset \tag{A.18}$$

$$F \cup F = F \tag{A.19}$$

$$F \cap F = F \tag{A.20}$$

Table A.1 *Set algebra*

seven axioms: From (A.4) and (A.6) we have that

$$A \cap B = ((A \cap B)^c)^c = (A^c \cup B^c)^c,$$

and using (A.1), (A.4), and (A.6), this becomes

$$(B^c \cup A^c)^c = (B^c)^c \cap (A^c)^c$$

as desired.

Relation (A.18). Set $F = \Omega$ in (A.5) to obtain $\Omega \cap \Omega^c = \emptyset$, which with (A.7) and (A.8) yields (A.18).

Relation (A.11). Complement (A.5), $(F^c \cap F)^c = \emptyset^c$, and hence, using (A.6), $(F^c \cup F) = \emptyset^c$, and finally, using (A.4) and (A.18), $F^c \cup F = \Omega$.

Relation (A.12). Using F^c in (A.7): $F^c \cap \Omega = F^c$. Complementing the result: $(F^c \cap \Omega)^c = (F^c)^c = F$ (by (A.4)). Using (A.6): $(F^c \cap \Omega)^c = F \cup \Omega^c = F$. From (A.18) $\Omega^c = \emptyset$, yielding (A.12).

Relation (A.20). Set $G = F$ and $H = F^c$ in (A.3) to obtain $F \cap (F \cup F^c) = (F \cap F) \cup (F \cap F^c) = F \cap F$ using (A.5) and (A.12). Applying (A.11) and (A.7) to the left-hand side of this relation yields $F \cap \Omega = F = F \cap F$.

Relation (A.19). Complement (A.20) using (A.6) and replace F^c by F.

The proofs for the examples were algebraic in nature, manipulating the operations based on the axioms. Proofs can also be constructed based on the definitions of the basic operations. For example, DeMorgan's law can be proved directly by considering individual points. To prove that $(F \cap G)^c = F^c \cup G^c$ it suffices to show separately that $(F \cap G)^c \subset F^c \cup G^c$ and $F^c \cup G^c \subset (F \cap G)^c$. Suppose that $\omega \in (F \cap G)^c$, then $\omega \notin F \cap G$ from the definition of complement and hence $\omega \notin F$ or $\omega \notin G$ from the definition of intersection (if ω were in both, it would be in the intersection). Thus either $\omega \in F^c$ or $\omega \in G^c$ and hence $\omega \in F^c \cup G^c$. Conversely, if $\omega \in F^c \cup G^c$, then $\omega \in F^c$ or $\omega \in G^c$, and hence either $\omega \neq F$ or $\omega \neq G$, which implies that $\omega \neq F \cap G$, which in turn implies that $\omega \in (F \cap G)^c$, completing the proof.

We will have occasion to deal with more general unions and intersections, that is, unions or intersections of more than two or three sets. As long as the number of unions and intersections is finite, the generalizations are obvious. The various set theoretic relations extend to unions and intersections of finite collections of sets. For example, DeMorgan's law for finite collections of sets is

$$\left(\bigcap_{i=1}^{n} F_i \right)^c = \bigcup_{i=1}^{n} F_i^c. \tag{A.21}$$

This result can be proved using the axioms or by induction. Point arguments are often more direct. Define the set on the left-hand side of the equation as G and that on the right-hand side as H and prove $G = H$ by considering individual points. This is done by separately showing that $G \subset H$ and $H \subset G$, which implies the two sets are the same. To show that $G \subset H$, let $\omega \in G = (\bigcap_{i=1}^{n} F_i)^c$, which means that $\omega \notin \bigcap_{i=1}^{n} F_i$, which means that $\omega \notin F_i$ for some i or, equivalently, that $\omega \in F_i^c$ for some i. This means that $\omega \in \bigcup_{i=1}^{n} F_i^c$ and hence that $\omega \in H$. Thus $G \subset H$ since we have shown that every point in G is also in H. The converse containment follows in a similar manner. If $\omega \in H = \bigcup_{i=1}^{n} F_i^c$, the $\omega \in F_i^c$ for some i and hence $\omega \notin F_i$ for some i. This implies that $\omega \notin \bigcap_{i=1}^{n} F_i$ and hence that $\omega \in G$, completing the proof.

The operations can also be defined for quite general infinite collections of sets as well. Say that we have an indexed collection of sets $\{A_i; i \in \mathcal{I}\}$, sometimes denoted $\{A_i\}_{i \in \mathcal{I}}$, for some index set $\mathcal{I}$. In other words, this collection is a set whose elements are sets — one set A_i for each possible value of an index i drawn from $\mathcal{I}$. We call such a collection a *family* or *class* of sets. (To avoid confusion we never say a "set of sets.") The index set $\mathcal{I}$ can be thought of as numbering the sets. Typical index sets are the set $\mathcal{Z}_+$ of all nonnegative integers, $\mathcal{Z} = \{\ldots, -1, 0, 1, \ldots\}$, or the real line $\Re$. The index set may be finite in that it has only a finite number of entries, say $\mathcal{I} = \mathcal{Z}_k = \{0, 1, \ldots, k-1\}$. The index set is said to be *countably infinite* if its elements can be counted, that is, can be put into a one-to-one correspondence with a subset of the nonnegative integers $\mathcal{Z}_+$; e.g., $\mathcal{Z}_+$ or $\mathcal{Z}$ itself. If an index set has an infinity of elements, but the elements cannot be counted, then it is said to be uncountably infinite, for example $\Re$ or the unit interval $[0, 1]$ (see Problem a.11).

The family of sets is said to be finite, countable, or uncountable if the respective index set is finite, countable, or uncountable. As an example, the family of sets $\{[0, 1/r); r \in \mathcal{I}\}$ is countable if $\mathcal{I} = \mathcal{Z}$ and uncountable if $\mathcal{I} = \Re$. Another way of describing countably infinite sets is that they can be put into one-to-one correspondence with the integers. For example, the set of rational numbers is countable because it can be enumerated, the set of irrational numbers is not.

The obvious extensions of the pairwise definitions of union and intersection will now be given. Given an indexed family of sets $\{A_i; i \in \mathcal{I}\}$, define the union by

$$\bigcup_{i \in \mathcal{I}} A_i = \{\omega : \omega \in A_i \text{ for at least one } i \in \mathcal{I}\}$$

and define the intersection by

$$\bigcap_{i \in \mathcal{I}} A_i = \{\omega : \omega \in A_i \text{ for all } i \in \mathcal{I}\}.$$

In certain special cases we make the notation more specific for particular index sets. For example, if $\mathcal{I} = \{0, \ldots, n-1\}$, then we write the union and intersection as

$$\bigcup_{i=0}^{n-1} A_i \text{ and } \bigcap_{i=0}^{n-1} A_i$$

respectively.

A collection of sets $\{F_i;\ i \in \mathcal{I}\}$ is said to be *disjoint* or *pairwise disjoint* or *mutually exclusive* if

$$F_i \cap F_j = \emptyset; \text{ all } i, j \in \mathcal{I}, i \neq j,$$

that is, if no sets in the collection contain points contained by other sets in the collection.

The class of sets is said to be *collectively exhaustive* or to *exhaust* the space if

$$\bigcup_{i \in \mathcal{I}} F_i = \Omega,$$

that is, together the F_i contain all the points of the space.

A collection of sets $\{F_j;\ i \in \mathcal{I}\}$ is called a *partition* of the space Ω if the collection is both disjoint and collectively exhaustive. A collection of sets $\{F_i;\ i \in \mathcal{I}\}$ is said to partition a set G if the collection is disjoint and the union of all of its members is identical to G.

A.3 Mappings and functions

We make much use of mappings of functions from one space to another. This is of importance in a number of applications. For example, the waveforms and sequences that we considered as members of an abstract space describing the outputs of a random process are just functions of time – for each value of time t in some continuous or discrete collection of possible times we assigned some output value to the function. As a more complicated example, consider a binary digit that is transmitted to a receiver at some destination by sending either plus or minus V volts through a noisy environment called a "channel." At the receiver a decision is made whether $+V$ or $-V$ was sent. The receiver puts out a 1 or a 0, depending on the decision. In this example three mappings are involved: the transmitter maps a binary symbol in $\{0, 1\}$ into either $+V$ or $-V$. During transmission, the channel has an input either

$+V$ or $-V$ and produces a real number, not usually equal to $0, 1, +V$, or $-V$. At the receiver, a real number is viewed and a binary number produced.

We will encounter a variety of functions or mappings, from simple arithmetic operations to general filtering operations. We now introduce some common terminology and notation for handling such functions. Given two abstract spaces Ω and A, an A-valued *function* or *mapping* F or, in more detail, $f : \Omega \to A$, is an assignment of a unique point in A to each point in Ω; that is, given any point $\omega \in \Omega$, $f(\omega)$ is some value in A. Ω is called the *domain* or *domain of definition* of the function f, and A is called the *range* of f. Given any sets $F \subset \Omega$ and $G \subset A$, define the *image* of F (under f) as the set

$$f(F) = \{a : a = f(\omega) \text{ for some } \omega \in F\}$$

and the *inverse image* (also called the *preimage*) of G (under f) as the set

$$f^{-1}(G) = \{\omega : f(\omega) \in G\}.$$

Thus $f(F)$ is the set of all points in A obtained by mapping points in F, and $f^{-1}(G)$ is the set of all points in Ω that map into G.

For example, let $\Omega = [-1, 1]$ and $A = [-10, 10]$. Given the function $f(\omega) = \omega^2$ with domain Ω and range A, define the sets $F = (-1/2, 1/2) \subset \Omega$ and $G = (-1/4, 1) \subset A$. Then $f(F) = [0, 1/4)$ and $f^{-1}(G) = [-1, 1]$. As you can see from this example, not all points in G have to correspond to points in F. In fact, the inverse image can be empty; e.g., continuing the same example, $f^{-1}((-1/4, 0)) = \emptyset$.

The image of the entire space Ω is called the *range space* of f, and it need not equal the range; the function f could map the whole input space into a single point in A. For example, $f : \Re \to \Re$ defined by $f(r) = 1$, all r, has a range space of a single point. If the range space equals the range, the mapping is said to be *onto*. (Is the mapping f of the preceding example onto? What is the range space? Is the range unique?)

A mapping f is called *one-to-one* if $x \neq y$ implies that $f(x) \neq f(y)$.

A real valued function f is said to be *continuous* at r_0 if given any $\epsilon > 0$ there is a $\delta > 0$ such that if $|r - r_0| \leq \delta$, then $|f(r) - f(r_0)| \leq \epsilon$.

A.4 Linear algebra

We collect a few definitions and results for vectors, matrices, and determinants.

There is a variety of notational styles for vectors. Historically a common form was to use boldface, as in $\mathbf{x} = (x_0, x_0, \ldots, x_{k-1})$, to denote a k-dimensional vector with k components x_i, $i = 0, 1, \ldots, k-1$. When dealing with linear algebra, however, it is most commonly the convention to assume that vectors are column vectors, e.g.,

$$\mathbf{x} = \begin{pmatrix} x_0 \\ x_1 \\ \vdots \\ x_{k-1} \end{pmatrix},$$

or, as an in-line equation, $\mathbf{x} = (x_0, x_0, \ldots, x_{k-1})^t$, where t denotes "transpose." We will often be lazy and write vectors inline without explicitly denoting the transpose unless it is needed, e.g., in vector/matrix equations. Although boldface makes it clear which symbols are vectors and which are scalars, in modern practice it is more common to drop the distinction and not use boldface, i.e., to write a vector as simply $x = (x_0, x_1, \ldots, x_{k-1})$ or, if it is desired to make clear it is a column vector, as $x = (x_0, x_1, \ldots, x_{k-1})^t$. Both boldface and nonboldface notations are used in this book. Generally, early on the boldface notation is used to clarify when vectors or scalars are being used while later in the book boldface is often dropped.

The inner product (or dot product) of two real-valued n-dimensional vectors y and n is defined by the scalar value

$$x^t y = \sum_{i=0}^{n-1} x_i y_i. \tag{A.22}$$

If the vectors are more generally complex-valued, then the transpose is replaced by a conjugate transpose

$$x^* y = \sum_{i=0}^{n-1} x_i^* y_i. \tag{A.23}$$

A matrix is a rectangular array of numbers

$$A = \begin{bmatrix} a_{0,0} & a_{0,1} & a_{0,2} & \cdots a_{0,n-1} \\ a_{1,0} & a_{1,1} & a_{1,2} & \cdots a_{1,n-1} \\ \cdot & \cdot & \cdot & \cdot \;\; \cdot \\ \cdot & \cdot & \cdot & \cdot\,\cdot \\ \cdot & \cdot & \cdot & \cdot\,\cdot \\ a_{m-1,0} & a_{m-1,1} & a_{m-1,2} & \cdots a_{m-1,n-1} \end{bmatrix}$$

with m rows and n columns. Boldface notation is also used for matrices. If $m = n$ the matrix is said to be *square*. A matrix is symmetric if $A^t = A$, where A^t is the transpose of the matrix A, that is, the $n \times m$ matrix whose k, jth element is $(A^t)_{k,j} = a_{j,k}$. If the matrix has complex elements and $A^* = A$, where $*$ denotes conjugate transpose so that $(A^*)_{k,j} = a^*_{j,k}$, then A is said to be *Hermitian*.

The product of an $m \times n$ matrix and an n-dimensional vector $y = Ax$ is an m dimensional vector with components

$$y_i = \sum_{k=0}^{n-1} a_{i,k} x_k,$$

that is, the inner product of x and the ith row of A.

The outer product of two n-dimensional vectors y and n is defined as the $n \times n$ matrix

$$xy^t = \begin{bmatrix} x_0y_0 & x_0y_1 & x_0y_2 & \cdots & x_0y_{n-1} \\ x_1y_0 & x_1y_1 & x_1y_2 & \cdots & x_1y_{n-1} \\ \cdot & \cdot & \cdot & \cdot & \cdot \\ \cdot & \cdot & \cdot & \cdot & \cdot \\ \cdot & \cdot & \cdot & & \cdot\cdot \\ x_{n-1}y_0 & x_{n-1}y_1 & x_{n-1}y_2 & \cdots & x_{n-1}y_{n-1} \end{bmatrix} \tag{A.24}$$

Given a square matrix A, a scalar λ is called an *eigenvalue* and a vector u is called an *eigenvector* if

$$Au = \lambda u. \tag{A.25}$$

An $n \times n$ matrix has n eigenvalues and eigenvectors, but they need not be distinct. Eigenvalues provide interesting formulas for two attributes of matrices, the trace defined by

$$\mathrm{Tr}(A) = \sum_{i=0}^{n-1} a_{i,i}$$

and the determinant of the matrix $\det(A)$:

$$\mathrm{Tr}(A) = \sum_{i=0}^{n-1} \lambda_i \tag{A.26}$$

$$\det(A) = \prod_{i=0}^{n-1} \lambda_i \tag{A.27}$$

The arithmetic–geometric means inequality says that the arithmetic mean of a collection of real positive numbers λ_i is bounded below by the geometric mean:

$$\frac{1}{n}\sum_{i=0}^{n-1}\lambda_i \geq \left(\prod_{i=0}^{n-1}\lambda_i\right)^{\frac{1}{n}} \tag{A.28}$$

with equality if and only if the λ are all the same. Application of the inequality to the eigenvalue representation of the determinant and trace in the case where the eigenvalues are real and positive (e.g., an Hermitian matrix) provides the inequality

$$\frac{1}{n}\operatorname{Tr}(A) \geq (\det(A))^{1/n} \tag{A.29}$$

with equality if and only if the eigenvalues are all constant.

A similar argument provides an inequality for the harmonic mean, the arithmetic mean of the reciprocals. If the matrix A has real and positive eigenvalues, then

$$\frac{1}{n}\operatorname{Tr}(A) \geq (\det(A))^{1/n} \geq \frac{n}{\operatorname{Tr}(A^{-1})}. \tag{A.30}$$

A square Hermitian matrix A can be diagonalized into the form

$$A = U\Lambda U^*, \tag{A.31}$$

where Λ is the diagonal matrix with diagonal entries $\Lambda(k,k) = \lambda_k$, the kth eigenvalue of the matrix, and where U is a unitary matrix, that is, $U^* = U^{-1}$.

The inner product and outer product of two vectors can be related as

$$x^t y = \operatorname{Tr}(xy^t). \tag{A.32}$$

Given an n-dimensional vector x and an $n \times n$ matrix A, the product

$$x^t Ax = \sum_{k=0}^{n-1}\sum_{j=0}^{n-1} x_k x_j a_{k.j}$$

is called a *quadratic form.* If the matrix A is such that $x^t Ax \geq 0$, the matrix is said to be *nonnegative definite.* If the matrix is such that $x^t Ax > 0$, then the matrix is said to be *positive definite.* These are the definitions for real-valued vectors and matrices. For complex vectors and matrices use the conjugate transpose instead of the transpose. If a matrix is positive definite, then its eigenvalues are all strictly positive and hence so is its determinant.

A quadratic form can also be written as

$$x^t Ax = \operatorname{Tr}(Axx^t). \tag{A.33}$$

If a matrix A is positive definite and Hermitian (e.g., real and symmetric), then its square root $A^{1/2}$ is well-defined as $U\Lambda^{1/2}U^*$. In particular, $A^{1/2}A^{1/2} = A$ and $(A^{1/2})^{-1} = (A^{-1})^{1/2}$.

A.5 Linear system fundamentals

In general, a *system* $\mathcal{L}$ is a mapping of an input time function or input *signal*, $x = \{x(t);\, t \in \mathcal{T}\}$ into an output time function or output signal, $\mathcal{L}(x) = y = \{y(t);\, t \in \mathcal{T}\}$. We now use $\mathcal{T}$ to denote the index set or domain of definition instead of $\mathcal{I}$ to emphasize that the members of the set correspond to "time." Usually the functions take on real or complex values of each value of time t in $\mathcal{T}$. The system is called a *discrete time system* if $\mathcal{T}$ is discrete, e.g., $\mathcal{Z}$ or $\mathcal{Z}_+$, and it is called a *continuous time system* if $\mathcal{T}$ is continuous, e.g., $\Re$ or $[0, \infty)$. If only nonnegative times are allowed, e.g., $\mathcal{T}$ is $\mathcal{Z}_+$ or $[0, \infty)$, the system is called a one-sided or single-sided system. If time can go on infinitely in both directions, then it is said to be a two-sided system.

A system $\mathcal{L}$ is said to be *linear* if the mapping is linear, that is, for all complex (or real) constants a and b and all input functions x_1 and x_2

$$\mathcal{L}(ax_1 + bx_2) = a\mathcal{L}(x_1) + b\mathcal{L}(x_2). \tag{A.34}$$

There are many ways to define or describe a particular linear system: one can provide a constructive rule for determining the output from the input; e.g., the output may be a weighted sum or integral of values of the input. Alternatively, one may provide a set of equations whose solution determines the output from the input, e.g., differential or difference equations involving the input and output at various times. Our emphasis is on the former constructive technique, but we shall occasionally consider examples of other techniques.

The most common and the most useful class of linear systems comprises systems that can be represented by a convolution, that is, where the output is described by a weighted integral or sum of input values. We first consider continuous time systems and then turn to discrete time systems.

For $t \in \mathcal{T} \subset \Re$, let $x(t)$ be a continuous time input to a system with output $y(t)$ defined by the superposition integral

$$y(t) = \int_{s:t-s\in\mathcal{T}} x(t-s)h_t(s)\,ds. \tag{A.35}$$

The function $h_t(t)$ is called the *impulse response* of the system since it can be considered the output of the system at time t which results from an input

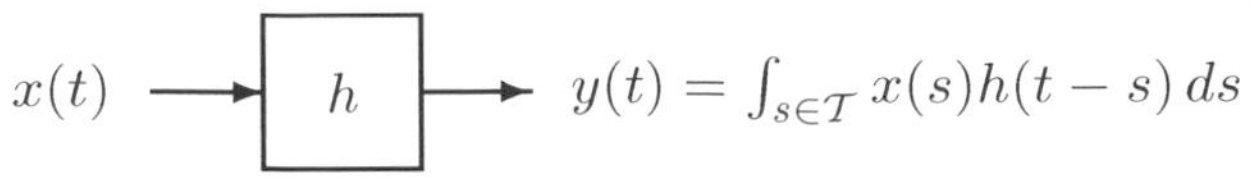

$$x(t) \longrightarrow \boxed{h} \longrightarrow y(t) = \int_{s \in \mathcal{T}} x(s)h(t-s)\,ds$$

Figure A.3 Linear filter

of a Dirac delta function $x(t) = \delta(t)$ at time 0. The index set is usually either $(-\infty, \infty)$ or $[0, \infty)$ for continuous time systems. The linearity of integration implies that the system defined by (A.35) is a linear system. A system of this type is called a *linear filter.* If the impulse response does not depend on time t, then the filter is said to be *time-invariant* and the superposition integral becomes a *convolution integral*:

$$y(t) = \int_{s:t-s \in \mathcal{T}} x(t-s)h(s)\,ds = \int_{s \in \mathcal{T}} x(s)h(t-s)\,ds. \tag{A.36}$$

We shall deal almost exclusively with time-invariant filters. Such a linear time-invariant system is often depicted using a block diagram as in Figure A.3.

If $x(t)$ and $h(t)$ are absolutely integrable, i.e.,

$$\int_{\mathcal{T}} |x(t)|dt, \quad \int_{\mathcal{T}} |h(t)|dt < \infty, \tag{A.37}$$

then their Fourier transforms exist:

$$X(f) = \int_{\mathcal{T}} x(t)e^{-j2\pi ft}dt, \quad H(f) = \int_{\mathcal{T}} h(t)e^{-j2\pi ft}dt. \tag{A.38}$$

Continuous time filters satisfying (A.37) are said to be *stable.* The function $H(f)$ is called the filter transfer function or the system function. We point out that (A.37) is a sufficient but not necessary condition for the existence of the transform. We are not usually concerned with the fine points of the existence of such transforms and their inverses. The inverse transforms that we require are accomplished either by inspection or by reference to a table.

A basic property of Fourier transforms is that convolution in the time domain corresponds to multiplication in the frequency domain, and hence the output transform is given by

$$Y(f) = H(f)X(f). \tag{A.39}$$

Even if a particular system has an input that does not have a Fourier transform, (A.39) can be used to find the transfer function of the system by using some other input that does have a Fourier transform.

As an example, consider Figure A.4, where two linear filters are concate-

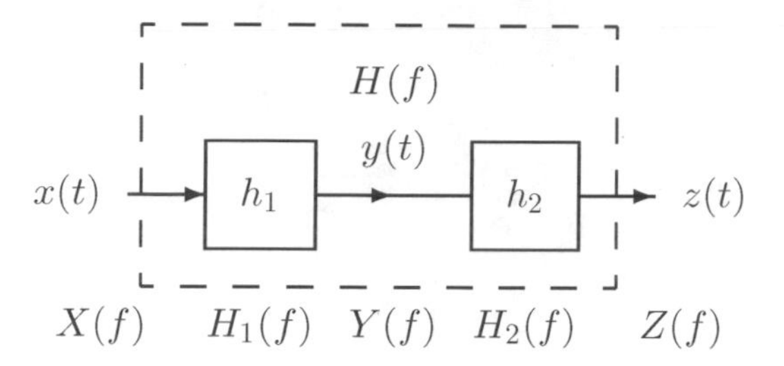

Figure A.4 Cascade filter

nated or cascaded: $x(t)$ is input to the first filter, and the output $y(t)$ is input to the second filter, with final output $z(t)$. If both filters are stable and $x(t)$ is absolutely integrable, the Fourier transforms satisfy

$$Y(f) = H_1(f)X(f), \quad Z(f) = H_2(f)Y(f), \tag{A.40}$$

or

$$Z(f) = H_2(f)H_1(f)X(f).$$

Obviously the overall filter transfer function is $H(f) = H_2(f)H_1(f)$. The overall impulse response is then the inverse transform of $H(f)$.

Frequently (but not necessarily) the output of a linear filter can also be represented by a finite-order differential equation in terms of the differential operator, $D = d/dt$:

$$\sum_{k=0}^{n} a_k D^k y(t) = \sum_{i=0}^{m} b_i D^i x(t). \tag{A.41}$$

The output is completely specified by the input, the differential equation, and appropriate initial conditions. Under suitable conditions on the differential equation, the linear filter is stable, and the transfer function can be obtained by transforming both sides of (A.41). However, we shall not pursue this approach further.

Turn now to Figure A.5. Here we show an idealized sampled data system to demonstrate the relationship between discrete and continuous time filters. The input function $x(t)$ is input to a mixer, which forms the product of $x(t)$ with a pulse train, $p(t) = \sum_{k \in \mathcal{T}} \delta(t - k)$, of Dirac delta functions spaced 1 second apart in time. $\mathcal{T}$ is a suitable subset of $\mathcal{Z}$. If we denote the sampled values $x(k)$ by x_k, the product is

$$x(t)p(t) = \sum_k x_k \delta(t - k),$$

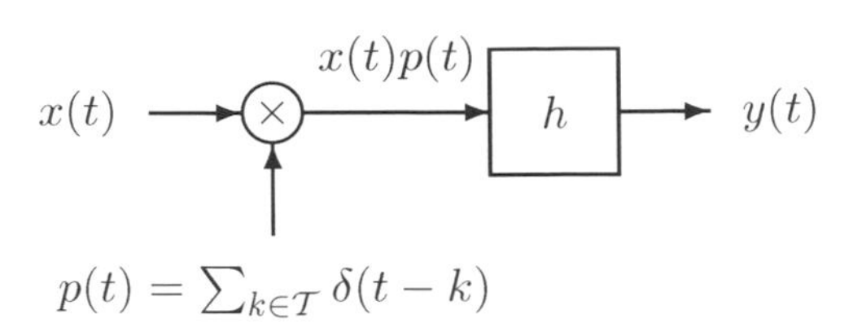

Figure A.5 Sampled data system

which is the input to a linear filter with impulse response $h(t)$. Applying the convolution integral of (A.36) and sampling the output with a switch at 1second intervals, we have as an output function at time n

$$\begin{aligned} y_n = y(n) &= \int x(t)p(t)h(t-n)\,dt \\ &= \int \sum_k x_k \delta(t-k)h(t-n)\,dt = \sum_{k:k\in\mathcal{T}} x_k h_{n-k} \\ &= \sum_{k:n-k\in\mathcal{T}} x_{n-k} h_k. \end{aligned} \tag{A.42}$$

Thus, macroscopically the filter is a discrete time linear filter with a discrete convolution sum in place of an integral. The sequence $\{h_k\}$ is called the *Kronecker* delta *response* or δ-response of the discrete time filter. Its name is derived from the fact that h_k is the output of the linear filter at time k when a Kronecker delta function is input at time zero. It is also sometimes referred to as the "discrete time impulse response" or "unit pulse response." If only a finite number of the h_k are nonzero, then the filter is sometimes referred to as an FIR (finite impulse response) filter. If a filter is not an FIR filter, then it is an IIR (infinite impulse response) filter.

Generally we shall use both δ-response and impulse response for both the discrete and continuous time cases.

If $\{h_k\}$ and $\{x_k\}$ are both absolutely summable,

$$\sum_k |h_k| < \infty, \ \sum_k |x_k| < \infty, \tag{A.43}$$

then their discrete Fourier transforms exist:

$$H(f) = \sum_k h_k e^{-j2\pi kf}, \ X(f) = \sum_k x_k e^{-j2\pi kf}. \tag{A.44}$$

Discrete time filters satisfying (A.43) are said to be *stable*. $H(f)$ is called the filter transfer function. The output transform is given by

$$Y(f) = H(f)X(f). \tag{A.45}$$

The example of Figure A.4 applies for discrete time as well as continuous time.

For convenience and brevity, we shall occasionally use a general notation $\mathcal{F}$ to denote both the discrete and continuous Fourier transforms; that is,

$$\mathcal{F}(x) = \begin{cases} \int_{\mathcal{T}} x(t)e^{-j2\pi ft}\,dt & \mathcal{T} \text{ continuous} \\ \\ \sum_{k\in\mathcal{T}} x_k e^{-j2\pi fk} & \mathcal{T} \text{ discrete} \end{cases}. \tag{A.46}$$

A more general discrete time linear system is described by a difference equation of the form

$$\sum_k a_k y_{n-k} = \sum_i b_i x_{n-k}. \tag{A.47}$$

Observe that the convolution of (A.42) is a special case of the above where only one of the a_k is not zero. Observe also that the difference equation (A.47) is a discrete time analog of the differential equation (A.41). As in that case, to describe an output completely one has to specify initial conditions.

A continuous time or discrete time filter is said to be *causal* if the pulse response or impulse response is zero for negative time; that is, if a discrete time pulse response h_k satisfies $h_k = 0$ for $k < 0$ or a continuous time impulse response $h(t)$ satisfies $h(t) = 0$ for $t < 0$.

A.6 Problems

1. Use the first seven relations to prove relations (A.10), (A.13), and (A.16).
2. Use relation (A.16) to obtain a partition $\{G_i;\, i = 1, 2, \ldots, k\}$ of Ω from an arbitrary finite class of collectively exhaustive sets $\{F_i;\, i = 1, 2, \ldots, k\}$ with the property that $G_i \subset F_i$ for all i and

$$\bigcup_{j=1}^{i} G_j = \bigcup_{j=1}^{i} F_j \text{ all } i.$$

   Repeat for a countable collection of sets $\{F_i\}$. (You must prove that the given collection of sets is indeed a partition.)
3. If $\{F_i\}$ partitions Ω, show that $\{G \cap F_i\}$ partitions G.
4. Show that $F \subset G$ implies that $F \cap G = F$, $F \cup G = G$, and $G^c \subset F^c$.

5. Show that if F and G are disjoint, then $F \subset G^c$.
6. Show that $F \cap G = (F \cup G) - (F \Delta G)$.
7. Let $F_r = [0, 1/r)$, $r \in (0, 1]$. Find $\bigcup_{r \in (0,1]} F_r$ and $\bigcap_{r \in (0,1]} F_r$.
8. Prove the countably infinite version of DeMorgan's "laws." For example, given a sequence of sets F_i; $i = 1, 2, \ldots$, then

$$\bigcap_{i=1}^{\infty} F_i = \left(\bigcup_{i=1}^{\infty} F_i^c \right)^c .$$

9. Define the subsets of the real line

$$F_n = \left\{ r : |r| > \frac{1}{n} \right\},$$

and

$$F = \{0\}.$$

Show that

$$F^c = \bigcup_{n=1}^{\infty} F_n .$$

10. Let F_i, $i = 1, 2, \ldots$ be a countable sequence of "nested" closed intervals whose length is not zero, but tends to zero; i.e., for every i, $F_i = [a_i, b_i] \subset F_{i-1} \subset F_{i-2} \ldots$ and $b_i - a_i \to 0$ and $i \to \infty$. What are the points in $\bigcap_{i=1}^{\infty} F_i$?
11. Prove that the interval $[0, 1]$ cannot be put into one-to-one correspondence with the set of integers as follows: suppose that there is such a correspondence so that $x_1, x_2, x_3, \ldots$ is a listing of all numbers in $[0, 1]$. Use Problem A.10 to construct a set that consists of a point not in this listing. This contradiction proves the statement.
12. Show that inverse images preserve set theoretic operations, that is, given $f : \Omega \to A$ and sets F and G in A, then

$$f^{-1}(F^c) = (f^{-1}(F))^c.$$

$$f^{-1}(F \cup G) = f^{-1}(F) \cup f^{-1}(G),$$

and

$$f^{-1}(F \cap G) = f^{-1}(F) \cap f^{-1}(G).$$

If $\{F_i,\ i \in \mathcal{I}\}$ is an indexed family of subsets of A that partitions A, show that $\{f^{-1}(F_i);\ i \in \mathcal{I}\}$ is a partition of Ω. Do images preserve set theoretic operations in general? (Prove that they do or provide a counterexample.)
13. An experiment consists of rolling two four-sided dice (each having faces labeled 1, 2, 3, 4) on a glass table. Depict the space Ω of possible outcomes. Define two functions on Ω: $X_1(\omega)$ = the sum of the two down faces and $X_2(\omega)$ = the product

of the two down faces. Let A_1 denote the range space of X_1, A_2 the range space of X_2, and A_{12} the range space of the vector-valued function $\mathcal{X} = (X_1, X_2)$, that is, $\mathcal{X}(\omega) = (X_1(\omega), X_2(\omega))$. Draw in both Ω and A_{12} the set $\{\omega : X_1(\omega) < X_2(\omega)\}$. The cartesian product $\prod_{i=1}^{2} A_i$ of two sets is defined as the collection of all pairs of elements, one from each set, that is

$$\prod_{i=1}^{2} A_i = \{\text{all } a, b : a \in A_1, b \in A_2\}.$$

Is it true above that $A_{12} = \prod_{i=1}^{2} A_i$?

14. Let $\Omega = [0, 1]$ and A be the set of all infinite binary vectors. Find a one-to-one mapping from Ω to A, being careful to note that some rational numbers have two infinite binary representations (e.g., $1/2 = 0.1000\ldots = 0.0111\ldots$ in binary).
15. Can you find a one-to-one mapping from:
    (a) $[0, 1]$ to $[0, 2)$?
    (b) $[0, 1]$ to the unit square $[0, 1]^2$?
    (c) $\mathcal{Z}$ to $\mathcal{Z}_+$? When is it possible to find a one-to-one mapping from one space to another?
16. Suppose that a voltage is measured that takes values in $\Omega = [0, 15]$. The voltage is mapped into the finite space $A = \{0, 1, \ldots, 15\}$ for transmission over a digital channel. A mapping of this type is called a *quantizer*. What is the best mapping in the sense that the maximum error is minimized?
17. Let A be as in Problem A.16, i.e., the space of 16 messages which is mapped into the space of 16 waveforms, $B = \{\cos nt, n = 0, 1, \ldots, 15; t \in [0, 2\pi]\}$. The selected waveform from B is transmitted on a *waveform channel,* which adds noise; i.e., B is mapped into $C = \{$set of all possible waveforms $\{y(t) = \cos nt + \text{noise}(t);\ t \in [0, 2\pi]\}\}$. (This is a random mapping in a sense that is explained in the text.) Find a good mapping from C into $D = A$. The set D is the *decision space* and the mapping is called a *decision rule.* (In other words, how would you perform this mapping knowing little of probability theory? Your mapping should at least give the correct decision if the noise is absent or small.)
18. A continuous time linear filter has impulse response $h(t)$ given by e^{-at} for $t \geq 0$ and 0 for $t < 0$, where a is a positive constant. Find the transfer function $H(f)$ of the filter. Is the filter stable? What happens if $a = 0$?
19. A discrete time linear filter has δ-response h_k given by r^k for $k \geq 0$ and 0 for $k < 0$, where r has magnitude strictly less than 1, find the transfer function $H(f)$. (Hint: use the geometric series formula.) Is the filter stable? What happens if $r = 1$? Assume that $|r| < 1$. Suppose that the input $x_k = 1$ for all nonnegative k and $x_k = 0$ for all negative k is put into the filter. Find a simple expression for the output as a function of time. Does the transform of the output exist?
20. A continuous time system is described by the following relation: an input $x =$

$\{x(t);\ t \in \Re\}$ is defined for each t by

$$y(t) = (a_0 + a_1 x(t)) \cos(2\pi f_0 t + \theta),$$

where a_0, a_1, f_0, and θ are fixed parameters. (This system is called an amplitude modulation (AM) system.) Under what conditions on the parameters is this system linear? Is it time-invariant?

21. Suppose that $x = \{x(t);\ t \in \mathcal{R}\}$, where $\mathcal{R} = (-\infty, \infty)$ is the real line, is a continuous time signal defined by

$$x(t) = \begin{cases} 1 & |t| \leq T \\ 0 & \text{otherwise} \end{cases},$$

where $T > 0$ is a fixed parameter. This signal is put into a linear, time-invariant (LTI) filter described by an impulse response $h = \{h(t);\ t \in \mathcal{R}\}$, where

$$h(t) = \begin{cases} e^{-t} & t \geq 0 \\ 0 & \text{otherwise} \end{cases}.$$

(a) Find the Fourier transform X of x, i.e.,

$$X(f) = \int_{-\infty}^{\infty} x(t) e^{-j2\pi f t}\, dt;\ f \in \mathcal{R},$$

where $j = \sqrt{-1}$. Find the Fourier transform H of h.

(b) Find y, the output signal of the LTI filter, and its Fourier transform Y.

22. Suppose that $x = \{x_n;\ n \in \mathcal{Z}\}$, where $\mathcal{Z}$ is the set of all integers $\{\ldots, -2, -1, 0, 1, 2, \ldots\}$, is a discrete time signal defined by

$$x_n = \begin{cases} r^n & n \geq 0 \\ 0 & \text{otherwise} \end{cases},$$

where r is a fixed parameter satisfying $|r| < 1$, is put into a linear, time-invariant (LTI) filter described by a Kronecker delta response $h = \{h_n;\ n \in \mathcal{Z}\}$, where

$$h_n = \begin{cases} 1 & n = 0, 1, \ldots, N-1 \\ 0 & \text{otherwise} \end{cases},$$

where $N > 0$ is a fixed integer. This filter is sometimes called a "comb filter." Note that the Kronecker delta response is the response to the filter when the input is δ_n, the Kronecker delta defined as

$$\delta_n = \begin{cases} 1 & n = 0 \\ 0 & \text{otherwise} \end{cases}.$$

(a) Find the (discrete time) Fourier transform X of x, i.e.,

$$X(f) = \sum_{n=-\infty}^{\infty} x_n e^{-j2\pi fn}\,;\; f \in (-1/2, 1/2).$$

Find the Fourier transform H of h.

(b) Find y, the output signal of the LTI filter, and its Fourier transform Y.

23. Evaluate the following integrals:

(a)

$$\int\int_{x,y:0\le x,y\le 2;\sqrt{(x-2)^2+(y-2)^2}\le 1} dx\,dy$$

(b)

$$\int_0^{\infty} dx e^{-x} \int_0^{x} dy e^{-y}.$$

24. Evaluate the following integrals:

(a)

$$\int\int_{x,y:|x|+|y|\le r} dx\,dy$$

(b)

$$\int_0^{\infty} dx \int_x^{\infty} dy e^{-y}$$

(c)

$$\int\int_{x,y:0\le x^2\le y\le 1} dx\,dy.$$

Appendix B
Sums and integrals

In this appendix a few useful definitions and results are gathered for reference.

B.1 Summation

The sum of consecutive integers

$$\sum_{k=1}^{n} k = \frac{n(n+1)}{2}. \tag{B.1}$$

Proof The result is easily proved by induction, which requires demonstrating the truth of the formula for $n = 1$ (which is obvious) and showing that if the formula is true for any positive integer n, then it must also be true for $n + 1$. This follows since if $S_n = \sum_{k=1}^{n} k$ and we assume that $S_n = n(n+1)/2$, then necessarily

$$\begin{aligned} S_{n+1} &= S_n + (n+1) \\ &= (n+1)(\frac{n}{2} + 1) \\ &= \frac{(n+1)(n+2)}{2}, \end{aligned}$$

proving the claim. □

The sum of consecutive squares of integers

$$\sum_{k=1}^{n} k^2 = \frac{1}{3}n^3 + \frac{1}{2}n^2 + \frac{1}{6}n. \tag{B.2}$$

The sum can also be expressed as

$$\sum_{k=1}^{n} k^2 = \frac{(2n+1)(n+1)n}{6}.$$

Proof This can also be proved by induction, but for practice we note another approach. Just as in solving differential or difference equations, one can guess a general form of solution and solve for unknowns. Since summing k up to n had a second-order solution in n, one might suspect that solving for a sum up to n of squares of k would have a third-order solution in n, that is, a solution of the form $f(n) = an^3 + bn^2 + cn + d$ for some real numbers a, b, c, d. Assume for the moment that this is the case, then if $f(n) = \sum_{k=1}^{n} k^2$, clearly $n^2 = f(n) - f(n-1)$ and hence with a little algebra

$$\begin{aligned} n^2 &= an^3 + bn^2 + cn + d - a(n-1)^3 + b(n-1)^2 + c(n-1) + d \\ &= 3an^2 + (2b - 3a)n + (a - b + c). \end{aligned}$$

This can only be true for all n however if $3a = 1$ so that $a = 1/3$, if $2b - 3a = 0$ so that $b = 3a/2 = 1/2$, and if $a - b + c = 0$ so that $c = b - a = 1/6$. This leaves d, but the initial condition that $f(1) = 1$ implies $d = 0$. □

The geometric progression

Given a complex constant a,

$$\sum_{k=0}^{n-1} a^k = \frac{1 - a^n}{1 - a}, \tag{B.3}$$

and if $|a| < 1$ this sum is convergent and

$$\sum_{k=0}^{\infty} a^k = \frac{1}{1 - a}. \tag{B.4}$$

Proof There are, in fact, many ways to prove this result. Perhaps the simplest is to define the sum with n terms $S_n = \sum_{k=0}^{n-1} a^k$ and observe that

$$\begin{aligned}(1-a)S_n &= \sum_{k=0}^{n-1} a^k - a\sum_{k=0}^{n-1} a^k \\ &= \sum_{k=0}^{n-1} a^k - \sum_{k=1}^{n} a^k \\ &= 1 - a^n,\end{aligned}$$

proving (B.3). Other methods of proof include induction and solving the difference equation $S_n = S_{n-1} + a^{n-1}$. Proving the finite n result gives the infinite sum since if $|a| < 1$,

$$\sum_{k=0}^{\infty} a^k = \lim_{n\to\infty} S_n = \frac{1}{1-a}. \tag{B.5}$$

For the reader who might be rusty with limiting arguments, this follows since

$$|S_n - \frac{1}{1-a}| = |\frac{a^n}{1-a}| = \frac{|a|^n}{|1-a|} \to 0$$

as $n \to \infty$ since by assumption $|a| < 1$. □

First moment of the geometric progression

Given $q \in (0,1)$,

$$\sum_{k=0}^{\infty} kq^{k-1} = \frac{1}{(1-q)^2}\ . \tag{B.6}$$

Proof Since $kq^{k-1} = \frac{d}{dq}q^k$ and interchanging differentiation and summation,

$$\sum_{k=0}^{\infty} kq^{k-1} = \frac{d}{dq}\sum_{k=0}^{\infty} q^k = \frac{d}{dq}(1-q)^{-1},$$

where we have used the geometric series sum formula. □

Second moment of the geometric progression

Given $q \in (0, 1)$,

$$\sum_{k=0}^{\infty} k^2 q^{k-1} = \frac{2q}{(1-q)^3} + \frac{1}{(1-q)^2}. \tag{B.7}$$

Proof Take a second derivative of a geometric progression to find

$$\begin{aligned}\frac{d^2}{dq^2} \sum_{k=0}^{\infty} q^k &= \sum_{k=0}^{\infty} k(k-1)q^{k-2} = \frac{1}{q} \sum_{k=0}^{\infty} k(k-1)q^{k-1} \\ &= \frac{1}{q} \sum_{k=0}^{\infty} k^2 q^{k-1} - \frac{1}{q} \sum_{k=0}^{\infty} k q^{k-1} = \frac{1}{q} \sum_{k=0}^{\infty} k^2 q^{k-1} - \frac{1}{(1-q)^2 q}\end{aligned}$$

and

$$\frac{d^2}{dq^2} \sum_{k=0}^{\infty} q^k = \frac{2}{(1-q)^3}$$

so that

$$\sum_{k=0}^{\infty} k^2 q^{k-1} = \frac{2q}{(1-q)^3} + \frac{1}{(1-q)^2},$$

proving the claim. □

B.2 ⋆Double sums

The following lemma provides a useful simplification of a double summation that crops up when considering sample averages and laws of large numbers.

Lemma B.1 *Given a sequence* $\{a_n\}$,

$$\sum_{k=0}^{N-1} \sum_{l=0}^{N-1} a_{k-l} = \sum_{n=-N+1}^{N-1} (N - |n|) a_n.$$

Proof This result can be thought in terms of summing the entries of a matrix $A = \{A_{k,l};\ k, l \in \mathcal{Z}_N\}$ which has the property that all elements along any diagonal are equal, i.e., $A_{k,l} = a_{k-l}$ for some sequence a. (As mentioned in the text, a matrix of this type is called a *Toeplitz* matrix. To sum up all of the elements in the matrix note that the main diagonal has N equal values of a_0, the next diagonal up has $N-1$ values of a_1, and so on with the nth diagonal having $N-n$ equal values of a_n. Note there is only one element a_{N-1} in the top diagonal. □

The next result is a limiting result for sums of the type considered in the previous lemma.

Lemma B.2 *Suppose that $\{a_n;\ n \in \mathcal{Z}\}$ is an absolutely summable sequence, i.e., that*

$$\sum_{n=-\infty}^{\infty} |a_n| < \infty.$$

Then

$$\lim_{N\to\infty} \sum_{n=-N+1}^{N-1} (1 - \frac{|n|}{N})a_n = \sum_{n=-\infty}^{\infty} a_n.$$

Comment The limit should be believable since the multiplier in the summand tends to 1 for each fixed n as $N \to \infty$.

Proof Absolute summability implies that the infinite sum exists and

$$\sum_{n=-\infty}^{\infty} a_n = \lim_{N\to\infty} \sum_{n=-N+1}^{N-1} a_n$$

so the result will follow if we show that

$$\lim_{N\to\infty} \sum_{n=-N+1}^{N-1} \frac{|n|}{N} a_n = 0.$$

Since the sequence is absolutely summable, given an arbitrarily small $\epsilon > 0$ we can choose an N_0 large enough to ensure that for any $N \geq N_0$ we have

$$\sum_{n:|n|\geq N} |a_n| < \epsilon.$$

For any $N \geq N_0$ we can then write

$$\begin{aligned}
|\sum_{n=-N+1}^{N-1} \frac{|n|}{N} a_n| \leq & \sum_{n=-N+1}^{N-1} \frac{|n|}{N} |a_n| \\
= & \sum_{n:|n|\leq N_0-1} \frac{|n|}{N}|a_n| + \sum_{n:N_0\leq|n|\leq N-1} \frac{|n|}{N}|a_n| \\
\leq & \sum_{n:|n|\leq N_0-1} \frac{|n|}{N}|a_n| + \sum_{n:|n|\geq N_0} |a_n| \\
\leq & \sum_{n:|n|\leq N_0-1} \frac{|n|}{N}|a_n| + \epsilon.
\end{aligned}$$

Letting $N \to \infty$ the remaining term can be made arbitrarily small, proving the result. □

B.3 Integration

A basic integral in calculus and engineering is the simple integral of an exponential, which corresponds to the sum of a "discrete time exponential," a geometric progression. This integral is most easily stated as

$$\int_0^\infty e^{-r}\,dr = 1. \tag{B.8}$$

If $\alpha > 0$, then making a linear change of variables as $r = \alpha x$ or $x = r/\alpha$ implies that $dr = \alpha\,dx$ and hence

$$\int_0^\infty e^{-\alpha x}\,dx = \frac{1}{\alpha}. \tag{B.9}$$

Integrals of the form

$$\int_0^\infty x^k e^{-\alpha x}\,dx$$

can be evaluated by parts, or by using the same trick that worked for the geometric progression. Take the kth derivative of both sides of B.9 with respect to α:

$$\begin{aligned}
\frac{d^k}{d\alpha^k}\int_0^\infty e^{-\alpha x}\,dx &= \frac{d^k}{d\alpha^k}\alpha^{-1} \\
\int_0^\infty (-x)^k e^{-\alpha x}\,dx &= (-1)^k k!\alpha^{-k-1} \\
\int_0^\infty x^k e^{-\alpha x}\,dx &= k!\alpha^{-k-1}.
\end{aligned} \tag{B.10}$$

Computations using a Gaussian pdf follow from the basic integral

$$I = \int_{-\infty}^\infty e^{-x^2}\,dx.$$

This integral is a bit trickier than the others considered. It can of course be found in a book of tables, but again a proof is provided to make it seem a bit less mysterious. The proof is not difficult, but the initial step may appear devious. Simplify things by considering the integral

$$\frac{I}{2} = \int_0^\infty e^{-x^2}\,dx$$

and note that this one-dimensional integral can also be written as a two-dimensional integral:

$$\begin{aligned}\frac{I}{2} &= \left(\left(\int_0^\infty e^{-x^2}\, dx\right)^2\right)^{1/2} \\ &= \left(\left(\int_0^\infty e^{-x^2}\, dx\right)\left(\int_0^\infty e^{-y^2}\, dy\right)\right)^{1/2} \\ &= \left(\int_0^\infty \int_0^\infty e^{-(x^2+y^2)}\, dx\, dy\right)^{1/2}.\end{aligned}$$

This subterfuge may appear to complicate matters, but it allows us to change to polar coordinates using $r = \sqrt{x^2 + y^2}$, $x = r\cos(\theta)$, $y = r\sin(\theta)$, and $dx\, dy = rdr\, d\theta$ to obtain

$$\left(\frac{I}{2}\right)^2 = \int_0^{\pi/2} \int_0^\infty re^{-r^2}\, dr\, d\theta = \frac{\pi}{2}\int_0^\infty re^{-r^2}\, dr.$$

Again this might appear to have complicated matters by introducing the extra factor of r, but now a change of variables of $u = r^2$ or $r = \sqrt{u}$ implies that $dr = du/2\sqrt{u}$ so that

$$\left(\frac{I}{2}\right)^2 = \frac{\pi}{2}\int_0^\infty \frac{1}{2}e^{-u}\, du = \frac{\pi}{4},$$

using (B.8). Thus

$$\int_{-\infty}^\infty e^{-x^2}\, dx = \sqrt{\pi}. \tag{B.11}$$

This is commonly expressed by changing variables to $r/\sqrt{2} = x$ so that $dx = dr/\sqrt{2}$ and the result becomes

$$\int_{-\infty}^\infty e^{-\frac{r^2}{2}}\, dr = \sqrt{2\pi}, \tag{B.12}$$

from which it follows that a zero-mean unit variance Gaussin pdf has unit integral. The general case is handled by a change of variables. In the following integral change variables by defining $r = (x - m)/\sigma$ so that $dx = \sigma dr$

$$\begin{aligned}\int_{-\infty}^\infty \frac{1}{\sqrt{2\pi\sigma^2}}e^{-(x-m)^2/2\sigma^2}\, dx &= \frac{1}{\sqrt{2\pi\sigma^2}}\int_{-\infty}^\infty e^{-r^2/2}\, \sigma\, dr \\ &= \frac{\sqrt{2\pi}\sigma}{\sqrt{2\pi\sigma^2}} = 1.\end{aligned} \tag{B.13}$$

B.4 ⋆The Lebesgue integral

This section provides a brief introduction to the Lebesgue integral, the calculus that underlies rigorous probability theory. In the authors' view the Lebesgue integral is not nearly as mysterious as is sometimes suggested in the engineering literature and that, in fact, it has a very intuitive engineering interpretation and avoids the rather clumsy limits required to study the Riemann integral. We present a few basic definitions and properties without proof. Details can be found in all books on measure theory or integration and in many books on advanced probability, including the first author's *Probability, Random Processes, and Ergodic Properties*[32].

Suppose that $(\Omega, \mathcal{F}, P)$ is a probability space as defined in Chapter-2. For simplicity we focus on real-valued random variables. The extensions to complex random variables and more general random vectors are straightforward. The integral or expectation of a random variable f defined on this probability space is defined in a sequence of steps treating random variables of increasing generality.

First suppose that f takes on only a finite number of values, for example

$$f(x) = \sum_{i=1}^{N} a_i 1_{F_i}(x); \; x \in \Re, \tag{B.14}$$

where $1_F(x) = 1$ if $x \in F$ and 0 otherwise and it is assumed that $F_i \in \mathcal{F}$ for all i. A discrete random variable of this form is sometimes called a *simple function*. The (Lebesgue) integral of f or expectation of f is then defined by

$$\int f\, dP = \sum_{i=1}^{N} a_i P(F_i). \tag{B.15}$$

The integral is also written as $\int f(x)\, dP(x)$ and is also denoted by $E(f)$. It is easy to see that this definition reduces to the Riemann integral.

The definition is next generalized to all nonnegative random variables by means of a sequence of *quantizers* which map the random variable into an ever better approximation with only a finite possible number of outputs.

Define for each real r and each positive integer n the quantizer

$$q_n(r) = \begin{cases} n & r \geq n \\ (k-1)2^{-n} & (k-1)2^{-n} \leq r < k2^{-n},\ k = 1, 2, \ldots, n2^n \\ -(k-1)2^{-n} & -(k-1)2^{-n} > r \geq k2^{-n},\ k = 1, 2, \ldots, n2^n \\ -n & r < -n \end{cases} . \tag{B.16}$$

The sequence of quantizers is asymptotically accurate in the sense that for all x

$$f(x) = \lim_{n\to\infty} q_n(f(x)) \tag{B.17}$$

It can be shown without much effort that thanks to the specific construction the sequence $q_n(x)$ is monotone increasing up to x. Given a general nonnegative random variable f, the integral is defined by

$$\int f\, dP = \lim_{n\to\infty} \int q_n(f)\, dP, \tag{B.18}$$

that is, as the limit of the simple integrals of the asymptotically accurate sequence of quantized versions of the random variable. The monotonicity of the quantizer sequence is enough to prove that the limit is well defined. Thus the expectation or integral of any nonnegative random variable exists, although it may be infinite.

For an arbitrary random variable f, the integral is defined by breaking f up into its positive and negative parts, defined by $f^+(x) = \max(f(x), 0) \geq 0$ and $f^-(x) = -\min(f(x), 0)$ so that $f(x) = f^+(x) - f^-(x) \geq 0$, and then defining

$$\int f\, dP = \int f^+\, dP - \int f^- - dP, \tag{B.19}$$

provided that this does not have the indeterminate form $\infty - \infty$, in which case the integral does not exist.

This is one of several equivalent ways to define the Lebesgue integral. A random variable f is said to be *integrable* or P-integrable if $E(f) = \int f\, dP$ exists and is finite. It can be shown that if f is integrable, then

$$\int f\, dP = \lim_{n\to\infty} \int q_n(f)\, dP, \tag{B.20}$$

that is, the form used to define the integral for nonnegative f gives the integral for integrable f.

A highly desirable property of integrals and one often taken for granted in engineering applications is that limits and integrations can be interchanged, e.g., if we are told we have a sequence of random variables $f_n;\ n = 1, 2, 3, \ldots$ which converge to a random variable f with probability one, that is, $F = \{\omega : \lim_{n\to\infty} f_n(\omega) = f(\omega)\}$ is an event with $P(F) = 1$, then

$$\lim_{n\to\infty} \int f_n\, dP \overset{(?)}{=} \int f\, dP. \tag{B.21}$$

Unfortunately this is not true in general and the Riemann integral in particular is poor when it comes to results along this line. There are two very useful such convergence theorems, however, for the Lebesgue integral, which we state next without proof. The first shows that this desirable property holds when the random variables are monotone, the second when they are dominated by an integrable random variable.

Theorem B.1 *If $f_n;\ n = 1, 2, \ldots$ is a sequence of nonnegative random variables that is monotone increasing up to f (with probability one) and $f_n \geq 0$ (with probability one) for all n, then*

$$\lim_{n\to\infty} \int f_n\, dP = \int f\, dP. \tag{B.22}$$

Theorem B.2 *If $f_n;\ n = 1, 2, \ldots$ is a sequence of random variables that converges to f (with probability one) and if there is an integrable function g which dominates the sequence in the sense that $|f_n| \leq g$ (with probability one) for all n, then*

$$\lim_{n\to\infty} \int f_n\, dP = \int f\, dP. \tag{B.23}$$

Appendix C

Common univariate distributions

Binary pmf. $\Omega = \{0, 1\}$; $p(0) = 1 - p$, $p(1) = p$, where p is a parameter in $(0, 1)$.
mean: p
variance: $p(1 - p)$

Uniform pmf. $\Omega = \mathcal{Z}_n = \{0, 1, \ldots, n-1\}$ and $p(k) = 1/n$; $k \in \mathcal{Z}_n$.
mean: $(n - 1)/2$
variance: $(2n-1)(n-1)/6 - [(n-1)/2]^2 = (n^2 - 1)/12$.

Binomial pmf. $\Omega = \mathcal{Z}_{n+1} = \{0, 1, \ldots, n\}$ and

$$p(k) = \binom{n}{k} p^k (1-p)^{n-k};\ k \in \mathcal{Z}_{n+1}\ ,$$

where

$$\binom{n}{k} = \frac{n!}{k!(n-k)!}$$

is the binomial coefficient.
mean: np
variance: $np(1 - p)$

Geometric pmf. $\Omega = \{1, 2, 3, \ldots\}$ and $p(k) = (1-p)^{k-1}p$; $k = 1, 2, \ldots$, where $p \in (0, 1)$ is a parameter.
mean: $1/p$
variance: $(1 - p)/p^2$

Poisson pmf. $\Omega = \mathcal{Z}_+ = \{0, 1, 2, \ldots\}$ and $p(k) = (\lambda^k e^{-\lambda})/k!$, where λ is a parameter in $(0, \infty)$. (Keep in mind that $0! \triangleq 1$.)

mean: λ
variance: λ

Uniform pdf. Given $b > a$, $f(r) = 1/(b-a)$ for $r \in [a,b]$.
mean: $(b+a)/2$
variance: $(b-a)^2/12$

Exponential pdf. $f(r) = \lambda e^{-\lambda r}$; $r \geq 0$.
mean: $1/\lambda$
variance: $1/\lambda^2$

Doubly exponential (or Laplacian) pdf. $f(r) = \dfrac{\lambda}{2} e^{-\lambda|r|}$; $r \in \Re$.
mean: 0
variance: $1/2\lambda^2$

Gaussian (or Normal) pdf. $f(r) = (2\pi\sigma^2)^{-1/2} \exp(-(r-m)^2/2\sigma^2)$; $r \in \Re$. Since the density is completely described by two parameters: the mean m and variance $\sigma^2 > 0$, it is common to denote it by $\mathcal{N}(m, \sigma^2)$.
mean: m
variance: σ^2

Gamma pdf. $f(r) = r^{b-1}e^{-r/a}/a^b\Gamma(b)$; $r > 0$, where $a > 0$ and $b > 0$, where

$$\Gamma(b) = \int_0^\infty e^{-r} r^{b-1}\, dr.$$

mean: ab
variance: $a^2 b$

Logistic pdf. $f(r) = e^{r/\lambda}/\lambda(1+e^{r/\lambda})^2$; $r \in \Re$, where $\lambda > 0$.
mean: 0
variance: $\lambda^2\pi^2/3$

Weibull pdf. $f(r) = br^{b-1}e^{-(r/a)^b}/a^b$; $r > 0$, where $a > 0$ and $b > 0$. If $b = 2$, this is called a *Rayleigh* distribution.
mean: $a\Gamma(1+(1/b))$
variance: $a^2(\Gamma(1+(2/b)) - \Gamma^2(1+(1/b)))$

Appendix D
Supplementary reading

In this appendix we provide some suggestions for supplementary reading. Our goal is to provide some leads for the reader interested in pursuing the topics treated in more depth. Admittedly we only scratch the surface of the large literature on probability and random processes. The books are selected based on our own tastes — they are books from which we have learned and from which we have drawn useful results, techniques, and ideas for our own research.

A good history of the theory of probability may be found in Maistrov [49], who details the development of probability theory from its gambling origins through its combinatorial and relative frequency theories to the development by Kolmogorov of its rigorous axiomatic foundations. A somewhat less serious historical development of elementary probability is given by Huff and Geis [40]. Several early papers on the application of probability are given in Newman [53]. Of particular interest are the papers by Bernoulli on the law of large numbers and the paper by George Bernard Shaw comparing the vice of gambling and the virtue of insurance.

An excellent general treatment of the theory of probability and random processes may be found in Ash [1], along with treatments of real analysis, functional analysis, and measure and integration theory. Ash is a former engineer turned mathematician, and his book is one of the best available for someone with an engineering background who wishes to pursue the mathematics beyond the level treated in this book. The only subject of this book completely absent in Ash is the second-order theory and linear systems material of Chapter 5 and the related examples of Chapter 6. Other good general texts on probability and random processes are those of Breiman [7] and Chung [10]. These books are mathematical treatments that are relatively accessible to engineers. All three books are a useful addition to any library, and most of the mathematical details avoided here can be found in these texts.

Wong's book [72] provides a mathematical treatment for engineers with a philosophy similar to ours but with an emphasis on continuous time rather than discrete time random processes. A further good addition to any library is Doob's classic text on stochastic processes [16]. Doob also forty years later a less well known but very nice little book on measure theory [17]. The first author found this book to have the best treatment available on the theory of signed measures, measures with the restrictions of nonnegativeness and normalization removed. Such generalizations of measures are encountered in the theory of quantization and Doob provides several useful extensions of the probability limit theorems.

Another general text of interest is the inexpensive paperback book by Sveshnikov [64], which contains a wealth of problems in most of the topics covered here as well as many others. While the notation and viewpoint often differ, this book is a useful source of applications, formulas, and general tidbits.

The set theory preliminaries of Appendix A can be found in all books on probability, elementary or otherwise, or in all books on elementary real analysis. In addition to the general books mentioned, more detailed treatments can be found in books on mathematical analysis such as those by Rudin [61], Royden [59], and Simmons [62]. These references also contain discussions of functions or mappings. A less mathematical text that treats set theory and provides an excellent introduction to basic applied probability is Drake [18]. This classic but out-of-print book has been replaced at MIT by an excellent text inheriting some of Drake's philosophy by Bertsekas and Tsitsiklis [3].

The linear systems fundamentals are typical of most electrical engineering linear systems courses. Good developments may be found in Chen [8], Kailath [41], Bose and Stevens [5], and Papoulis [55], among others. A treatment emphasizing discrete time may be found in Stieglitz [63]. A minimal treatment of the linear systems aspects used in this book may also be found in Gray and Goodman [33].

Detailed treatments of Fourier techniques may be found in Bracewell [6], Papoulis [54], Gray and Goodman [33], and the early classic Wiener [68]. This background is useful both for the system theory applications and for the manipulation of characteristic functions or moment-generating functions of probability distributions.

Although the development of probability theory is self-contained, elementary probability is best viewed as a prerequisite. An introductory text on the subject for review (or for the brave attempting the course without such experience) can be a useful source of intuition, applications, and practice of some of the basic ideas. Two books that admirably fill this function are Drake [18] and the classic introductory text by two of the primary contributors to the early development of probability theory, Gnedenko and Khinchine [30]. The more complete text by Gnedenko [29] also provides a useful backup text. A virtual encyclopedia of basic probability, including a wealth of examples, distributions, and computations, may be found in Feller [22].

The axiomatic foundations of probability theory presented in Chapter 2 were developed by Kolmogorov and first published in 1933. (See the English translation [44].) Although not the only theory of probability (see, e.g., Fine [23] for a survey of other approaches), it has become the standard approach to the analysis of random systems. The general references cited previously provide good additional material for the basic development of probability spaces, measures, Lebesgue integration, and expectation. The reader interested in probing more deeply into the mathematics is referred to the classics by Halmos [37] and Loeve [47].

As observed in Chapter 4, instead of beginning with axioms of probability and deriving the properties of expectation, one can go the other way and begin with axioms of expectation or integration and derive the properties of probability. Some texts treat measure and integration theory in this order, e.g., Asplund and Bungart [2]. A nice paperback book treating probability and random processes from this viewpoint in a manner accessible for engineers is that by Whittle [67].

A detailed and quite general development of the Kolmogorov extension theorem of Chapter 3 may be found in Parthasarathy [56], who treats probability theory for general metric spaces instead of just Euclidean spaces. The mathematical level of this book is high, though, and the going can be rough. It is useful, however, as a reference for very general results of this variety and for detailed statements of the theorem. A treatment may also be found in Gray [32].

Good background reading for Chapters 4 and 6 are the book on convergence of random variables by Lukacs [48] and the book on ergodic theory by Billingsley [4]. The Billingsley book is a real gem for engineers interested in learning more about the varieties and proofs of ergodic theorems for discrete time processes. The book also provides nice tutorial reviews on advanced conditional probability and a variety of other topics. Several proofs are given for the mean and pointwise ergodic theorems. Most are accessible given a knowledge of the material of this book plus a knowledge of the projection theorem of Hilbert space theory. The book also provides insight into applications of the general formulation of ergodic theory to areas other than random process theory. Another nice survey of ergodic theory is that of Halmos [38].

As discussed in Chapter 6, stationarity and ergodicity are sufficient but not necessary conditions for the ergodic theorem to hold, that is, for sample averages to converge. A natural question, then, is what conditions are both necessary and sufficient. The answer is known for discrete time processes in the following sense: A process is said to be *asymptotically mean stationary* or ams if its process distribution, say m, is such that the limits

$$\lim_{n\to\infty} \frac{1}{n}\sum_{i=0}^{n-1} m(T^{-i}F)$$

exist for all process events F, where T is the left-shift operation. The limits trivially exist if the process is stationary. They also exist when they die out with time and in a variety of other cases. It is known that a process will have an ergodic theorem in the sense of having all sample averages of bounded measurements converge if *and only if* the process is ams [34, 32]. The sample averages of an ams process will converge to constants with probability one if and only the process is also ergodic.

Second-order theory of random processes and its application to filtering and estimation form a bread-and-butter topic for engineering applications and are the subject of numerous good books such as Grenander and Rosenblatt [35], Cramér and Leadbetter [13], Rozanov [60], Yaglom [73], and Liptster and Shiryayev [46]. It was pointed out that the theory of weakly stationary processes is intimately related to the theory of Toeplitz forms and Toeplitz matrices. An excellent treatment of the topic and its applications to random processes is given by Grenander and Szego [36]. A more informal engineering-oriented treatment of Toeplitz matrices can be found in Gray [31].

It is emphasized in our book that the focus is on discrete time random processes because of their simplicity. While many of the basic ideas generalize, the details can become far more complicated, and much additional mathematical power becomes required. For example, the simple product sigma fields used here to generate process events are not sufficiently large to be useful. A simple integral of the process over a finite time window will not be measurable with respect to the resulting event spaces. Most of the added difficulties are technical — that is, the natural analogs to the discrete time results may hold, but the technical details of their proof can be far more complicated. Many excellent texts emphasizing continuous time random processes are available, but most require a solid foundation in functional analysis and in measure and integration theory. Perhaps the most famous and complete treatment is that of Doob [16]. Several of the references for second-order theory focus on continuous time random processes, as do Gikhman and Skorokhod [27], Hida [39], and McKean [51]. Lamperti [45] presents a clear summary of many facets of continuous time and discrete time random processes, including second-order theory, ergodic theorems, and prediction theory.

In Chapter 5 we sketch briefly some basic ideas of Wiener and Kalman filters as an application of second-order theory. A detailed general development of the fundamentals and recent results in this area may be found in Kailath [42] and the references listed therein. In particular, the classic development of Wiener [69] is an excellent treatment of the fundamentals of Wiener filtering.

Of the menagerie of processes considered in the book, most may be found in the various references already mentioned. The communication modulation examples may also be found in Gagliardi [24], among others. Compound Poisson processes are treated in detail in Parzen [57]. There is an extensive literature on Markov processes and their applications, as examples we cite Kemeny and Snell [43], Chung [9], Rosenblatt [58], and Dynkin [21].

Perhaps the most notable beast absent from our menagerie of processes is the class of martingales. If the book and the target class length were longer, martingales would be the next topic to be added. They were not included simply because we felt the current content already filled a semester, and we did not want to expand the book past that goal. An excellent mathematical treatment for the discrete time case may be found in Neveu [52], and a readable description of the applications of martingale theory to gambling may be found in the classic by Dubins and Savage [20].

References

[1] R. B. Ash. *Real Analysis and Probability.* Academic Press, New York, 1972.

[2] E. Asplund and L. Bungart. *A First Course in Integration.* Holt,Rinehart and Winston, New York, 1966.

[3] D. T. Bertsekas and J. N. Tsitsiklis. *Introduction to Probability.* Athena Scientific, Boston, 2002.

[4] P. Billingsley. *Ergodic Theory and Information.* Wiley, New York, 1965.

[5] A. G. Bose and K. N. Stevens. *Introductory Network Theory.* Harper & Row, New York, 1965.

[6] R. Bracewell. *The Fourier Transform and Its Applications.* McGraw-Hill, New York, 1965.

[7] L. Breiman. *Probability.* Addison-Wesley, Menlo Park, CA, 1968.

[8] C. T. Chen. *Introduction to Linear System Theory.* Holt, Rinehart and Winston, New York, 1970.

[9] K. L. Chung. *Markov Chains with Stationary Transition Probabilities.* Springer-Verlag, New York, 1967.

[10] K. L. Chung. *A Course in Probability Theory.* Academic Press, New York, 1974.

[11] T. M. Cover, P. Gacs, and R. M. Gray. Kolmogorov's contributions to information theory and algorithmic complexity. *Ann. Probab.*, 17:840–865, 1989.

[12] T. M. Cover and J. A. Thomas. *Elements of Information Theory.* Wiley, New York, 1991.

[13] H. Cramér and M. R. Leadbetter. *Stationary and Related Stochastic Processes.* Wiley, New York, 1967.

[14] W. B. Davenport and W. L Root. *An Introduction to the Theory of Random Signals and Noise.* McGraw-Hill, New York, 1958.

[15] R. L. Dobrushin. The description of random field by means of condi-

tional probabilities and conditions of its regularity. *Theory Prob. Appl.*, 13:197–224, 1968.

[16] J. L. Doob. *Stochastic Processes.* Wiley, New York, 1953.

[17] Joseph L. Doob. *Measure Theory.* Springer-Verlag, New York, 1994.

[18] A. W. Drake. *Fundamentals of Applied Probability Theory.* McGraw-Hill, San Francisco, 1967.

[19] R. C. Dubes and A. K. Jain. Random field model in image analysis. *Journal of Applied Statistics*, 16:131–164, 1989.

[20] L. E. Dubins and L. J. Savage. *Inequalities for Stochastic Processes: How to Gamble If You Must.* Dover, New York, 1976.

[21] E. B. Dynkin. *Markov Processes.* Springer-Verlag, New York, 1965.

[22] W. Feller. *An Introduction to Probability Theory and its Applications*, volume 2. Wiley, New York, 1960. 3rd ed.

[23] T. Fine. Properties of an optimal digital system and applications. *IEEE Trans. Inform. Theory*, 10:287–296, Oct 1964.

[24] R. Gagliardi. *Introduction to Communications Engineering.* Wiley, New York, 1978.

[25] Karl F. Gauss. *Theoria Motus Corporum Coelestium (Theory of the Motion of the Heavenly Bodies Moving about the Sun in Conic Sections).* Little, Brown, and Co.,, 1987. republished by Dover, 1963.

[26] A. Gersho and R. M. Gray. *Vector Quantization and Signal Compression.* Kluwer Academic Publishers, Boston, 1992.

[27] I. I. Gikhman and A. V. Skorokhod. *Introduction to the Theory of Random Processes.* Saunders, Philadelphia, 1965.

[28] James Gleick. *Chaos: making a new science.* Penguin Books, 1987.

[29] B. V. Gnedenko. *The Theory of Probability.* Chelsea, New York, 1963. Translated from the Russian by B. D. Seckler.

[30] B. V. Gnedenko and A. Ya. Khinchine. *An Elementary Introduction to the Theory of Probability.* Dover, New York, 1962. Translated from the 5th Russian edition by L. F. Boron.

[31] R. M. Gray. Toeplitz and Circulent Matrices. ISL technical report 6504–1, Stanford University Information Systems Laboratory, 1977. Revised version is available online at http:/ee.stanford.edu/~gray/toeplitz.html.

[32] R. M. Gray. *Probability, Random Processes, and Ergodic Properties.* Springer-Verlag, New York, 1988. http://ee.stanford.edu/arp.html.

[33] R. M. Gray and J. G. Goodman. *Fourier Transforms.* Kluwer Academic Publishers, Boston, Mass., 1995.

[34] R. M. Gray and J. C. Kieffer. Asymptotically mean stationary measures. *Ann. Probab.*, 8:962–973, 1980.

[35] U. Grenander and M. Rosenblatt. *Statistical Analysis of Stationary Time Series.* Wiley, New York, 1957.

[36] U. Grenander and G. Szego. *Toeplitz Forms and Their Applications.* University of California Press, Berkeley and Los Angeles, 1958.

[37] P. R. Halmos. *Measure Theory.* Van Nostrand Reinhold, New York, 1950.

[38] P. R. Halmos. *Lectures on Ergodic Theory.* Chelsea, New York, 1956.

[39] T. Hida. *Stationary Stochastic Processes.* Princeton University Press, Princeton, NJ, 1970.

[40] D. Huff and I. Geis. *How to Take a Chance.* W. W. Norton, New York, 1959.

[41] T. Kailath. *Linear Systems.* Prentice-Hall, Englewood Cliffs, NJ, 1980.

[42] T. Kailath. *Lectures on Wiener and Kalman Filtering.* CISM Courses and Lectures No. 140. Springer-Verlag, New York, 1981.

[43] J. G. Kemeny and J. L. Snell. *Finite Markov Chains.* D. Van Nostrand, Princeton, NJ, 1960.

[44] A. N. Kolmogorov. *Foundations of the Theory of Probability.* Chelsea, New York, 1950.

[45] J. Lamperti. *Stochastic Processes: A Survey of the Mathematical Theory.* Springer-Verlag, New York, 1977.

[46] R. S. Liptser and A. N. Shiryayev. *Statistics of Random Processes.* Springer-Verlag, New York, 1977. Translated by A. B. Aries.

[47] M. Loeve. *Probability Theory.* D. Van Nostrand, Princeton, NJ, 1963. Third Edition.

[48] E. Lukacs. *Stochastic Convergence.* Heath, Lexington, MA, 1968.

[49] L. E. Maistrov. *Probability Theory: A Historical Sketch.* Academic Press, New York, 1974. Translated by S. Kotz.

[50] J. D. Markel and A. H. Gray, Jr. *Linear Prediction of Speech.* Springer-Verlag, New York, 1976.

[51] H. P. McKean, Jr. *Stochastic Integrals.* Academic Press, New York, 1969.

[52] J. Neveu. *Discrete-Parameter Martingales.* North-Holland, New York, 1975. Translated by T. P. Speed.

[53] J. R. Newman. *The World of Mathematics*, volume 3. Simon & Schuster, New York, 1956.

[54] A. Papoulis. *The Fourier Integral and Its Applications.* McGraw-Hill, New York, 1962.

[55] A. Papoulis. *Signal Analysis.* McGraw-Hill, New York, 1977.

[56] K. R. Parthasarathy. *Probability Measures on Metric Spaces.* Academic

Press, New York, 1967.

[57] E. Parzen. *Stochastic Processes.* Holden Day, San Francisco, 1962.

[58] M. Rosenblatt. *Markov Processes: Structure and Asymptotic Behavior.* Springer-Verlag, New York, 1971.

[59] H. L. Royden. *Real Analysis.* Macmillan, London, 1968.

[60] Yu. A. Rozanov. *Stationary Random Processes.* Holden Day, San Francisco, 1967. Translated by A. Feinstein.

[61] W. Rudin. *Principles of Mathematical Analysis.* McGraw-Hill, New York, 1964.

[62] G. F. Simmons. *Introduction to Topology and Modern Analysis.* McGraw-Hill, New York, 1963.

[63] K. Steiglitz. *An Introduction to Discrete Systems.* Wiley, New York, 1974.

[64] A. A. Sveshnikov. *Problems in Probability Theory, Mathematical Statistics,and Theory of Random Functions.* Dover, New York, 1968.

[65] Peter Swerling. A proposed stagewise differential correction procedure for satellite tracking and prediction. Technical Report P-1292, Rand Corporation, January 1958.

[66] P. Whittle. On stationary processes in the plane,. *Biometrika*, 41,:434–449, 1954.

[67] P. Whittle. *Probability.* Penguin Books, Middlesex, 1970.

[68] N. Wiener. *The Fourier Integral and Certain of Its Applications.* Cambridge University Press, New York, 1933.

[69] N. Wiener. *Time Series: Extrapolation,Interpolation,and Smoothing of Stationary Time Series with Engineering Applications.* MIT Press, Cambridge, MA, 1966.

[70] N. Wiener and R. E. A. C. Paley. *Fourier Transforms in the Complex Domain.* American Mathematics Society Coll. Pub., Providence, RI, 1934.

[71] Chee Son Won and Robert M. Gray. *Stochastic Image Processing.* Information Technology: Transmission, Processing, and Storage. Kluwer/Plenum, New York, 2004.

[72] E. Wong. *Introduction to Random Processes.* Springer-Verlag, New York, 1983.

[73] A. M. Yaglom. *An Introduction to the Theory of Stationary Random Functions.* Prentice-Hall, Englewood Cliffs, NJ, 1962. Translated by R. A. Silverman.

Index